HIGHWAY 99

HIGHWAY 99

A Literary Journey through California's Great Central Valley

Edited by Stan Yogi

HEYDAY BOOKS
IN CONJUNCTION WITH
CALIFORNIA COUNCIL FOR THE HUMANITIES

nia Council for the Humanities

. conjunction with the California Council for the
ort by the National Endowment for the Humanities.

Publisher's Cataloging in Publication
(Prepared by Quality Books Inc.)

Highway 99 : a literary journey through California's Great Central Valley / edited by Stan Yogi.
p. cm.
Includes bibliographical references.
ISBN 0-930588-82-7

1. Central Valley (Calif. : Valley)--In literature. 2. Central Valley (Calif. : Valley)--Literary collections. 3. Central Valley (Calif. : Valley)--History. 4. Central Valley (Calif. : Valley)--Description and travel. I. Yogi, Stan.

PS283.C2H54 1996 810'.9'97945
QBI96-40143

Interior Design and Production Coordination: Wendy Low
Cover Design: Jack Myers, DesignSite, Berkeley
Printing and Binding: Publishers Press, Salt Lake City

Please address orders, inquiries, and correspondence to
Heyday Books
Box 9145
Berkeley, CA 94709
(510) 549-3564

Printed in the United States of America

10 9 8 7 6 5 4 3 2 1

Cover photo of road by Roman Loranc, courtesy of the photographer.
Photo of Mexican railroad workers courtesy of City of Sacramento History and Science Division, Sacramento Archives and Museum Collection Center.

CONTENTS

ACKNOWLEDGMENTS

The seed for this book was planted by Susan Gordon, who deserves credit for nurturing it through its germination.

The valley's unofficial prose laureate Gerry Haslam and Jim Houston provided critical advice and guidance during the early conceptual stages of this project. Gerry was especially helpful, kind, and generous in helping to shape the themes that recur throughout this collection.

Special thanks go to the staff of Heyday Books. In particular, Malcolm Margolin, Heyday's inspirational publisher, recognized the need for this anthology, enthusiastically supported it since its inception, and provided insightful editorial help with the introduction. Wendy Low, Heyday's project manager, handled all of the complicated and sometimes tedious administrative and production tasks, from securing permissions for reprinting to layout, with organized grace and amazing calm.

Thanks to Stan Morita for his helpful comments about the introduction and to Bonnie Akimoto and Sam Blazer for help with proofreading.

Numerous individuals assisted with the collection of literary and photographic material:

Mike Cole, C.G. Hanzlicek (CSU Fresno), Michelle Hester-Reyes (College of the Sequoias), Lawson Inada, William Rintoul, Jon Veinberg, Lillian Vallee, and John Walden (Kern County Library) helped in the process of selecting literary material by alerting me to authors and works that they felt were appropriate for this collection and also by helping me contact writers.

The gathering of photos for this project involved visits to numerous museums, libraries and historical societies where I received expert and very kind help from: Marianne Bailey (California State University, Chico), Pam Bush (California State University, Chico), Terry Brazil (Tulare City Historical Society/Tulare Historical Museum), Gail Cismowski (Merced Courthouse Museum), Michael Clifton (Fresno City and County Historical Society), Leslie Crow (Bank of Stockton), Carola Enriquez (Kern County Museum,

Bakersfield), Jeanine Ford (Holt Atherton Center, University of the Pacific), Ellen Gorelick (Tulare City Historical Society/Tulare Historical Museum), Joyce Hall (Fresno County Library), Tom Hennion (Tulare City Historical Society/Tulare Historical Museum), Linda Johnson (Sacramento Archives and Museum Collection Center), Jacqueline Lowe (County Museum, Yuba City), Ronald Mahoney (California State University, Fresno), Charlene Noyes (Sacramento Archives and Museum Collection Center), William Rintoul, Julie Stark (County Museum, Yuba City), Lillian Vallee, and Heidi Warner (McHenry Museum, Modesto).

Special thanks go to Rondal Partridge, Andrea Metz (Merced Courthouse Museum), and especially to Roman Loranc for the generous use of their photographs.

And deep gratitude to the National Endowment for the Humanities for their generous support.

INTRODUCTION

Passing through cities such as Sacramento and Fresno, through the orchards of Yuba City and the vineyards of Selma, past rice fields, pastures, cotton fields, wetlands and scrub brush, Highway 99 spans almost the entire length of California's Great Central Valley, one of the most fertile agricultural regions in the world. Bathed by ground-kissing tule fog in the winter and smothered by weighty summer heat, the area from Bakersfield to Redding, bordered by the Sierra Nevada to the east and the coastal ranges to the west, is famous for its abundant crops.

The area is less well known, however, for its bountiful literary products. Yet the cities, towns, farms, and fields that hug Highway 99 have produced some of the most celebrated American writers of the 20th century: novelist, playwright, and short-story writer William Saroyan and the poet William Everson (aka Brother Antoninus) were born and raised in the Valley. One of the country's most respected living poets, Philip Levine, has since 1958 lived in Fresno, where along with C.G. Hanzlicek and Peter Everwine, he has mentored some of the country's leading contemporary poets. Among them are Lawson Inada, Larry Levis, Gary Soto, Roberta Spear, David St. John, and Sherley Anne Williams.

Joan Didion and Richard Rodriguez, two of America's premier essayists, grew up in Sacramento. Stockton has inspired the memoirs of Maxine Hong Kingston and the plays of Philip Kan Gotanda. The landscapes and people of Kern County inhabit the poems of Wilma Elizabeth McDaniel, the stories of Gerald Haslam, and the dramas of Luis Valdez.

Other writers—John Muir, Frank Norris, Mary Austin, John Steinbeck, Cherríe Moraga and Gary Snyder, to name a handful—are among those who have spent time in the Central Valley and have contributed greatly to its remarkable literary legacy.

Bringing together the works of some seventy authors who have written about the Great Central Valley, this anthology contains many vibrant cross-currents. One recurring theme is the century-long metamorphosis of the Valley's landscape. John Muir's description of the Valley floor as "one smooth, continuous bed of honey-bloom" might strike many contemporary Valley residents as a fantasy. Similarly, Tulare Lake (as described by George Derby in this anthology) might seem like a dream: once the most extensive freshwater lake west of the Mississippi, this one-time Valley wonder no longer exists. Many pieces in this collection, sometimes overtly sometimes beneath the surface, convey a poignant lament for a landscape that has slipped away, perhaps forever.

For the last 150 years, farming has been fundamental to the Central Valley, the source not only of livelihood but of cultural identity. People work the land, cajole and caress it into productivity, live with it on terms of greatest intimacy, as William Everson expressed in these lines from "San Joaquin":

> I in the vineyard, in green-time and dead-time, come to it dearly,
> And take nature neither freaked nor amazing
> But the secret shining, the soft indeterminate wonder.
> I watch it morning and noon, the unutterable sundowns;
> And love as the leaf does the bough.

Several other authors in this anthology write with similar intensity about the relationships people living in the Valley have established with the land.

Hard work is very much a part of living in the Central Valley. Farmers, field laborers, ranchers, oil field workers, and dairy men and women have over generations developed physical and spiritual toughness to survive. Many of the writers included in this collection know firsthand the aching muscles and sweat-soaked clothes that result from picking grapes, climbing oil rigs, and irrigating fields. They do not glamorize what city folk might romanticize as back-to-the-land farm work. From the oil field roustabouts in William Rintoul's story "The Gangpusher" to the cotton field laborer in Gary Soto's poem "Hoeing," the hard reality of labor is evident in many of the works in this anthology.

Still other selections in this collection depict the experiences of various groups of people from all over the globe who have come to call the Valley home. William Saroyan writes of Armenian farmers, Sherley Anne Williams of African American migrants. Wilma Elizabeth McDaniel and John Steinbeck portray the lives of "Okie" refugees, Carlos Bulosan of Filipino laborers; Maxine Hong Kingston recalls Chinese pioneers, while Jose Montoya and Ernesto Galarza tell of Mexican workers. Art Coelho describes Azorean immigrants; Thiphavanah Louangrath brings us into the lives of Laotian newcomers. As each wave of new settlers makes a home in the Valley, the social and economic relations of the area become more complex—rich ground for writers to till—as witnessed in stories like Gerald Haslam's "The Doll" and David Mas Masumoto's "Firedance."

Modern life has brought still more complications. In many parts of the Valley, the suburban is crashing in on the rural. Looking east, for instance, down Fresno's Shaw Avenue (or a major street in another Valley town) at the ribbons of strip malls, it is increasingly difficult to remember that the Valley's cities are any different from hundreds of other suburban towns anywhere in the U.S. Yet raise your eyes and, at least on a clear day, the Sierras are still humbling. Drive just a few miles west and the chain stores and fast food restaurants give way to vineyards, which in the winter resemble crops of ancient menorahs, and fields spread out like vast fans. Such complexity and contradictions nourish the Valley's ongoing literary community.

The material in this anthology is in a general chronological order according to the events and eras depicted in each of the selections, not by dates of initial publication. In a few instances, however, I have departed from chronology for reasons of thematic resonance and interplay among authors.

It would be too simplistic to argue that there is a monolithic Valley character or culture, some theme or mood that unites the various authors represented in this anthology. It would similarly be an oversimplification to argue that all of the writers in this collection identify themselves regionally. A few never lived in the Valley, while many who did have moved away. Some of them look back on the Valley through the veil of nostalgia, others through hard-edged critical lenses. Yet read together, the poems, stories, essays and drama included here work in concert to make California's Great Central Valley imaginable in ways that might otherwise not be possible—for people who are decades-long residents and for those who have never set foot on Valley soil.

Welcome to a literary journey through the Great Central Valley. It's full of excitement, some surprises, and far from over.

Stan Yogi
Oakland, 1996

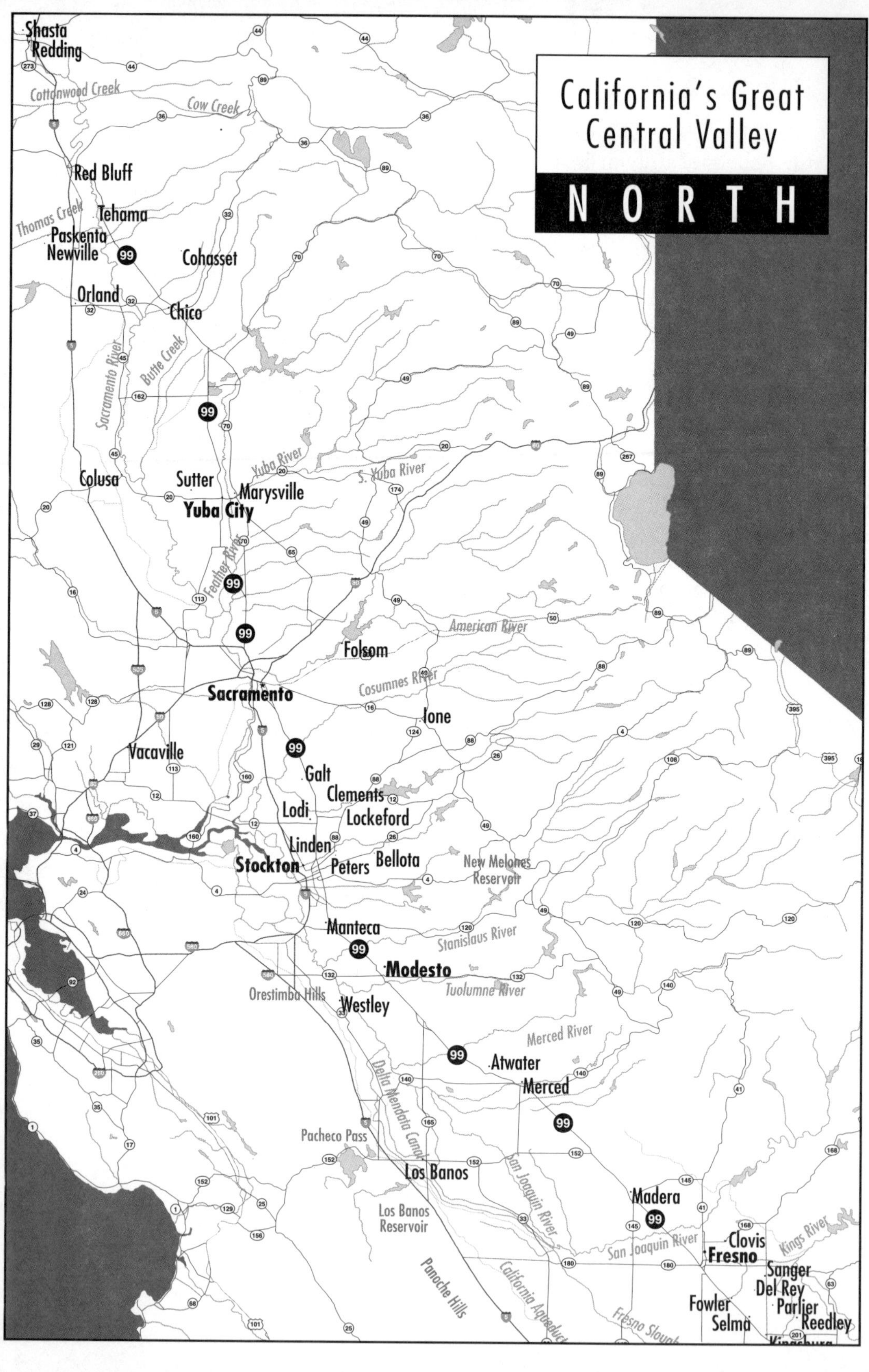
California's Great Central Valley
NORTH
Shasta
Redding
Cottonwood Creek
Cow Creek
Red Bluff
Thomas Creek
Tehama
Paskenta
Newville
Cohasset
Orland
Chico
Sacramento River
Butte Creek
Yuba River
S. Yuba River
Colusa
Sutter
Marysville
Yuba City
Feather River
American River
Folsom
Sacramento
Cosumnes River
Ione
Vacaville
Galt
Clements
Lodi
Lockeford
Linden
Stockton
Peters
Bellota
New Melones Reservoir
Manteca
Stanislaus River
Modesto
Tuolumne River
Orestimba Hills
Westley
Merced River
Atwater
Merced
Delta Mendota Canal
Pacheco Pass
Los Banos
San Joaquin River
Los Banos Reservoir
Madera
Clovis
Fresno
Kings River
Panoche Hills
California Aqueduct
Sanger
Del Rey
Fowler
Parlier
Selma
Reedley
Fresno Slough

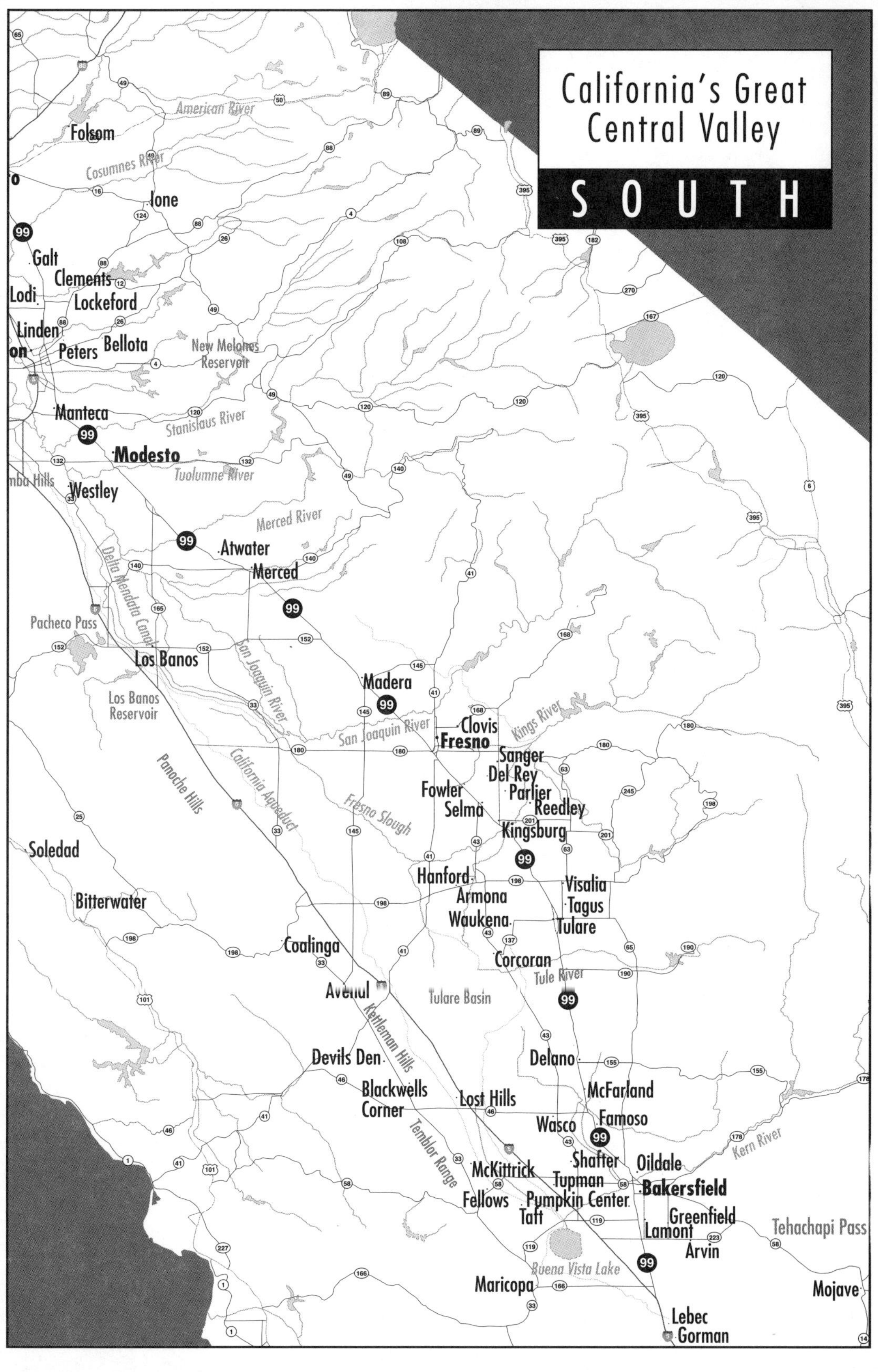
California's Great Central Valley
SOUTH
Folsom
American River
Cosumnes River
Ione
Galt
Clements
Lodi
Lockeford
Linden
Peters
Bellota
New Melones Reservoir
Manteca
Stanislaus River
Modesto
Tuolumne River
Westley
Merced River
Atwater
Merced
Delta Mendota Canal
Pacheco Pass
Los Banos
San Joaquin River
Los Banos Reservoir
Madera
Clovis
Fresno
Kings River
Sanger
Del Rey
Fowler
Parlier
Selma
Reedley
Kingsburg
Panoche Hills
California Aqueduct
Fresno Slough
Soledad
Hanford
Visalia
Armona
Tagus
Bitterwater
Waukena
Tulare
Coalinga
Corcoran
Tule River
Tulare Basin
Kettleman Hills
Devils Den
Delano
Blackwells Corner
Lost Hills
McFarland
Famoso
Wasco
Temblor Range
Shafter
Oildale
Kern River
McKittrick
Tupman
Bakersfield
Fellows
Pumpkin Center
Taft
Greenfield
Lamont
Tehachapi Pass
Arvin
Buena Vista Lake
Maricopa
Mojave
Lebec
Gorman

Yokuts Tale

THE MAN AND THE OWLS

A MAN AND HIS WIFE were traveling. They camped overnight in a cave. They had a fire burning. Then they heard a horned owl *(hutulu)* hoot. The woman said to her husband: "Call in the same way. He will come and you can shoot him. Then we will eat him for supper." The man got his bow and arrows ready and called. The owl answered, coming nearer. At last it sat on a tree near the fire. The man shot. He killed it. Then his wife told him: "Do it again. Another one will come." Again he called and brought an owl and shot it. He said: "It is enough now." But his wife said: "No. Call again. If you call them in the morning they will not come. We have had no meat for a long time. We shall want something to eat tomorrow as well as now." Then the man called. More owls came. There were more and more of them. He shot, but more came. The air was full of owls. All his arrows were gone. The owls came closer and attacked them. The man took sticks from the fire and fought them off. He covered the woman with a basket and kept on fighting. More and more owls came. At last they killed both the man and the woman.

 First appeared in "Myths of South Central California" by Alfred Kroeber, *University of California Publications in American Archaeology and Ethnology* 4, no. 4 (1907). The story was collected from an anonymous storyteller of the Yawdanchi Yokuts tribal group which inhabited the Tule River drainage in the foothills northeast of present-day Bakersfield.

Pedro Fages

IN THE SOUTH SAN JOAQUIN

FROM THE CAMP OF THE RETURN or Bad Water [a great valley extends] to a village called Buena Vista. This village, because it is on a fair sized elevation, overlooks a great plain along the course of the River San Francisco, which will exceed one hundred and twenty leagues in length and in parts is twenty, fifteen, and even less in width. It is all a labyrinth of lakes and tulares, and the River San Francisco, divided into several branches, winding in the middle of the plain, now enters and now flows out of the lakes until very near to the place where it empties into the estuary of the river. In the midst of the winding of the river and on the sides there are large rises of land of good soil where, with ease, irrigating ditches could be made.

All this plain is very thickly settled with many and large villages. They have many seeds, and especially one which the soldiers call "heathen rice," which grows larger than our rice and also has better flavor. There are two species, one white and the other yellow, and the grain is like mustard. There is also plentiful game, such as deer, antelope, mule deer, bear, geese, cranes, ducks, and many other species of animals, both terrestrial and wingéd.

In their villages the natives live in the winter in very large squares, the families divided from each other, and outside they have very large houses in the form of hemispheres, where they keep their seeds and utensils. They are people of very good features and of a superior height, and are very frank and liberal. It has not been noted that they have committed the least theft. They have some large stones like *metates* on which they grind their seeds.

From the village of Buena Vista the plain continues toward the south for seven leagues more, over good lands with some water. And at the end of

---

From "In the South San Joaquin Ahead of Garcés" by Herbert E. Bolton, *California Historical Society Quarterly* 10, no. 3 (1931). Reprinted by permission of the California Historical Society. This account from 1773 records the impressions of one of the first Europeans to see the Valley.

these seven leagues one goes towards the south through a pass, partly of valleys and arroyos, very thickly grown with groves of live oaks, as are also the hills and sierras which form these valleys. Going now three leagues more in the same direction, one comes to a very large plain, which keeps getting wider and wider, both toward the east and toward the south, leaving to the north and northwest many sierras.

Last year, coming from San Diego in pursuit of the deserters, I went and struck the plain fifty leagues toward the east. Lack of water forced us into the sierra but, when we were parallel with the mission of San Gabriel, we went about fifteen leagues to strike the plain again; and we went along the plain toward the north, keeping close to the sierra, on account of the water, travelling for about twenty-five leagues until we reached the pass of Buena Vista. Most of these twenty-five leagues we were passing through groves of date palms, the land both to the east and the south having more and more palm groves. But the country appeared to be very short of water. We saw many smokes all along the plain.

Andrew Freeman

THE ARRIVAL OF WHITES

WE HAD A MAN at Thomas Creek that had power given to him. He was young. He sang all the time. He drank water and ate once a month. He ate a little of everything, then took one swallow of water and smoked. He stayed in the sweat-house all the time.

Now our captain [chief] used to get out early every morning on top of the sweat-house and, calling everybody by name, would tell them what to do.

This fortuneteller from Thomas Creek would tell the people just how much game they would get and whether any mishaps would fall. He lived across from our present reservation at Paskenta. One day he said, "There are some people from across the ocean who are going to come to this country." He looked for them for three years. "They have some kind of boat *[tco 'ltci]* with which they can cross, and they will make it. They are on the way." Finally he said that they were on the land and that they were coming now. He said that they had fire at night and lots to eat. "They cook the same as we do; they smoke after meals, and they have a language of their own. They talk, laugh, and sing, just as we do. Besides, they have five fingers and toes; they are built like we are, only they are light." He said their blood was awfully light.

"They have a four-legged animal which some are riding and some are packing. They haven't any wives, any of them. They all are single. They are bringing some kind of sickness."

So everybody was notified. The night watch and day watch were kept. He said that they had something long which shoots little round things a long distance. They have something short that shoots just the same.

From *The Way We Lived: California Indian Reminiscences, Stories, and Songs: California Indian Reminiscences, Stories, and Songs* copyright 1981, 1993 by Malcolm Margolin, published by Heyday Books. First appeared in "Nomlaki Ethnography" by Walter Goldschmidt, *University of California Publications in American Archaeology and Ethnology* 42, no. 4 (1951). This is an account of the mid-19th century that Freeman must have heard from people of his parents' and grandparents' generation.

Finally the whites came in at Orland; many of them. When they came in they started shooting. There were thousands of Indians in the hills who went to fighting the whites. The Indians went after them but they couldn't do anything to them. Finally they got to Newville, and the man who was telling these fortunes said the whites were going to be there. The Indians were ready for them. The whites came by Oakes' place and down the flat at one o'clock in the morning. They killed the first Indian that showed himself. The captain told the others to stay in the house and get their bows and arrows ready.

The captain yelled to the whites that he was ready inside the house. He told his men, "When you get ready, run out and crowd into it." The captain sent them to fight at close range. He said, "We are dead anyway." The whites couldn't load their muzzle-loaders, so they used revolvers. The captain told his men to spear them. They fought from morning till afternoon. The Indians had come all the way from Colusa. They killed all those whites. The Indians were afraid of gray horses. They killed the horses. They examined everything. They divided everything up. One old man from south of the Tapscott place took away a lot of their money. His children used to take the money and play with it. Finally he took it up the canyon and hid it. The whites are looking for that money today but can't find it.

Another group of whites came to Mountain House *[lopom]*. They killed many of the Indians. White people hit women and children in the head. One Indian shouted from a rock when the white man started back. The whites came up there, and that Indian went into the rock cave, and they shot one white man from there. But the whites threw fire into the cave and killed all the Indians in there.

They had been hiding in the hills. Indians couldn't get to the salt. They got very weak—they say salt keeps a person fit. There was no rain for three years, and fighting going on every day. No clover, no acorn, juniper berries, or peppergrass. Nothing for three years. Very little rain.

Finally the Indians got smallpox, and the Indian doctor couldn't cure them. They died by the thousands. Gonorrhea came amongst the Indians. That killed a lot of them. My grandfather said that if he had fought he would have been killed too. But he went up to Yolla Bolly Mountain with about six hundred others and stayed three years. On the third winter there was a heavy snowstorm. The snow was over his head. He said women can stand more starvation than men. They singed the hair off a deerhide shoulder strap and ate it.

Men died every day from starvation. That was in Camp of Dark Canyon in the winter. Women would find a little bunch of grass and eat it and would bring a handful back for their husbands. The women would have to chew it for the men. The man was too weak to swallow it. She would take a mouthful of water and pour it into his mouth. That was the way they saved a lot of them.

One man and his brother were lying among the rocks. They wanted to steal something. They saw six riders with forty head of cattle. The Indians lay there and watched them. The riders left after they corralled the cattle. The two Indians got a long willow sprout and made a whip. They tore the fence down and drove the cattle up the creek. They killed one of the steers. First they shot it with an arrow and then hit it in the head with a rock. Then they cut its throat. They skinned it and divided the meat among the people. They had a big feast and they had to make soup for the weak people. They heated water in a basket. Then they put meat in the basket. That saved the people.

After that the whites began to gather up the Indians. They made the Nome Lackee Reservation in Tehama County. They take a tame Indian along when they bring Indians together on a reservation. They worked the Indians on the reservation. Old Martin was given a saddle mule and clothes. He wouldn't wear anything but the shirt—the overalls hurt his legs. He was a kind of foreman. Every Saturday they killed four or five beef and divided it among the Indians. They ground wheat and made biscuits. The women shocked hay. They had to examine all the men and women for disease.

Garland on the present Oakes' place wouldn't let them take the Indians off of his land, and that's what saved them. When they took the Indians to Covelo [in Round Valley, on the Nome Cult Reserve] they drove them like stock. Indians had to carry their own food. Some of the old people began to give out when they got to the hills. They shot the old people who couldn't make the trip. They would shoot children who were getting tired. Finally they got the Indians to Covelo. They killed all who tried to get away and wouldn't return to Covelo.

Jedediah S. Smith

from THE SOUTHWEST EXPEDITION OF JEDEDIAH S. SMITH, 1826–1827

I ARRIVED AT A LAKE called by the Spaniards Too Larree or Flag Lake. I arrived at the Lake quite late and found the bank so muddy that it was impossible for my horses to get any water yet I was obliged to encamp. From what I could learn of the Indians the Spaniards had named it from report but be that as it may the name was quite appropriate. Too Larre Lake is about 12 miles in circumference and is in a fine large valley which commences about 12 miles South of it. Coming into the valley from the South East I had passed over a range of hills which in their course a little East of North appeared to increase in height. On the declivity of these hills there was some Oak timber. I observed the trees had many holes made in their trunks in which an acorn was pressed so tight that it was difficult to get it out. By watching I found this to be the work of a bird of the woodpecker kind who takes this method to lay up his stock of provision for the winter. The bird is of a seal color and somewhat larger than the red head woodpecker. I called this bird the Provident woodpecker. The following day in moving along the bank of the Lake I surprised some Indians who immediately pushed out into the lake in canoes or rather rafts made of flag. My guide succeeded in getting them to return to the shore. One of them could talk some Spanish and I engaged him for a guide. I watered my horses and got some fish from the Indians (who I observed had some horses stolen no doubt from the Spaniards) and moved on about 3 miles along the Lake and then up an inlet about 10 miles crossed over and encamped. On this inlet was some timber Cotton wood and willow. Where I crossed it was 8 or 10 yard wide rapid current 2 feet deep and comes from the East. Several Indians, some of them having horses, visited the encampment. The principal characters brought with them each a small sack of down and sprinkled me from head to foot. To this I submitted knowing it was a custom among them and wishing to avoid giving offence. The[y] told me of a river to the north that had an animal which I supposed from their description to be the Beaver although they had no name for the animal by which it was known to me. These Indians call themselves Wa-ya-la-ma. The Indian that spoke

Spanish and the same I engaged at the Flag Lake told me he would go on with me and my other guide returned. On the following day I moved Northwardly 15 miles across low hills which were spurs of the mountain on the East. This mountain had been gradually increasing in elevation and had now attained a considerable heighth. The next day I moved nearly North West 30 miles over a level country the ground being so completely undermined by the paths of an animal like the Lizard that the horses were continually sinking in the Earth frequently up to the knees. I encamped on the bank of a Lake. Since leaving the Wa-ya-la-ma the country has been dry and destitute of water and grass. I found water in but one place in the bed of a stream which was nearly dry. East of my route at the foot of the Mountain there was some timber and plenty of grass and water. The Lake on which I encamped was apparently large extending to the N W so far that the shore was not visible. But as I supposed it not more than 80 miles to the Ocean. I did not think it of very great extent. It appeared shallow from the number of Boggy Islands seen in many parts of it. Near where I encamped was an Indian village of two or three hundred inhabitants. Their lodges were built of willows and mats. The willows were placed in the ground in rows at the distance of ten feet apart and bent over and joined together at the top and then covered with the mats forming a lodge in exterior appearance like a line of barracks and about 100 yards in length with a door at proper intervals for each family. My provision being nearly exhausted I visited the village for the purpose of trading for some provision. My interpreter having gone before to inform them of my approach, when I arrived some mats were spread in front of the Lodges and I was invited to sit down. Grass seed was then brought and poured on my head until I was nearly covered. This seed which was gathered during the summer formed at this time the principal subsistence of these Indians. I gave them some presents and after some conversation with the chiefs made arrangement to have my grass seed formed into meal. At night I was invited to attend a dance and went to the Lodge at 8 O' Clock and found a seat prepared for me. I was immediately treated with some roasted fish and a mush made of the grass seed. After supper the dancers came in 10 or 12 in number and seated themselves in a cluster. They were painted and some of them had head dresses made of feathers and a skin around the waist. Having remained at this encampment two days I moved north along the beach of the Lake and again encamped, on a low spot of ground on the beach. During the night a high north west wind raised the water of the Lake and drove it into my encampment so that I was obliged to remove to higher ground. East of this encampment a level country with but little vegetation extends to the foot of the mountain a distance of about 20 miles. On the following day I left the beach of the Lake as my guide said there was a stream putting in which I could not cross near its mouth. I therefore traveled North Eastwardly and at 18 miles encamped on a small River 20 yds wide deep and muddy with low Oak timber along its banks. During the day I saw several antelope and some Elk sign and passed a country like the last described. The next day I crossed the stream carrying my goods over on a log and swimming my horses and traveled North Eastwardly 12 miles crossing several small streams having oak timber and a plenty of grass on their banks. The soil was very fine and although somewhat wet yet it produced most excellent grass. The prairie and woodland was mingled in pleasing variety and my encampment was on a small stream in a fine little grove of timber. My guide informed me that in the neighborhood was a plenty of Elk. I therefore sent some men hunting. They killed 2 antelope and found Elk but killed none.

In company with my guide I visited some Indians that were up near the foot of the Mountain and at the distance of about 15 miles. Their lodges were built like those before described at the Chin-ta-che Lake. The country appeared populous but the soil gravelly and not as rich as at the encampment. Near the Mountain there is a good deal of Oak timber the trees having large trunks but low and spreading tops so that it is not of the most valuable kind for building or fencing. My guide saying he was unacquainted with the country further north I engaged another who told me that in one day I could travel to where I could find Beaver—Having remained two days at the last mentioned encampment I moved on N Westwardly 25 miles crossing in the course of the day 2 small streams and encamping near a large Indian village on the bank of a river 80 yds wide where I found some Beaver sign. These Indians called themselves Wim-mil-che and this name I applied to the river which comes from E N E. In the vicinity was considerable timber (Oak) and a plenty of grass. The game of the country was principally Elk and Antelope. On the 28th of February I commenced trapping on the Wim-mil-che and during 10 days I moved up the river 25 or 30 miles. I was then near the foot of the Mountain and finding no further inducement for trapping and the Indians telling me of a river they called the Peticutry [San Joaquin] in which there was beaver I traveled north along near the foot of the Mountain about 15 miles and encamped on the bank of the Peticutry (at the place where I first struck the Peticutry were a great number of small artificial mounds) running at this place west and not quite as large as the Wim-mil-che. In this vicinity the plains are generally clothed with grass and were at that time covered with blossoms. Along the river there is some timber. At the foot of the Mountain the timber is Oak and far up the Mountain Pine.

George H. Derby

from REPORT OF THE TULARE VALLEY

THE ESCORT HAVING ARRIVED and encamped in our vicinity, I started accordingly at about 10 A.M., and, crossing two ranges of low hills over a broad and smooth trail, arrived on the shore of the great Taché lake [Tulare Lake, now entirely gone] about 1 P.M. We were unable to get close to the water, in consequence of the tulé which environed it extending into the lake from two hundred yards to one-fourth of a mile, as far as the eye could reach. With a glass I could distinguish the timber at the north and the tulé at the south ends of the lake, the length of which I estimated at about twenty miles, but we could not distinctly make out the opposite or eastern shore. The peaks of the Sierra Nevada, at this place twelve thousand feet above the level of the sea and covered with perpetual snow, appeared in close proximity, and, rising far above the horizon, seemed to us to come down precipitously to the very edge of the water. The distance from our encampment to the lake we estimated at eighteen miles, or nearly a day's march, and as the country passed over was a perfect desert, and I found here no forage for the animals but wire grass, the water standing in the tulé marshes blackish, and no wood at all, I concluded to return immediately to camp, and in the morning to make a reconnaissance to the south of our position, for the purpose of finding a road to the southern extremity of the lake, which point I hoped to be able to reach in one day's march. An examination was accordingly made on the 2d, (a portion of the party being left in camp to cut wild oats, which I purposed to transport for forage, as it was evident we would find none upon the shores of the lake,) which terminated favorably, a good path being found through the southern extremity of the valley, and a trail leading apparently around the south of the lake. On the 3d we broke up our pleasant encampment at *Dick's creek* [probably Avenal Creek] and succeeded in reaching the southern part of the lake,

---

where we encamped upon the sand for the night, having marched twenty-four and a quarter miles nearly in an easterly direction from the termination of the pass. We found here a ridge of sand about one hundred yards in width, and twelve feet above the level of the lake, which divides the water of the northern or Taché from the bed (now nearly dry,) of the southern or Ton Taché lake. This last is little more than a very extensive swamp, covering the plain for fifteen miles in a southerly direction, and is about ten in width. It is filled with sloughs and small tulé lakes, and is of course impassable, except with the assistance of boats or rafts. The gradual receding of the water is distinctly marked by the ridges of decayed tulé upon its shore, and I was informed, and see no reason to disbelieve, that ten years ago it was nearly as extensive a sheet of water as the northern lake, having been gradually drained by the connecting sloughs, and its bed filled by the encroachments of the tulé.

• • •

On the 20th we marched 24.92 miles in a devious course with the object of striking the "Sanjon de San Jose," which is a slough connecting the river San Joaquin with the Taché lake, but were unable to get further on account of the mire, the ground between the lake and the San Joaquin being entirely cut up by small sloughs which had overflown in every direction, making the country a perfect swamp, which I found it a matter of great difficulty to cross. We saw numerous bands of wild horses, numbering in all more than a thousand; they were at some distance, but their appearance rendered me extremely anxious for the safety of my animals, which would infallibly have been lost if they could have broken loose and joined them, as they appeared much inclined to do. We were engaged on the 21st, 22d and 23d in getting through the mire, crossing no less than eight distinct sloughs, one of which we were obliged to raft over, before arriving at the Sanjon. In all of these sloughs a strong current was running southwest, or from the San Joaquin river to the lake. The country over which we passed between these sloughs was miserable in the extreme, and our animals suffered terribly for want of grass. There being no wood upon the plain except an occasional willow on the largest slough we could make no fires, and were consequently obliged to return to bread and water, a diet which, though simple in the extreme, I somewhat preferred to the raw salt pork on which the men luxuriated. The "Sanjon de San Jose" is a large and deep slough about forty miles in length, connecting the waters of the Taché lake with the San Joaquin river, with which it unites at its great southern bend. At this time it was about two hundred and forty feet in width, and with an extremely slow current setting towards the river. I do not think it possible to communicate directly with the lake through this slough. An attempt had been made a week or two previous to our arrival by a party of men in a whale-boat, who examined it for twenty or thirty miles, and found it branching off into innumerable smaller sloughs, which intersected the Tulé swamp in every direction. They also reported that there was a fall of water about six or eight feet at this point. I think it highly probable that there may be a rapid near the mouth of the slough, but should find it difficult to account for a direct fall of that height; and as I could hear of no one else who had seen it, although I met with many who had crossed the Sanjon in every direction during the dry season, I am inclined to disbelieve in its existence. The whole country for forty miles in extent in a southerly direction by ten in width, between the San Joaquin river and the Taché lake, is, during the rainy season and succeeding months until the middle of July, a vast swamp everywhere intersected by sloughs, which are deep, miry and dangerous. A wagon road therefore crossing the Sanjon at any point is impracticable, as it could not by any possibility be travelled more than three months in the year; I regret that we were unable to follow up the Sanjon to the lake, but this was utterly impossible. We found that we could not proceed a quarter of a mile in either direction without getting hopelessly mired, and as the water was fast rising, we could not get back over the swamp which we had already with great exertion and difficulty crossed.

Thomas Jefferson Mayfield

from TAILHOLT TALES

LEAVING THE STREAM, we started across the plains in an easterly direction. We had been told at Rancho San Luis that we would in this way arrive at Río San Joaquín where there was a ford.

By this time we could see what had caused the mass of color so noticeable from the mountain the day before. The entire plain, as far as we could see, was covered with wild flowers. Almost all of the flowers were new to us.

Along the creek were many lupines, some of them growing on bushes six and eight feet high. The low foothills were covered with two pretty, lily-like flowers, one tall and straight-stemmed with a cluster of lavender, bell-shaped flowers at the top *[brodiaea laxia]* and the other a purple, ball-shaped blossom on a similar stem *[brodiaea capitata]*.

As we passed below the hills the whole plain was covered with great patches of rose, yellow, scarlet, orange, and blue. The colors did not seem to mix to any great extent. Each kind of flower liked a certain kind of soil best, and some of the patches of one color were a mile or more across.

I believe that we were more excited out there on the plains among the wild flowers than we had been on the mountain the day before when we saw the valley for the first time. Several times we stopped to pick the different kinds of flowers, and soon we had our horses and packs decorated with masses of all colors.

My daddy had travelled a great deal, and it was not easy to get him excited about wild flowers or pretty scenery. But he said that he would not have believed that such a place existed if he had not seen it himself. Mother cried with joy and wanted to make a home right there in the midst of it all.

For my own part, I have never seen anything to equal the virgin San Joaquin Valley before there was a plow or a fence within it. I have always loved nature and have liked to live close to her.

Many times, travelling alone, when night has overtaken me, I have tied my horse and rolled up in my saddle blanket and slept under a bank, or

---

 The experiences in these excerpts were recounted to Latta by Mayfield in 1928 and describe events that took place during the mid- to late 19th century.

among the wild flowers, or on the desert under a bush. I remember those experiences as the greatest in my life.

The two most beautiful remembrances I have are of the virgin San Joaquin and of my mother.

• • •

A large group of Indians, numbering probably forty, were bathing in the south edge of the stream. As I remember, the group was composed mostly of young people. They all appeared very much excited at seeing our party but did not seem afraid of us.

We sat on our horses awhile and watched the bathers. Some of them were washing their hair. They would lean over, allowing their long hair to fall forward into the water. Then they would comb it with their fingers.

They soon became too deeply interested in us to go on with their bathing, and several of them swam to our side of the river. By this time we had dismounted, and were trying to decide whether to attempt a crossing there or to follow the north bank of the river toward the hills.

My daddy found that one of the Indians could understand a little Spanish, and this Indian encouraged him to cross the river, saying that his people would help him. So they again made preparations to ferry the river, as we had done before when crossing the valley from Rancho San Luis to Las Mariposas.

While the animals were being unpacked, I stayed near my mother. She kept close watch over me, as eight or ten of the Indians, many of them wearing no clothing at all, were crowding about, looking at me and excitedly talking to one another.

Finally a young girl about sixteen years of age offered to take me on her back and swim the river with me.

At this, Mother took me in her arms and held me close to her, motioning for the Indians to go away. She called to my daddy, and he ran to us, thinking that some hostile move had been made. He had fought the Mescalero Apaches for years and knew Indians well.

He soon saw that the Indians were all right and told my mother to let the girl carry me across, as I would be safer with her than I would on one of our horses.

So Mother took off my clothes and put me on the back of the Indian girl. I clasped my hands around her neck, and she took my feet under her arms and waded into the water.

Soon she started swimming with a long overhand stroke. She was as slippery as could be, and I was afraid of being carried away by the current and clung to her neck so close that she could not breathe.

Several times she stopped swimming and reached up and pulled my arms down until she recovered her breath. Then she started on again after each stop, until we arrived at the south edge of the stream in shallow water. She stood me in the water, which was about a foot deep, near a sand bar.

During this time all of the Indian women and children from the ranchería had accumulated where we were about to land, and they crowded around me and laughed and talked to each other about me and called me "chólo wé-chep" (little white boy). I did not know their language then, but I have always felt sure that they were also telling each other how cute I was.

The girl who had carried me across the river was very proud of me and, holding to my hand, kept the rest of the Indians at a distance of several feet. She would talk to me and laugh, but, of course, I understood nothing she said and remembered only the words I mentioned before.

• • •

After the reservation was formed and the white settlers began to come into the valley in great numbers, the Indians had a hard time of it. I used to hear all of their troubles discussed at the ranchería where I lived, and I know how they felt about the way they were treated.

The elk and antelope were all gone in just a few years. Fences were built, and the Indians were not allowed to roam about and gather plants and seeds and hunt as they had done before. They were forced into contact with the Monaches and other mountain Indians. I believe that this had as much to do with their disappearance as anything else.

They were finally crowded into small camps, and had to shear sheep and wash clothes for the white settlers about them. White men were always furnishing them liquor, and many of them were killed by whisky.

Whisky made devils of them. Quarrels and fights would take place between persons who had always been friends, and when such quarrels arose, someone was sure to be badly knifed or shot. The older men of the tribe tried to stop them but could do nothing with them.

The Indians were quite honest among themselves and never stole from one another. I believe they were more truthful among themselves than the white people.

The early settlers used to be scandalized when the white man married a mokee [Indian woman]. I believe that in nine cases out of ten the mokee was the loser in the bargain and was better than the man she married.

The Indians realized that the white people were smarter and cleverer than they were, and they looked to the whites many times for advice until they found they could not trust them at all.

They would do many things when advised to by the whites that they would not think of doing upon the advice of their own people. In their own life marriage consisted simply of providing a home for the bride. Many white scamps would fix up a cabin for one of the mokees and then, when they wanted to, would go away and leave her. She would provide for and raise the children. Many times she was better off when he did leave, as she otherwise had to provide for him also.

Then, too, some of the Indians started living in houses. They could not stand an indoor life, and many died of consumption and measles. When they had these diseases, they would go in the sweat house and sweat for an hour or more and then jump in the cold streams. This killed them by the hundreds.

They had never known about fevers. The sweat house was a good remedy for rheumatism, but it was deadly when used as a treatment for fevers.

When I left the Indians for the last time in 1862, there were not more than forty left of a group that had numbered more than three hundred when I went to them in 1850 or 1851. Battles with the whites accounted for very few of those missing, for the Indians I was with, as a group, had no battles with the whites. I believe that a few of the young men joined in some of the difficulties for the excitement they would get from it, but only a very few individuals.

• • •

"You remember one time you cross the river on one mokee's back?" "Yes," I almost yelled, "I'll never forget that as long as I live." "Well," she said, "Me that mokee."

This mokee was the girl who had carried me across the San Joaquin River on her back when we first came to the San Joaquin plains forty-three years before.

After we had left the San Joaquin River that year, I had seen very few of the Indians we had known there and had never known who had carried me across the river.

But she had remembered me. She had known of my life with the Indians on Kings River and on Tulare Lake. But she had recognized me largely through my conversation with her husband and my resemblance to my daddy.

I stayed with the Medleys for a week. Each day I would make preparations to leave the next morning. Mrs. Medley would pay no attention to my preparations but would go on talking or working. After breakfast she would walk over to where I was packing the burro and say, "Ugh, you no go today, you stay." Then nothing would do but I would stay another day.

I believe that the week with the Medleys was one of the most pleasant I have ever spent. We used to talk about the old times. With tears in her eyes and streaming down her face, Mrs. Medley would tell how the white man had ruined the country.

He had built fences so that the Indians could not roam about and gather their food. He had plowed up the fields and killed the wild flowers. He had killed most of the Indians and had crowded the rest of them off their land.

With all of my experience with Indians I had never until then really appreciated how badly they felt about the way they had been treated by the white people.

Of course, Mrs. Medley was an exceptionally fine type of Indian. As I remember the group we met when she carried me across the river, I would say that she was easily the most intelligent of the lot.

Her action in offering to carry me across the river was not what one would expect from a strange Indian. She was a far stronger character than her husband. He drank heavily and squandered everything they might have saved. In fact, he got gloriously drunk the week I visited with them.

William Henry Brewer

from UP AND DOWN CALIFORNIA, 1860–1864

Red Bluff, on the Upper Sacramento.
Sunday, August 17.

BUSY, ALTHOUGH WEAK and out of sorts, I got ready and at 4 P.M. of Tuesday, August 12, left San Francisco by steamer for Sacramento. It was a most lovely afternoon—the beautiful bay was crossed, the sun set, gilding in the most golden colors the bare hills, now either brown or a rich straw color. Mount Diablo stood up, a most majestic object, until shut out by the shades of evening. We were in the "sloughs," as the many mouths of the Sacramento and San Joaquin rivers are called, when the moon rose from the plain as from the sea. The illusion was heightened by its blood-red color and distorted shape as it rose from the low horizon. The river was low, although still eight feet above the low water of previous years. The bed has been so raised, however, that we stuck on a bar for seven hours, and only arrived at Sacramento at eight o'clock in the morning.

Wednesday, August 13, I took a short stroll through that city before going on board the steamer bound for Red Bluff. Everywhere one sees the effects of the flood in that unfortunate city, and, indeed, the water was still over a part of it. The morning was intensely hot, in strong contrast with the San Francisco weather.

Although the channel of the Sacramento is insufficient to carry off all the water of wet winters, yet it is rapidly filling up, each year increasing the difficulty. Previous to 1848 the river was noted for the purity of its waters, flowing from the mountains as clear as crystal; but, since the discovery of gold, the "washings" render it as muddy and turbid as is the Ohio at spring flood—in fact it is perfectly "riley," discoloring even the waters of the great bay into which it empties. A man pointed out to me a spot at the mouth of the American River at Sacramento where, in 1849, he had sounded the river and found it fifty feet deep. He had seen a schooner sink there, that

From the Yale University Press edition (1930). Later reprinted by University of California Press as *Up and Down California, 1860–1864: The Journal of William Henry Brewer* (1949, 1966).

only a little of her masts stuck out a short distance. Now there is a luxuriant growth of young willows on the mud bank that occupies the spot. Last winter's floods alone are supposed to have raised the bed of the river at Sacramento six or seven feet at least—that is, in spots, so as to raise the water that much.

Red Bluff is at the head of navigation on the river—three hundred miles above Sacramento by the river, but only half that distance by land. Sternwheel steamers, drawing but eighteen or nineteen inches of water run up. Our boat was the *Gem*, and we towed a barge with two hundred tons of freight, quite an impediment to rapid progress. We got off at 11 A.M. I had plenty of books along, and although it was very hot, 90° to 96° every day, yet I enjoyed that trip much.

Thursday, August 14, we kept on our slow and winding way, often on bars and shoals that took long to get over. A wide plain borders the river on each side. We caught distant views of the mountains, but generally we saw only the river and its banks, which were more or less covered with trees—willows, cottonwoods, oaks, and sycamores—with wild grapevines trailing from them. Some of the views were pretty indeed. When it got dark, we tied up, it being impossible to run in the night, owing to snags, bars, and rapids.

On Friday, August 15, . . . our progress this day was slower than before. Many bars and rapids in the crooked river were but slowly surmounted. During the day Lassen's Buttes stood out in clear outlines in the east—two majestic sharp peaks, their sides streaked with snow. Before night we saw Mount Shasta rising above the horizon, clear in outline, although its snowy crest was 150 miles distant in air line.

We tied up that night at Tehama, a little village on the bank. A circus was the excitement of the time and I attended. Such an audience! At least two-thirds were Digger Indians, who enjoyed the riding much, but were decidedly undemonstrative as to the rest. After it, there was the usual excitement of "side shows"—the bearded woman, the stone eater, etc.—the agent of each yelling for custom. There were gambling tables in saloons, where *monte* was in progress—the usual music, women, liquor, piles of gold and silver on the tables, etc. It was decidedly a scene to be remembered.

Saturday we again went on and arrived here in the afternoon. My party had arrived just before me and had encamped near town. This is a stirring little town of a few hundred inhabitants—saloons, taverns ("hotels"), and corrals being the chief features, for here pack trains and teams start for the whole northern country, Oregon, etc. But, oh, how hot it is! I am now writing at eleven o'clock at night, and it is 94° in my room—it has been

100° to 102° most of the day. I went to church this morning—an audience of about twenty-five only—in the schoolhouse.

Here the low hills close in on the river, and here begins a most interesting country to visit. I went out to camp a little while this afternoon, but I shall stay at the hotel until we leave here, then take to camp again.

Camp 94, North Fork of Cottonwood Creek.
Sunday, August 24.

Tuesday, August 19, the camp went on, but I rode with a gentleman to visit the Tuscan Springs, some eight miles east, in a wild region of low, desolate, forbidding hills. These springs have some repute—a house, bathhouses, etc., have been erected—but they are not quite a Saratoga or an Avon in either surroundings or accommodations, for they are hot, dry, dreary, and desolate. The waters are varied in the different springs, but all are sulphury, stinking, with salt, alkali, iron, etc. They have much repute for the cure of certain diseases. Not the least interesting feature was the evolution of gas along with the water. It is so abundant, and burns so well, that it is collected, purified, and burned to heat the baths. The whole region around shows strong marks of volcanic action.

Wednesday, August 20, I got up at 4 A.M. to take the stage, but had to wait three hours before it came along, heavily loaded with mail matter and boot full. Only two men could ride inside, the rest on top. A hot ride of forty miles brought us into Shasta about two o'clock—over hills and through gulches, very dusty. The last four miles was through a mining region.

• • •

Yesterday, August 23, we came here, where we will stop about three days, then return to Shasta and wait for Professor Whitney. The whole aspect of the scenery is changed on arriving in this region. Across the head of the Sacramento Valley stretches a great table-land, perhaps 1,200 feet above the sea, and north of this a chain of mountains stretches across and unites the Sierra Nevada with the Coast Range. This table-land stretches south from the foot of these mountains. Passing through, it would not be noticed that it is a table, so cut and furrowed is it with canyons and gulches, but from any considerable height it seems like an immense plain, if one is high enough so as not to notice its furrowed ravines.

From all the more elevated points, Mount Shasta is a most sublime object. Distant about sixty to seventy miles in an air line, so clear does it seem in its sharp outline that to an eye unaccustomed to our scenery here it would seem scarcely ten miles off. Its whole top is white with snow, save where cragged rocks peep through. The snow line is a well-defined, perfectly level line all around.

• • •

Marysville.
October 20.

October 2 we left Shasta City. First, six miles southwest to the river, over the table-land before described—a table furrowed entirely into hills and gulches—then crossed the river, then rose to another table. This was unlike the first; it is in reality the head of the great Sacramento Valley plain, which at its northern end rises into level tables, perhaps six hundred to eight hundred feet above the sea, often for many miles without streams or deep gulches, bearing no gold, covered at this place with scattered oaks and pines, the soil dry, but barren because of its dryness only. The surface is as hard as a paved road, the trees and shrubs have a dry aspect, although they are mostly evergreen, and all the herbage is long since dried up and gone.

We camped on Cow Creek, about twelve miles from old Fort Reading. We had heard of coal mines discovered up this creek, about fifteen miles from our camp, so on October 3 Rémond and I started for them, intending to be gone two days. Our ride was a gentle ascent, sometimes passing over tables sloping to the west and elevated perhaps a hundred feet above the surrounding country. These were remarkable, being made up of strata of volcanic ashes, sometimes mixed with bowlders of lava. This had covered the entire region, but up about fourteen miles from camp we found where the stream had cut through these strata into the sandstone beneath, which is rich in fossils—shells of many species were as thickly imbedded in the rocks as if the sea had but lately left these shores.

It was a solitary region, with houses only at intervals of several miles. We did not find the coal mines, did not find the men who were to show us where they were, could find no place where we could get shelter or feed for our animals, so returned the same night. It rained in the night and drove us into the tent.

San Francisco.
October 26.

Safely back here again I will go on with my journal, but first, a few more words on the great interior valley of the state, that of the San Joaquin and the Sacramento.

This great feature is a vast valley, often thirty to forty miles wide, a perfect plain enclosed by high mountains on both sides, its only opening, the Straits of Carquinez, being less than a mile wide. One can start on this plain, near Shasta, and travel southeast *four hundred and fifty miles,* in a nearly straight line, without crossing any hill of any considerable height—that is, if a road is run near the rivers.

The extreme north end rises in a table-land a few hundred feet high, but the valley does not taper to a point—it is cut off nearly square, where it is at least thirty miles wide, by the mountains that extend across the north part of the state. But the eastern edge is modified by a range of hills that stretches east from Lassen's Peak, down into the valley, way to the river, near Red Bluff, so that the upper end of the plain spreads out above it, something like a letter T. Now these hills mentioned are mostly of lava and need more than a passing notice, for they impart features to the landscape so unlike anything else that I must make these preliminary explanations so that you may understand what I will have to write for some time to come.

Lassen's Peak, and in fact, that whole part of that chain, like Mount Shasta, is a gigantic extinct volcano, perhaps about twelve thousand feet high—a volcano not only much higher, but vastly greater in every respect of magnitude and effect than Etna. It is flanked by a considerable number of smaller cones, old volcanoes, from one thousand feet high, up to that of the main peak itself, many of these cones being much higher and greater than Mount Vesuvius.

Here, in a former age of the world, was a scene of volcanic activity vastly surpassing anything existing now on the earth. The materials from these volcanoes not only formed the mountains themselves and covered the foothills, but also came down on the plain for more than a hundred miles. Sometimes volcanic ashes covered the whole region many feet thick, then sheets of molten lava would flow over it, hardening into the hardest rock, then ashes and lava again. Thus were formed beds of enormous thickness, regularly stratified, descending with a gentle slope toward the Sacramento River, and even crossing it in one place near Red Bluff.

But all volcanic action ceased ages ago, and the snows and rains falling on the high lands about Lassen's Peak formed streams which radiate from it. They have worn deep canyons, channels, in this lava, often a thousand feet deep, but generally less. Between these are table-lands, sometimes strewn with loose bowlders of lava, at others showing a surface of nearly naked lava with only enough soil to support, here and there, low cragged shrubs and a few herbs during the wet parts of the year.

• • •

October 10 we were off in good season and came down the valley twenty miles and camped at Chico, on Chico Creek. As we come south the valley becomes more fertile, and more highly cultivated. Here it is ten miles from the river to the hills, of which about eight is most excellent land and produces immense crops.

Chico is a thriving little place. We camped in the private yard of Major Bidwell, the principal citizen of the place, and while there ate at his house. We had a pleasant time.

By the way, I have forgotten whether I have given you the height of Mount Shasta—it is 14,440 feet, *the highest land yet measured in the United States.* I feel proud that I took *first* accurate barometrical observations to measure the highest point over which the Stars and Stripes hold jurisdiction.

San Francisco.
November 2.

My last letter brought me up to Chico, and here I will begin again. Here, as elsewhere, one man is often the town in heart and soul. Major Bidwell left the "States" in 1840, and arrived here in 1841. He *is* Chico. He is very wealthy now, very public spirited—owns a ranch of five leagues (over twenty-two thousand acres) of fine land in the Sacramento Valley, a large mill, store, etc., and is the spirit of the growing town of Chico. Unfortunately, he was not at home; but, knowing that we were coming, he had left orders for our entertainment.

Around his yard cottonwood trees had been planted ten years ago—these trees have now an average diameter of two feet—showing how trees grow here with care. In this shady yard we pitched our tent, the most pleasant place we have seen for a long time. Back of us was a fruit garden of some thirty acres, teeming with peaches, figs, grapes, etc., of which we were invited to partake *ad libitum—ad nauseam* if we chose.

Chico Creek seems on the map a little short stream. It is not so—it heads back in the hills many miles, in the volcanic table-land so often spoken of before.

Gary Thompson

OLD COHASSET ROAD

If John and Annie Bidwell rode
this road alone,
did they rein in the team,
climb down and try to count
the poppies burning dots
of flame up through the valley floor,
and did they think the earth
was burning up inside
with wanton love,
or did they choose to see
the stars of heaven scattered
in the field, like eyes, the many eyes
of God who sees and counts
the things we do in spring
when poppies bloom and winter cloaks
come off, and John and Annie
left alone with love
of God and with their own
stern passion?

John Muir

from THE BEE-PASTURES

THE GREAT CENTRAL PLAIN of California, during the months of March, April, and May, was one smooth, continuous bed of honey-bloom, so marvelously rich that, in walking from one end of it to the other, a distance of more than 400 miles, your foot would press about a hundred flowers at every step. Mints, gilias, nemophilas, castilleias, and innumerable compositæ were so crowded together that, had ninety-nine per cent of them been taken away, the plain would still have seemed to any but Californians extravagantly flowery. The radiant, honeyful corollas, touching and overlapping, and rising above one another, glowed in the living light like a sunset sky—one sheet of purple and gold, with the bright Sacramento pouring through the midst of it from the north, the San Joaquin from the south, and their many tributaries sweeping in at right angles from the mountains, dividing the plain into sections fringed with trees.

Along the rivers there is a strip of bottom-land, countersunk beneath the general level, and wider toward the foot-hills, where magnificent oaks, from three to eight feet in diameter, cast grateful masses of shade over the open, prairie-like levels. And close along the water's edge there was a fine jungle of tropical luxuriance, composed of wild-rose and bramble bushes and a great variety of climbing vines, wreathing and interlacing the branches and trunks of willows and alders, and swinging across from summit to summit in heavy festoons. Here the wild bees reveled in fresh bloom long after the flowers of the drier plain had withered and gone to seed. And in midsummer, when the "blackberries" were ripe, the Indians came from the mountains to feast—men, women, and babies in long, noisy trains, often joined by the farmers of the neighborhood, who gathered this wild fruit with commendable appreciation of its superior flavor, while their home orchards were full of ripe peaches, apricots, nectarines, and figs, and their vineyards were laden with grapes. But, though these luxuriant, shaggy river-beds were

---

From *The Mountains of California* by John Muir, published by The Century Co. (1894).

thus distinct from the smooth, treeless plain, they made no heavy dividing lines in general views. The whole appeared as one continuous sheet of bloom bounded only by the mountains.

When I first saw this central garden, the most extensive and regular of all the bee-pastures of the State, it seemed all one sheet of plant gold, hazy and vanishing in the distance, distinct as a new map along the foot-hills at my feet.

Descending the eastern slopes of the Coast Range through beds of gilias and lupines, and around many a breezy hillock and bush-crowned headland, I at length waded out into the midst of it. All the ground was covered, not with grass and green leaves, but with radiant corollas, about ankle-deep next the foot-hills, knee-deep or more five or six miles out. Here were bahia, madia, madaria, burrielia, chrysopsis, corethrogyne, grindelia, etc., growing in close social congregations of various shades of yellow, blending finely with the purples of clarkia, orthocarpus, and œnothera, whose delicate petals were drinking the vital sunbeams without giving back any sparkling glow.

Because so long a period of extreme drought succeeds the rainy season, most of the vegetation is composed of annuals, which spring up simultaneously, and bloom together at about the same height above the ground, the general surface being but slightly ruffled by the taller phacelias, pentstemons, and groups of *Salvia carduacea*, the king of the mints.

Sauntering in any direction, hundreds of these happy sun-plants brushed against my feet at every step, and closed over them as if I were wading in liquid gold. The air was sweet with fragrance, the larks sang their blessed songs, rising on the wing as I advanced, then sinking out of sight in the polleny sod, while myriads of wild bees stirred the lower air with their monotonous hum—monotonous, yet forever fresh and sweet as every-day sunshine. Hares and spermophiles showed themselves in considerable numbers in shallow places, and small bands of antelopes were almost constantly in sight, gazing curiously from some slight elevation, and then bounding swiftly away with unrivaled grace of motion. Yet I could discover no crushed flowers to mark their track, nor, indeed, any destructive action of any wild foot or tooth whatever.

The great yellow days circled by uncounted, while I drifted toward the north, observing the countless forms of life thronging about me, lying down almost anywhere on the approach of night. And what glorious botanical beds I had! Oftentimes on awaking I would find several new species leaning over me and looking me full in the face, so that my studies would begin before rising.

All the seasons of the great plain are warm or temperate, and bee-flowers are never wholly wanting; but the grand springtime—the annual resurrection—is governed by the rains, which usually set in about the middle of November or the beginning of December. Then the seeds, that for six months have lain on the ground dry and fresh as if they had been gathered into barns, at once unfold their treasured life. The general brown and purple of the ground, and the dead vegetation of the preceding year, give place to the green of mosses and liverworts and myriads of young leaves. Then one species after another comes into flower, gradually overspreading the green with yellow and purple, which lasts until May.

The "rainy season" is by no means a gloomy, soggy period of constant cloudiness and rain. Perhaps nowhere else in North America, perhaps in the world, are the months of December, January, February, and March so full of bland, plant-building sunshine. Referring to my notes of the winter and spring of 1868-69, every day of which I spent out of doors, on that section of the plain lying between the Tuolumne and Merced rivers, I find that the first rain of the season fell on December 18th. January had only six rainy days—that is, days on which rain fell; February three, March five, April three, and May three, completing the so-called rainy season, which was about an average one. The ordinary rain-storm of this region is seldom very cold or violent. The winds, which in settled weather come from the northwest, veer round into the opposite direction, the sky fills gradually and evenly with one general cloud, from which the rain falls steadily, often for days in succession, at a temperature of about 45° or 50°.

• • •

In 1855, two years after the time of the first arrivals from New York, a single swarm was brought over from San Jose, and let fly in the Great Central Plain. Bee-culture, however, has never gained much attention here, notwithstanding the extraordinary abundance of honey-bloom, and the high price of honey during the early years. A few hives are found here and there among settlers who chanced to have learned something about the business before coming to the State. But sheep, cattle, grain, and fruit raising are the chief industries, as they require less skill and care, while the profits thus far have been greater. In 1856 honey sold here at from one and a half to two dollars per pound. Twelve years later the price had fallen to twelve and a half cents. In 1868 I sat down to dinner with a band of ravenous sheep-shearers at a ranch on the San Joaquin, where fifteen or twenty hives were kept, and our host advised us not to spare the large pan of honey he had placed on the table, as it was the cheapest article he had to offer. In all my walks, however, I have never come upon a regular bee-ranch in the Central Valley like those so common and so skilfully managed in the southern counties of the State. The few pounds of honey and wax produced are consumed at home, and are scarcely taken into account among the coarser products of the farm. The swarms that escape from their careless owners have a weary, perplexing time of it in seeking suitable homes. Most of them make their way to the foot-hills of the mountains, or to the trees that line the banks of the rivers, where some hollow log or trunk may be found. A friend of mine, while out hunting on the San Joaquin, came upon an old coon trap, hidden among some tall grass, near the edge of the river, upon which he sat down to rest. Shortly afterward his attention was attracted to a crowd of angry bees that were flying excitedly about his head, when he discovered that he was sitting upon their hive, which was found to contain more than 200 pounds of honey. Out in the broad, swampy delta of the Sacramento and San Joaquin rivers, the little wanderers have been known to build their combs in a bunch of rushes, or stiff, wiry grass, only slightly protected from the weather, and in danger every spring of being carried away by floods. They have the advantage, however, of a vast extent of fresh pasture, accessible only to themselves.

The present condition of the Grand Central Garden is very different from that we have sketched. About twenty years ago, when the gold placers had been pretty thoroughly exhausted, the attention of fortune-seekers—not home-seekers—was, in great part, turned away from the mines to the fertile plains, and many began experiments in a kind of restless, wild agriculture. A load of lumber would be hauled to some spot on the free wilderness, where water could be easily found, and a rude box-cabin built. Then a gang-plow was procured, and a dozen mustang ponies, worth ten or fifteen dollars apiece, and with these hundreds of acres were stirred as easily as if the land had been under cultivation for years, tough, perennial roots being almost wholly absent. Thus a ranch was established, and from these bare wooden huts, as centers of desolation, the wild flora vanished in ever-widening circles. But the arch destroyers are the shepherds, with their flocks of hoofed locusts, sweeping over the ground like a fire, and trampling down every rod that escapes the plow as completely as if the whole plain were a cottage garden-plot without a fence. But notwithstanding these destroyers, a thousand swarms of bees may be pastured here for every one now gathering honey. The greater portion is still covered every season with a repressed growth of bee-flowers, for most of the species are annuals, and many of them are not relished by sheep or cattle, while the rapidity of their growth enables them to develop and mature their seeds before any foot has time to crush them. The ground is, therefore, kept sweet, and the race is perpetuated, though only as a suggestive shadow of the magnificence of its wildness.

The time will undoubtedly come when the entire area of this noble valley will be tilled like a garden, when the fertilizing waters of the mountains, now flowing to the sea, will be distributed to every acre, giving rise to prosperous towns, wealth, arts, etc. Then, I suppose, there will be few left, even among botanists, to deplore the vanished primeval flora. In the mean time, the pure waste going on—the wanton destruction of the innocents—is a sad sight to see, and the sun may well be pitied in being compelled to look on.

Gary Snyder

COVERS THE GROUND

"When California was wild, it was one sweet bee-garden..."
John Muir

Down the Great Central Valley's
blossoming almond orchard acres
lines of tree-trunks shoot a glance through
            as the rows flash by—

And the ground is covered with
cement culverts standing on end,
house-high & six feet wide
culvert after culvert far as you can see
      covered with
mobile homes, pint size portable housing, johnny-on-the-spots,
concrete freeway, overpass, underpass,
      exit floreals, entrance curtsies, railroad bridge,
long straight miles of divider oleanders;
scrappy ratty grass and thistle, tumbled barn, another age,

yards of tractors, combines lined up—
new bright-painted units down at one end,
old stuff broke and smashed down at the other,
cypress tree spires, frizzy lonely palm tree,
steep and gleaming
fertilizer tank towers fine-line catwalk in the sky—

      covered with walnut orchard acreage
irrigated, pruned and trimmed;
with palleted stacks of cement bricks
      waiting for yellow fork trucks;

---

quarter acre stacks of wornout car tires,
dust clouds blowing off the new plowed fields,
taut-strung vineyards trimmed out even on the top,

cubic blocks of fresh fruit loading boxes,
long aluminum automated chicken feeder houses,
        spring fur of green weed
        comes on last fall's hard-baked ground,
                beyond "Blue Diamond Almonds"
come the rows of red-roofed houses
& the tower that holds catfood
with a red / white checkered sign

crows whuff over almond blossoms
beehives sit tight between fruit tree ranks
eucalyptus boughs shimmer in the wind—a pale blue hip-roof
house   behind a weathered fence—
crows in the almonds
        trucks on the freeways,
                Kenworth, Peterbilt, Mack,
                rumble diesel depths,
like boulders bumping in an outwash glacial river

        drumming to a not-so ancient text

        "The Great Central Plain of California
        was one smooth bed of honey-bloom
                400 miles, your foot would press
        hundred flowers at every step
        it seemed one sheet of plant gold;

        all the ground was covered
        with radiant corollas ankle-deep:
        bahia, madia, madaria, burielia,
        chrysopsis, grindelia,
                wherever a bee might fly—"

us and our stuff just covering the ground.

Frank Norris

from THE OCTOPUS

MAGNUS CLIMBED into the buggy, helping himself with Harran's outstretched hand which he still held. The two were immensely fond of each other, proud of each other. They were constantly together, and Magnus kept no secrets from his favorite son.

"Well, boy."

"Well, Governor."

"I am very pleased you came yourself, Harran. I feared that you might be too busy and send Phelps. It was thoughtful."

Harran was about to reply, but at that moment Magnus caught sight of the three flatcars loaded with bright-painted farming machines which still remained on the siding above the station. He laid his hands on the reins, and Harran checked the team.

"Harran," observed Magnus, fixing the machinery with a judicial frown, "Harran, those look singularly like our plows. Drive over, boy."

The train had by this time gone on its way, and Harran brought the team up to the siding.

"Ah, I was right," said the Governor. "'Magnus Derrick, Los Muertos, Bonneville, from Ditson & Co., Rochester.' These are ours, boy."

Harran breathed a sigh of relief.

"At last," he answered, "and just in time, too. We'll have rain before the week is out. I think now that I am here I will telephone Phelps to send the wagon right down for these. I started bluestoning today."

Magnus nodded a grave approval.

"That was shrewd, boy. As to the rain, I think you are well informed; we will have an early season. The plows have arrived at a happy moment."

"It means money to us, Governor," remarked Harran.

But as he turned the horses to allow his father to get into the buggy again, the two were surprised to hear a thick throaty voice wishing them good-morning and turning about were aware of S. Behrman, who had come

---

From the New American Library edition (1981). First published by Doubleday (1901).

up while they were examining the plows. Harran's eyes flashed on the instant and through his nostrils he drew a sharp, quick breath, while a certain rigor of carriage stiffened the set of Magnus Derrick's shoulders and back. Magnus had not yet got into the buggy, but stood with the team between him and S. Behrman, eyeing him calmly across the horses' backs. S. Behrman came around to the other side of the buggy and faced Magnus.

He was a large, fat man with a great stomach; his cheek and the upper part of his thick neck ran together to form a great, tremulous jowl, shaven and blue-gray in color; a roll of fat, sprinkled with sparse hair, moist with perspiration, protruded over the back of his collar. He wore a heavy black moustache. On his head was a round-topped hat of stiff brown straw, highly varnished. A light brown linen vest, stamped with innumerable interlocked horseshoes, covered his protuberant stomach, upon which a heavy watch chain of hollow links rose and fell with his difficult breathing, clinking against the vest buttons of imitation mother-of-pearl.

S. Behrman was the banker of Bonneville. But besides this he was many other things. He was a real estate agent. He bought grain; he dealt in mortgages. He was one of the local political bosses, but more important than all this, he was the representative of the Pacific and Southwestern Railroad in that section of Tulare County. The railroad did little business in that part of the country that S. Behrman did not supervise, from the consignment of a shipment of wheat to the management of a damage suit, or even to the repair and maintenance of the right of way. During the time when the ranchers of the county were fighting the grain-rate case, S. Behrman had been much in evidence in and about the San Francisco courtrooms and the lobby of the legislature in Sacramento. He had returned to Bonneville only recently, a decision adverse to the ranchers being foreseen. The position he occupied on the salary list of the Pacific and Southwestern could not readily be defined, for he was neither freight agent, passenger agent, attorney, real estate broker, nor political servant, though his influence in all these offices was undoubted and enormous. But for all that, the ranchers about Bonneville knew whom to look to as a source of trouble. There was no denying the fact that for Osterman, Broderson, Annixter, and Derrick, S. Behrman was the railroad.

"Mr. Derrick, good morning," he cried as he came up. "Good morning, Harran. Glad to see you back, Mr. Derrick." He held out a thick hand.

Magnus, head and shoulders above the other, tall, thin, erect, looked down upon S. Behrman, inclining his head, failing to see his extended hand.

"Good morning, sir," he observed, and waited for S. Behrman's further speech.

"Well, Mr. Derrick," continued S. Behrman, wiping the back of his neck with his handkerchief, "I saw in the city papers yesterday that our case had gone against you."

"I guess it wasn't any great news to *you*," commented Harran, his face scarlet. "I guess you knew which way Ulsteen was going to jump after your very first interview with him. You don't like to be surprised in this sort of thing, S. Behrman."

"Now, you know better than that, Harran," remonstrated S. Behrman blandly. "I know what you mean to imply, but I ain't going to let it make me get mad. I wanted to say to your Governor—I wanted to say to you, Mr. Derrick—as one man to another—letting alone for the minute that we were on opposite sides of the case—that I'm sorry you didn't win. Your side made a good fight, but it was in a mistaken cause. That's the whole trouble. Why, you could have figured out before you ever went into the case that such rates are confiscation of property. You must allow us—must allow the railroad—a fair interest on the investment. You don't want us to go into the receiver's hands, do you now, Mr. Derrick?"

"The Board of Railroad Commissioners was bought," remarked Magnus sharply, a keen, brisk flash glinting in his eye.

"It was part of the game," put in Harran, "for the Railroad Commission to cut rates to a ridiculous figure, far below a *reasonable* figure, just so that it *would* be confiscation. Whether Ulsteen is a tool of yours or not, he had to put the rates back to what they were originally."

"If you enforced those rates, Mr. Harran," returned S. Behrman calmly, "we wouldn't be able to earn sufficient money to meet operating expenses or fixed charges, to say nothing of a surplus left over to pay dividends—"

"Tell me when the P. and S. W. ever paid dividends."

"The lowest rates," continued S. Behrman, "that the legislature can establish must be such as will secure us a fair interest on our investment."

"Well, what's your standard? Come, let's hear it. Who is to say what's a fair rate? The railroad has its own notions of fairness sometimes."

"The laws of the state," returned S. Behrman, "fix the rate of interest at seven per cent. That's a good enough standard for us. There is no reason, Mr. Harran, why a dollar invested in a railroad should not earn as much as a dollar represented by a promissory note—seven per cent. By applying your schedule of rates we would not earn a cent; we would be bankrupt."

"Interest on your investment!" cried Harran, furious. "It's fine to talk about fair interest. *I* know and *you* know that the total earnings of the P. and S. W.—their main, branch, and leased lines for last year—were between nineteen and twenty millions of dollars. Do you mean to say that twenty million dollars is seven percent of the original cost of the road?"

S. Behrman spread out his hands, smiling.

"That was the gross, not the net, figure—and how can you tell what was the original cost of the road?"

"Ah, that's just it," shouted Harran, emphasizing each word with a blow of his fist upon his knee, his eyes sparkling, "you take cursed good care that we don't know anything about the original cost of the road. But we know you are bonded for treble your value; and we know this: that the road *could* have been built for fifty-four thousand dollars per mile and that you *say* it cost you eighty-seven thousand. It makes a difference, S. Behrman, on which of these two figures you are basing your seven per cent."

"That all may show obstinacy, Harran," observed S. Behrman vaguely, "but it don't show common sense."

"We are threshing out old straw, I believe, gentlemen," remarked Magnus. "The question was thoroughly sifted in the courts."

"Quite right," assented S. Behrman. "The best way is that the railroad and the farmer understand each other and get along peaceably. We are both dependent on each other. Your plows, I believe, Mr. Derrick." S. Behrman nodded toward the flatcars.

"They are consigned to me," admitted Magnus.

"It looks a trifle like rain," observed S. Behrman, easing his neck and jowl in his limp collar. "I suppose you will want to begin plowing next week."

"Possibly," said Magnus.

"I'll see that your plows are hurried through for you, then, Mr. Derrick. We will route them by fast freight for you and it won't cost you anything extra."

"What do you mean?" demanded Harran. "The plows are here. We have nothing more to do with the railroad. I am going to have my wagons down here this afternoon."

"I am sorry," answered S. Behrman, "but the cars are going north, not, as you thought, coming *from* the north. They have not been to San Francisco yet."

Magnus made a slight movement of the head as one who remembers a fact hitherto forgotten. But Harran was as yet unenlightened.

"To San Francisco!" he answered. "We want them here—what are you talking about?"

"Well you know, of course, the regulations," answered S. Behrman. "Freight of this kind coming from the eastern points into the state must go first to one of our common points and be reshipped from there."

Harran did remember now, but never before had the matter so struck home. He leaned back in his seat in dumb amazement for the instant. Even Magnus had turned a little pale. Then, abruptly, Harran broke out, violent and raging.

"What next? My God, why don't you break into our houses at night? Why don't you steal the watch out of my pocket, steal the horses out of the harness, hold us up with a shotgun; yes, 'Stand and deliver; your money or your life.' Here we bring our plows from the East over your lines, but you're not content with your long-haul rate between eastern points and Bonneville. You want to get us under your ruinous short-haul rate between Bonneville and San Francisco, *and return*. Think of it! Here's a load of stuff for Bonneville that can't stop at Bonneville, where it is consigned, but has got to go up to San Francisco first *by way of* Bonneville, at forty cents per ton, and then be reshipped from San Francisco back to Bonneville again at *fifty-one* cents per ton, the short-haul rate. And we have to pay it all or go without. Here are the plows right here, in sight of the land they have got to be used on, the season just ready for them, and we can't touch them. Oh," he exclaimed in deep disgust, "isn't it a pretty mess! Isn't it a farce! The whole dirty business!"

S. Behrman listened to him unmoved, his little eyes blinking under his fat forehead, the gold chain of hollow links clicking against the pearl buttons of his waistcoat as he breathed.

"It don't do any good to let loose like that, Harran," he said at length. "I am willing to do what I can for you. I'll hurry the plows through, but I can't change the freight regulation of the road."

"What's your blackmail for this?" vociferated Harran. "How much do you want to let us go? How much have we got to pay you to be *allowed* to use our own plows—what's your figure? Come, spit it out."

"I see you are trying to make me angry, Harran," returned S. Behrman, "but you won't succeed. Better give up trying, my boy. As I said, the best way is to have the railroad and the farmer get along amicably. It is the only way we can do business. Well, s'long, Governor, I must trot along. S'long, Harran." He took himself off.

Mary Austin

THE WALKING WOMAN

THE FIRST TIME of my hearing of her was at Temblor. We had come all one day between blunt, whitish bluffs rising from mirage water, with a thick, pale wake of dust billowing from the wheels, all the dead wall of the foothills sliding and shimmering with heat, to learn that the Walking Woman had passed us somewhere in the dizzying dimness, going down to the Tulares on her own feet. We heard of her again in the Carrisal, and again at Adobe Station, where she had passed a week before the shearing, and at last I had a glimpse of her at the Eighteen-Mile House as I went hurriedly northward on the Mojave stage; and afterward, sheepherders at whose camps she slept, and cowboys at rodeos, told me as much of her way of life as they could understand. Like enough they told her as much of mine. That was very little. She was the Walking Woman, and no one knew her name, but because she was a sort of whom men speak respectfully, they called her to her face Mrs. Walker, and she answered to it if she was so inclined. She came and went about our western world on no discoverable errand, and whether she had some place of refuge where she lay by in the interim, or whether between her seldom, unaccountable appearances in our quarter she went on steadily walking, was never learned. She came and went, oftenest in a kind of muse of travel which the untrammelled space begets, or at rare intervals flooding wondrously with talk, never of herself, but of things she had known and seen. She must have seen some rare happenings, too—by report. She was at Maverick the time of the Big Snow, and at Tres Piños when they brought home the body of Morena; and if anybody could have told whether De Borba killed Mariana for spite or defence, it would have been she, only she could not be found when most wanted. She was at Tunawai at the time of the cloud-burst, and if she had cared for it could

From *Lost Borders* by Mary Austin (1909). Reprinted in *Western Trails: A Collection of Short Stories by Mary Austin,* selected and edited by Melody Graulich, copyright 1987 by Melody Graulich. Reprinted by permission of the University of Nevada Press.

have known most desirable things of the ways of trail-making, burrow-habiting small things.

All of which should have made her worth meeting, though it was not, in fact, for such things I was wishful to meet her; and as it turned out, it was not of these things we talked when at last we came together. For one thing, she was a woman, not old, who had gone about alone in a country where the number of women is as one in fifteen. She had eaten and slept at the herder's camps, and laid by for days at one-man stations whose masters had no other touch of human kind than the passing of chance prospectors, or the halting of the tri-weekly stage. She had been set on her way by teamsters who lifted her out of white, hot desertness and put her down at the crossing of unnamed ways, days distant from anywhere. And through all this she passed unarmed and unoffended. I had the best testimony to this, the witness of the men themselves. I think they talked of it because they were so much surprised at it. It was not, on the whole, what they expected of themselves.

Well I understand that nature which wastes its borders with too eager burning, beyond which rim of desolation it flares forever quick and white, and have had some inkling of the isolating calm of a desire too high to stoop to satisfaction. But you could not think of these things pertaining to the Walking Woman; and if there were ever any truth in the exemption from offence residing in a frame of behavior called ladylike, it should have been inoperative here. What this really means is that you get no affront so long as your behavior in the estimate of the particular audience invites none. In the estimate of the immediate audience—conduct which affords protection in Mayfair gets you no consideration in Maverick. And by no canon could it be considered ladylike to go about on your own feet, with a blanket and a black bag and almost no money in your purse, in and about the haunts of rude and solitary men.

There were other things that pointed the wish for a personal encounter with the Walking Woman. One of them was the contradiction of reports of her—as to whether she was comely, for example. Report said yes, and again, plain to the point of deformity. She had a twist to her face, some said; a hitch to one shoulder; they averred she limped as she walked. But by the distance she covered she should have been straight and young. As to sanity, equal incertitude. On the mere evidence of her way of life she was cracked; not quite broken, but unserviceable. Yet in her talk there was both wisdom and information, and the word she brought about trails and water-holes was as reliable as an Indian's.

By her own account she had begun by walking off an illness. There had been an invalid to be taken care of for years, leaving her at last broken in body, and with no recourse but her own feet to carry her out of that predicament. It seemed there had been, besides the death of her invalid, some other worrying affairs, upon which, and the nature of her illness, she was never quite clear, so that it might very well have been an unsoundness of mind which drove her to the open, sobered and healed at last by the large soundness of nature. It must have been about that time that she lost her name. I am convinced that she never told it because she did not know it herself. She was the Walking Woman, and the country people called her Mrs. Walker. At the time I knew her, though she wore short hair and a man's boots, and had a fine down over all her face from exposure to the weather, she was perfectly sweet and sane.

I had met her occasionally at ranch-houses and road-stations, and had got as much acquaintance as the place allowed; but for the things I wished to know there wanted a time of leisure and isolation. And when the occasion came we talked altogether of other things.

It was at Warm Spring in the Little Antelope I came upon her in the heart of a clear forenoon. The spring lies off a mile from the main trail, and has the only trees about it known in that country. First you come upon a pool of waste full of weeds of a poisonous dark green, every reed ringed about the water-level with a muddy white incrustation. Then the three oaks appear staggering on the slope, and the spring sobs and blubbers below them in ashy-colored mud. All the hills of that country have the down plunge toward the desert and back abruptly toward the Sierra. The grass is thick and brittle and bleached straw-color toward the end of the season. As I rode up the swale of the spring I saw the Walking Woman sitting where the grass was deepest, with her black bag and blanket, which she carried on a stick, beside her. It was one of those days when the genius of talk flows as smoothly as the rivers of mirage through the blue hot desert morning.

You are not to suppose that in my report of a Borderer I give you the words only, but the full meaning of the speech. Very often the words are merely the punctuation of thought; rather, the crests of the long waves of intercommunicative silences. Yet the speech of the Walking Woman was fuller than most.

The best of our talk that day began in some dropped word of hers from which I inferred that she had had a child. I was surprised at that, and then wondered why I should have been surprised, for it is the most natural of all experiences to have children. I said something of that purport, and also that it was one of the perquisites of living I should be least willing to do without. And that led to the Walking Woman saying that there were three things which if you had known you could cut out all the rest, and they were good any way you got them, but best if, as in her case, they were related to and grew each one out of the others. It was while she talked that I decided that she really did have a twist to her face, a sort of natural warp or skew into which it fell when it was worn merely as a countenance, but which disappeared the moment it became the vehicle of thought or feeling.

The first of the experiences the Walking Woman had found most worth while had come to her in a sand-storm on the south slope of Tehachapi in a dateless spring. I judged it should have been about the time she began to find herself, after the period of worry and loss in which her wandering began. She had come, in a day pricked full of intimations of a storm, to the camp of Filon Geraud, whose companion shepherd had gone a three days' *pasear* to Mojave for supplies. Geraud was of great hardihood, red-blooded, of a full laughing eye, and an indubitable spark for women. It was the season of the year when there is a soft bloom on the days, but the nights are cowering cold and the lambs tender, not yet flockwise. At such times a sand-storm works incalculable disaster. The lift of the wind is so great that the whole surface of the ground appears to travel upon it slant-wise, thinning out miles high in air. In the intolerable smother the lambs are lost from the ewes; neither dogs nor man make headway against it.

The morning through a horizon of yellow smudge, and by mid-forenoon the flock broke.

"There were but the two of us to deal with the trouble," said the Walking Woman. "Until that time I had not known how strong I was, nor how good it is to run when running is worth while. The flock travelled down the wind, the sand bit our faces; we called, and after a time heard the words broken and beaten small by the wind. But after a little we had not to call. All the time of our running in the yellow dusk of day and the black dark of night, I knew where Filon was. A flock-length away, I knew him. Feel? What should I feel? I knew. I ran with the flock and turned it this way and that as Filon would have.

"Such was the force of the wind that when we came together we held by one another and talked a little between partings. We snatched and ate what we could as we ran. All that day and night until the next afternoon the camp kit was not out of the cayaques. But we held the flock. We herded them under a butte when the wind fell off a little, and the lambs sucked; when the storm rose they broke, but we kept upon their track and brought them together again. At night the wind quieted, and we slept by turns; at least Filon slept. I lay on the ground when my turn was and beat with the storm. I was no more tired than the earth was. The sand filled in the creases of the blanket, and where I turned, dripped back upon the ground. But we saved the sheep. Some ewes there were that would not give down their milk because of the worry of the storm, and the lambs died. But we kept the flock together. And I was not tired."

The Walking Woman stretched out her arms and clasped herself, rocking in them as if she would have hugged the recollection to her breast.

"For you see," said she, "I worked with a man, without excusing, without any burden on me of looking or seeming. Not fiddling or fumbling as women work, and hoping it will all turn out for the best. It was not for Filon to ask, Can you, or Will you. He said, Do, and I did. And my work was good. We held the flock. And that," said the Walking Woman, the twists coming in her face again, "is one of the things that make you able to do without the others."

"Yes," I said; and then, "What others?"

"Oh," she said, as if it pricked her, "the looking and the seeming."

And I had not thought until that time that one who had the courage to be the Walking Woman would have cared! We sat and looked at the pattern of the thick crushed grass on the slope, wavering in the fierce noon like the waterings in the coat of a tranquil beast; the ache of a world-old bitterness sobbed and whispered in the spring. At last—

"It is by the looking and the seeming," said I, "that the opportunity finds you out."

"Filon found out," said the Walking Woman. She smiled; and went on from that to tell me how, when the wind went down about four o'clock and left the afternoon clear and tender, the flock began to feed, and they had out the kit from the cayaques, and cooked a meal. When it was over, and Filon had his pipe between his teeth, he came over from his side of the fire, of his own notion, and stretched himself on the ground beside her. Of his own notion. There was that in the way she said it that made it seem as if nothing of the sort had happened before to the Walking Woman, and for a moment I thought she was about to tell me one of the things I wished to know; but she went on to say what Filon had said to her of her work with the flock. Obvious, kindly things, such as any man in sheer decency would have said, so that there must have something more gone with the words to make them so treasured of the Walking Woman.

"We were very comfortable," said she, "and not so tired as we expected to be. Filon leaned up on his elbow. I had not noticed until then how broad he was in the shoulders, and how strong in the arms. And we had saved the flock together. We felt that. There was something that said together, in the slope of his shoulders toward me. It was around his mouth and on the cheek high up under the shine of his eyes. And under the shine the look—the look that said, 'We are of one sort and one mind'—his eyes that were the color of the flat water in the tulares—do you know the look?"

"I know it."

"The wind was stopped and all the earth smelled of dust, and Filon understood very well that what I had done with him I could not have done so well with another. And the look—the look in the eyes—"

"Ah-ah—!"

I have always said, I will say again, I do not know why at this point the Walking Woman touched me. If it were merely a response to my unconscious throb of sympathy, or the unpremeditated way of her heart to declare that this, after all, was the best of all indispensable experiences; or if in some flash of forward vision, encompassing the unimpassioned years, the stir, the movement of tenderness were for *me*—but no; as often as I have thought of it, I have thought of a different reason, but no conclusive one, why the Walking Woman should have put out her hand and laid it on my arm.

"To work together, to love together," said the Walking Woman, withdrawing her hand again; "there you have two of the things; the other you know."

"The mouth at the breast," said I.

"The lips and the hands," said the Walking Woman. "The little, pushing hands and the small cry." There ensued a pause of fullest understanding, while the land before us swam in the noon, and a dove in the oaks behind the spring began to call. A little red fox came out of the hills and lapped delicately at the pool.

"I stayed with Filon until the fall," said she. "All that summer in the Sierra, until it was time to turn south on the trail. It was a good time, and longer than he could be expected to have loved one like me. And besides, I was no longer able to keep the trail. My baby was born in October."

Whatever more there was to say to this, the Walking Woman's hand said it, straying with remembering gesture to her breast. There are so many ways of loving and working, but only one way of the first-born. She added after an interval, that she did not know if she would have given up her walking to keep at home and tend him, or whether the thought of her son's small feet running beside her in the trails would have driven her to the open again. The baby had not stayed long enough for that. "And whenever the wind blows in the night," said the Walking Woman, "I wake and wonder if he is well covered."

She took up her black bag and her blanket; there was the ranch-house of

Dos Palos to be made before night, and she went as outliers do, without a hope expressed of another meeting and no word of good-bye. She was the Walking Woman. That was it. She had walked off all sense of society-made values, and, knowing the best when the best came to her, was able to take it. Work—as I believed; love—as the Walking Woman had proved it; a child—as you subscribe to it. But look you: it was the naked thing the Walking Woman grasped, not dressed and tricked out, for instance, by prejudices in favor of certain occupations; and love, man love, taken as it came, not picked over and rejected if it carried no obligation of permanency; and a child; *any* way you get it, a child is good to have, say nature and the Walking Woman; to have it and not to wait upon a proper concurrence of so many decorations that the event may not come at all.

At least one of us is wrong. To work and to love and to bear children. That sounds easy enough. But the way we live establishes so many things of much more importance.

Far down the dim, hot valley I could see the Walking Woman with her blanket and black bag over her shoulder. She had a queer, sidelong gait, as if in fact she had a twist all through her.

Recollecting suddenly that people called her lame, I ran down to the open space below the spring where she had passed. There in the bare, hot sand the track of her two feet bore evenly and white.

Ernesto Galarza

from BARRIO BOY

ON SIXTH NEAR K there was the Lyric Theater with a sign that we easily translated into Lírico. It was next to a handsome red stone house with high turrets, like a castle. Navigating by these key points and following the rows of towering elms along L Street, one by one we found the post office on 7th and K; the cathedral, four blocks farther east; and the state capitol with its golden dome.

It wasn't long before we ventured on walks around Capitol Park which reminded me of the charm and the serenity of the Alameda in Tepic. In some fashion Mrs. Dodson had got over to us that the capitol was the house of the government. To us it became El Capitolio or, as more formally, the Palacio de Gobierno. Through the park we walked into the building itself, staring spellbound at the marble statue of Queen Isabel and Christopher Columbus. It was awesome, standing in the presence of that gigantic admiral, the one who had discovered America and Mexico and Jalcocotán, as Doña Henriqueta assured me.

After we had thoroughly learned our way around in the daytime we found signs that did not fail us at night. From the window of the projection room of the Lyric Theater a brilliant purple light shone after dark. A snake of electric lights kept whipping round and round a sign over the Albert Elkus store. K Street on both sides was a double row of bright show windows that led up to the Land Hotel and back to Breuner's, thence down one block to the lumber yard, the grocery store, and our house. We had no fear of getting lost.

These were the boundaries of the lower part of town, for that was what everyone called the section of the city between Fifth Street and the river and from the railway yards to the Y-street levee. Nobody ever mentioned an upper part of town; at least, no one could see the difference because the

---

 This excerpt describes Sacramento in the early part of the 20th century.

K Street looking west near 10th, Sacramento, ca. 1920s. Courtesy of David Joslyn Collection, City of Sacramento History and Science Division, Sacramento Archives and Museum Collection Center.

whole city was built on level land. We were not lower topographically, but in other ways that distinguished between Them, the uppers, and Us, the lowers. Lower Sacramento was the quarter that people who made money moved away from. Those of us who lived in it stayed there because our problem was to make a living and not to make money. A long while back, Mr. Howard, the business agent of the union told me, there had been stores and shops, fancy residences, and smart hotels in this neighborhood. The crippled old gentleman who lived in the next room down the hall from us, explained to me that our house, like the others in the neighborhood, had been the home of rich people who had stables in the back yards, with back entrances by way of the alleys. Mr. Hansen, the Dutch carpenter, had helped build such residences. When the owners moved uptown, the back yards had been fenced off and subdivided, and small rental cottages had been built in the alleys in place of the stables. Handsome private homes were turned into flop-houses for men who stayed one night, hotels for working people, and rooming houses, like ours.

Among the saloons, pool halls, lunch counters, pawnshops, and poker parlors was skid row, where drunk men with black eyes and unshaven faces lay down in the alleys to sleep.

The lower quarter was not exclusively a Mexican *barrio* but a mix of many nationalities. Between L and N Streets two blocks from us, the Japanese had taken over. Their homes were in the alleys behind shops, which they advertised with signs covered with black scribbles. The women walked on the street in kimonos, wooden sandals, and white stockings, carrying neat black bundles on their backs and wearing their hair in puffs with long ivory needles stuck through them. When they met they bowed, walked a couple of steps, and turned and bowed again, repeating this several times. They carried babies on their backs, not in their arms, never laughed or went into the saloons. On Sundays the men sat in front of their shops, dressed in gowns, like priests.

Chinatown was on the other side of K Street, toward the Southern Pacific shops. Our houses were old, but those in which the Chinese kept stores, laundries, and restaurants were older still. In black jackets and skullcaps the older merchants smoked long pipes with a tiny brass cup on the end. In their dusty store windows there was always the same assortment of tea packages, rice bowls, saucers, and pots decorated with blue temples and dragons.

In the hotels and rooming houses scattered about the *barrio* the Filipino farm workers, riverboat stewards, and houseboys made their homes. Like the Mexicans they had their own poolhalls, which they called clubs. Hindus from the rice and fruit country north of the city stayed in the rooming houses when they were in town, keeping to themselves. The Portuguese and Italian families gathered in their own neighborhoods along Fourth and Fifth Streets southward toward the Y-street levee. The Poles, Yugo-Slavs, and Koreans, too few to take over any particular part of it, were scattered throughout the *barrio.* Black men drifted in and out of town, working the waterfront. It was a kaleidoscope of colors and languages and customs that surprised and absorbed me at every turn.

Although we, the foreigners, made up the majority of the population of that quarter of Sacramento, the Americans had by no means given it up to us. Not all of them had moved above Fifth Street as the *barrio* became more crowded. The bartenders, the rent collectors, the insurance salesmen, the mates on the riverboats, the landladies, and most importantly, the police—these were all gringos. So were the craftsmen, like the barbers and printers, who did not move their shops uptown as the city grew. The teachers of our one public school were all Americans. On skid row we rarely saw a drunk wino who was not a gringo. The operators of the pawnshops and secondhand stores were white and mostly Jewish.

For the Mexicans the *barrio* was a colony of refugees. We came to know families from Chihuahua, Sonora, Jalisco, and Durango. Some had come to the United States even before the revolution, living in Texas before migrating to California. Like ourselves, our Mexican neighbors had come this far moving step by step, working and waiting, as if they were feeling their way up a ladder. They talked of relatives who had been left behind in Mexico, or in some far-off city like Los Angeles or San Diego. From whatever place they had come, and however short or long the time they had lived in the United States, together they formed the *colonia mexicana*. In the years between our arrival and the First World War, the *colonia* grew and spilled out from the lower part of town. Some families moved into the alley shacks east of the Southern Pacific tracks, close to the canneries and warehouses and across the river among the orchards and rice mills.

The *colonia* was like a sponge that was beginning to leak along the edges, squeezed between the levee, the railroad tracks, and the river front. But it wasn't squeezed dry, because it kept filling with newcomers who found families who took in boarders: basements, alleys, shanties, run-down rooming houses and flop joints where they could live.

Crowded as it was, the *colonia* found a place for these *chicanos*, the name by which we called an unskilled worker born in Mexico and just arrived in the United States. The *chicanos* were fond of identifying themselves by saying they had just arrived from *el macizo*, by which they meant the solid Mexican homeland, the good native earth. Although they spoke of *el macizo* like homesick persons, they didn't go back. They remained, as they said of themselves, *pura raza. So* it happened that José and Gustavo would bring home for a meal and for conversation workingmen who were *chicanos* fresh from *el macizo* and like ourselves, *pura raza*. Like us, they had come straight to the *barrio* where they could order a meal, buy a pair of overalls, and look for work in Spanish. They brought us vague news about the revolution, in which many of them had fought as *villistas, huertistas, maderistas*, or *zapatistas*. As an old *maderista*, I imagined our *chicano* guests as battle-tested revolutionaries, like myself.

As poor refugees, their first concern was to find a place to sleep, then to eat and find work. In the *barrio* they were most likely to find all three, for not knowing English, they needed something that was even more urgent than a room, a meal, or a job, and that was information in a language they could understand. This information had to be picked up in bits and pieces from families like ours, from the conversation groups in the poolrooms and the saloons.

Beds and meals, if the newcomers had no money at all, were provided—in one way or another—on trust, until the new *chicano* found a job. On trust and not on credit, for trust was something between people who had plenty of nothing, and credit was between people who had something of plenty. It was not charity or social welfare but something my mother called *asistencia*, a helping given and received on trust, to be repaid because those who had given it were themselves in need of what they had given. *Chicanos* who had found work on farms or in railroad camps came back to pay us a few dollars for *asistencia* we had provided weeks or months before.

Because the *barrio* was a grapevine of job information, the transient *chicanos* were able to find work and repay their obligations. The password of the barrio was *trabajo* and the community was divided in two—the many who were looking for it and the few who had it to offer. Pickers, foremen, contractors, drivers, field hands, pick and shovel men on the railroad and in construction came back to the *barrio* when work was slack, to tell one another of the places they had been, the kind of *patrón* they had, the wages paid, the food, the living quarters, and other important details. Along Second Street, labor recruiters hung blackboards on their shop fronts, scrawling in chalk offers of work. The grapevine was a mesh of rumors and gossip, and men often walked long distances or paid bus fares or a contractor's fee only to find that the work was over or all the jobs were filled. Even the chalked signs could not always be relied on. Yet the search for *trabajo,* or the *chanza,* as we also called it, went on because it had to.

We in the *barrio* considered that there were two kinds of *trabajo.* There were the seasonal jobs, some of them a hundred miles or more from Sacramento. And there were the closer *chanzas* to which you could walk or ride on a bicycle. These were the best ones, in the railway shops, the canneries, the waterfront warehouses, the lumber yards, the produce markets, the brick kilns, and the rice mills. To be able to move from the seasonal jobs to the close-in work was a step up the ladder. Men who had made it passed the word along to their relatives or their friends when there was a *chanza* of this kind.

It was all done by word of mouth, this delicate wiring of the grapevine. The exchange points of the network were the places where men gathered in small groups, apparently to loaf and chat to no purpose. One of these points was our kitchen, where my uncles and their friends sat and talked of *el macizo* and of the revolution but above all of the *chanzas* they had heard of.

There was not only the everlasting talk about *trabajo,* but also the never-ending action of the *barrio* itself. If work was action the *barrio* was where the action was. Every morning a parade of men in oily work clothes and carrying lunch buckets went up Fourth Street toward the railroad shops, and every evening they walked back, grimy and silent. Horse drawn drays with low platforms rumbled up and down our street carrying the goods the city traded in, from kegs of beer to sacks of grain. Within a few blocks of our house there were smithies, hand laundries, a macaroni factory, and all manner of places where wagons and buggies were repaired, horses stabled, bicycles fixed, chickens dressed, clothes washed and ironed, furniture repaired, candy mixed, tents sewed, wine grapes pressed, bottles washed, lumber sawed, suits fitted and tailored, watches and clocks taken apart and put together again, vegetables sorted, railroad cars unloaded, boxcars iced, barges freighted, ice cream cones molded, soda pop bottled, fish scaled, salami stuffed, corn ground for masa, and bread ovened. To those who knew where these were located in the alleys, as I did, the whole *barrio* was an open workshop. The people who worked there came to know you, let you look in at the door, made jokes, and occasionally gave you an odd job.

This was the business district of the *barrio.* Around it and through it moved a constant traffic of drays, carts, bicycles, pushcarts, trucks, and high-wheeled automobiles with black canvas tops and honking horns. On the tailgates of drays and wagons, I nipped rides when I was going home with a gunnysack full of empty beer bottles or my gleanings around the packing sheds.

Once we had work, the next most important thing was to find a place to live we could afford. Ours was a neighborhood of leftover houses. The cheapest rents were in the back quarters of the rooming houses, the basements, and the run-down clapboard rentals in the alleys. Clammy and dank as they were, they were nevertheless one level up from the barns and tents where many of our *chicano* friends lived, or the shanties and lean-to's of the migrants who squatted in the "jungles" along the levees of the Sacramento and the American rivers.

Barrio people, when they first came to town, had no furniture of their own. They rented it with their quarters or bought a piece at a time from the secondhand stores, the *segundas,* where we traded. We cut out the ends of tin cans to make collars and plates for the pipes and floor moldings where the rats had gnawed holes. Stoops and porches that sagged we propped with bricks and fat stones. To plug the drafts around the windows in winter, we cut strips of corrugated cardboard and wedged them into the frames. With squares of cheesecloth neatly cut and sewed to screen doors holes were covered and rents in the wire mesh mended. Such repairs, which landlords never paid any attention to, were made *por mientras,* for the time being or temporarily. It would have been a word equally suitable for the house itself, or for the *barrio.* We lived in run-down places furnished with seconds in a hand-me-down neighborhood all of which were *por mientras.*

We found the Americans as strange in their customs as they probably found us. Immediately we discovered that there were no *mercados* and that when shopping you did not put the groceries in a *chiquihuite.* Instead everything was in cans or in cardboard boxes or each item was put in a brown paper bag. There were neighborhood grocery stores at the corners and some big ones uptown, but no *mercado.* The grocers did not give children a *pilón,* they did not stand at the door and coax you to come in and buy, as they did in Mazatlán. The fruits and vegetables were displayed on counters instead of being piled up on the floor. The stores smelled of fly spray and oiled floors, not of fresh pineapple and limes.

Neither was there a plaza, only parks which had no bandstands, no concerts every Thursday, no Judases exploding on Holy Week, and no promenades of boys going one way and girls the other. There were no parks in the *barrio;* and the ones uptown were cold and rainy in winter, and in summer there was no place to sit except on the grass. When there were celebrations nobody set off rockets in the parks, much less on the street in front of your house to announce to the neighborhood that a wedding or a baptism was taking place. Sacramento did not have a *mercado* and a plaza with the cathedral to one side and the Palacio de Gobierno on another to make it obvious that there and nowhere else was the center of the town.

It was just as puzzling that the Americans did not live in *vecindades,* like our block on Leandro Valle. Even in the alleys, where people knew one another better, the houses were fenced apart, without central courts to wash clothes, talk and play with the other children. Like the city, the Sacramento *barrio* did not have a place which was the middle of things for everyone.

In more personal ways we had to get used to the Americans. They did not listen if you did not speak loudly, as they always did. In the Mexican style, people would know that you were enjoying their jokes tremendously if you merely smiled and shook a little, as if you were trying to swallow your mirth. In the American style there was little difference between a laugh and a roar, and until you got used to them you could hardly tell whether the boisterous Americans were roaring mad or roaring happy.

It was Doña Henriqueta more than Gustavo or José who talked of these oddities and classified them as agreeable or deplorable. It was she also who pointed out the pleasant surprises of the American way. When a box of rolled oats with a picture of red carnations on the side was emptied, there was a plate or a bowl or a cup with blue designs. We ate the strange stuff regularly for breakfast and we soon had a set of the beautiful dishes. Rice and beans we bought in cotton bags of colored prints. The bags were unsewed, washed, ironed, and made into gaily designed towels, napkins, and handkerchiefs. The American stores also gave small green stamps which were pasted in a book to exchange for prizes. We didn't have to run to the corner with the garbage; a collector came for it.

With remarkable fairness and never-ending wonder we kept adding to our list the pleasant and the repulsive in the ways of the Americans. It was my second acculturation.

The older people of the *barrio,* except in those things which they had to do like the Americans because they had no choice, remained Mexican. Their language at home was Spanish. They were continuously taking up collections to pay somebody's funeral expenses or to help someone who had had a serious accident. Cards were sent to you to attend a burial where you would throw a handful of dirt on top of the coffin and listen to tearful speeches at the graveside. At every baptism a new *compadre* and a new *comadre* joined the family circle. New Year greeting cards were exchanged, showing angels and cherubs in bright colors sprinkled with grains of mica so that they glistened like gold dust. At the family parties the huge pot of steaming tamales was still the center of attention, the *atole* served on the side with chunks of brown sugar for sucking and crunching. If the party lasted long enough, someone produced a guitar, the men took over and the singing of *corridos* began.

In the *barrio* there were no individuals who had official titles or who were otherwise recognized by everybody as important people. The reason must have been that there was no place in the public business of the city of Sacramento for the Mexican immigrants. We only rented a corner of the city and as long as we paid the rent on time everything else was decided at City Hall or the County Court House, where Mexicans went only when they were in trouble. Nobody from the *barrio* ever ran for mayor or city councilman. For us the most important public officials were the policemen who walked their beats, stopped fights, and hauled drunks to jail in a paddy wagon we called *La Julia.*

The one institution we had that gave the *colonia* some kind of image was the *Comisión Honorífica,* a committee picked by the Mexican Consul in San Francisco to organize the celebration of the *Cinco de Mayo* and the Sixteenth of September, the anniversaries of the battle of Puebla and the beginning of our War of Independence. These were the two events which stirred everyone in the *barrio,* for what we were celebrating was not only the heroes of Mexico but also the feeling that we were still Mexicans ourselves. On these occasions there was a dance preceded by speeches and a concert. For both the *cinco* and the sixteenth queens were elected to preside over the ceremonies.

Between celebrations neither the politicians uptown nor the *Comisión*

Honorífica attended to the daily needs of the *barrio.* This was done by volunteers—the ones who knew enough English to interpret in court, on a visit to the doctor, a call at the county hospital, and who could help make out a postal money order. By the time I had finished the third grade at the Lincoln School I was one of these volunteers. My services were not professional but they were free, except for the IOU's I accumulated from families who always thanked me with "God will pay you for it."

My clients were not *pochos,* Mexicans who had grown up in California, probably had even been born in the United States. They had learned to speak English of sorts and could still speak Spanish, also of sorts. They knew much more about the Americans than we did, and much less about us. The *chicanos* and the *pochos* had certain feelings about one another. Concerning the *pochos,* the *chicanos* suspected that they considered themselves too good for the *barrio* but were not, for some reason, good enough for the Americans. Toward the *chicanos,* the *pochos* acted superior, amused at our confusions but not especially interested in explaining them to us. In our family when I forgot my manners, my mother would ask me if I was turning *pochito.*

Turning *pocho* was a half-step toward turning American. And America was all around us, in and out of the *barrio.* Abruptly we had to forget the ways of shopping in a *mercado* and learn those of shopping in a corner grocery or in a department store. The Americans paid no attention to the Sixteenth of September, but they made a great commotion about the Fourth of July. In Mazatlán Don Salvador had told us, saluting and marching as he talked to our class, that the *Cinco de Mayo* was the most glorious date in human history. The Americans had not even heard about it.

William Saroyan

THE SUMMER OF THE BEAUTIFUL WHITE HORSE

ONE DAY BACK THERE in the good old days when I was nine and the world was full of every imaginable kind of magnificence, and life was still a delightful and mysterious dream, my cousin Mourad, who was considered crazy by everybody who knew him except me, came to my house at four in the morning and woke me up by tapping on the window of my room.

Aram, he said.

I jumped out of bed and looked out the window.

I couldn't believe what I saw.

It wasn't morning yet, but it was summer and with daybreak not many minutes around the corner of the world it was light enough for me to know I wasn't dreaming.

My cousin Mourad was sitting on a beautiful white horse.

I stuck my head out of the window and rubbed my eyes.

Yes, he said in Armenian. It's a horse. You're not dreaming. Make it quick if you want to ride.

I knew my cousin Mourad enjoyed being alive more than anybody else who had ever fallen into the world by mistake, but this was more than even I could believe.

In the first place, my earliest memories had been memories of horses and my first longings had been longings to ride.

This was the wonderful part.

In the second place, we were poor.

This was the part that wouldn't permit me to believe what I saw.

We were poor. We had no money. Our whole tribe was poverty-stricken. Every branch of the Garoghlanian family was living in the most amazing and comical poverty in the world. Nobody could understand where we ever got money enough to keep us with food in our bellies, not even the

---

old men of the family. Most important of all, though, we were famous for our honesty. We had been famous for our honesty for something like eleven centuries, even when we had been the wealthiest family in what we liked to think was the world. We were proud first, honest next, and after that we believed in right and wrong. None of us would take advantage of anybody in the world, let alone steal.

Consequently, even though I could see the horse, so magnificent; even though I could *smell* it, so lovely; even though I could *hear* it breathing, so exciting; I couldn't *believe* the horse had anything to do with my cousin Mourad or with me or with any of the other members of our family, asleep or awake, because I *knew* my cousin Mourad couldn't have *bought* the horse, and if he couldn't have bought it he must have *stolen* it, and I refused to believe he had stolen it.

No member of the Garoghlanian family could be a thief.

I stared first at my cousin and then at the horse. There was a pious stillness and humor in each of them which on the one hand delighted me and on the other frightened me.

Mourad, I said, where did you steal this horse?

Leap out of the window, he said, if you want to ride.

It was true, then. He *had* stolen the horse. There was no question about it. He had come to invite me to ride or not, as I chose.

Well, it seemed to me stealing a horse for a ride was not the same thing as stealing something else, such as money. For all I knew, maybe it wasn't stealing at all. If you were crazy about horses the way my cousin Mourad and I were, it wasn't stealing. It wouldn't become stealing until we offered to sell the horse, which of course I knew we would never do.

Let me put on some clothes, I said.

All right, he said, but hurry.

I leaped into my clothes.

I jumped down to the yard from the window and leaped up onto the horse behind my cousin Mourad.

That year we lived at the edge of town, on Walnut Avenue. Behind our house was the country: vineyards, orchards, irrigation ditches, and country roads. In less than three minutes we were on Olive Avenue, and then the horse began to trot. The air was new and lovely to breathe. The feel of the horse running was wonderful. My cousin Mourad who was considered one of the craziest members of our family began to sing. I mean, he began to roar.

Every family has a crazy streak in it somewhere, and my cousin Mourad was considered the natural descendant of the crazy streak in our tribe.

Before him was our uncle Khosrove, an enormous man with a powerful head of black hair and the largest mustache in the San Joaquin Valley, a man so furious in temper, so irritable, so impatient that he stopped anyone from talking by roaring, *It is no harm; pay no attention to it.*

That was all, no matter what anybody happened to be talking about. Once it was his own son Arak running eight blocks to the barber shop where his father was having his mustache trimmed to tell him their house was on fire. This man Khosrove sat up in the chair and roared, It is no harm; pay no attention to it. The barber said, But the boy says your house is on fire. So Khosrove roared, Enough, it is no harm, I say.

My cousin Mourad was considered the natural descendant of this man, although Mourad's father was Zorab, who was practical and nothing else. That's how it was in our tribe. A man could be the father of his son's flesh, but that did not mean that he was also the father of his spirit. The distribution of the various kinds of spirit of our tribe had been from the beginning capricious and vagrant.

We rode and my cousin Mourad sang. For all anybody knew we were still in the old country where, at least according to some of our neighbors, we belonged. We let the horse run as long as it felt like running.

At last my cousin Mourad said, Get down. I want to ride alone.

Will you let me ride alone? I said.

That is up to the horse, my cousin said. Get down.

The *horse* will let me ride, I said.

We shall see, he said. Don't forget that I have a way with a horse.

Well, I said, any way you have with a horse, I have also.

For the sake of your safety, he said, let us hope so. Get down.

All right, I said, but remember you've got to let me try to ride alone.

I got down and my cousin Mourad kicked his heels into the horse and shouted, *Vazire,* run. The horse stood on its hind legs, snorted, and burst into a fury of speed that was the loveliest thing I had ever seen. My cousin Mourad raced the horse across a field of dry grass to an irrigation ditch, crossed the ditch on the horse, and five minutes later returned, dripping wet.

The sun was coming up.

Now it's my turn to ride, I said.

My cousin Mourad got off the horse.

Ride, he said.

I leaped to the back of the horse and for a moment knew the awfulest fear imaginable. The horse did not move.

Kick into his muscles, my cousin Mourad said. What are you waiting for? We've got to take him back before everybody in the world is up and about.

I kicked into the muscles of the horse. Once again it reared and snorted. Then it began to run. I didn't know what to do. Instead of running across the field to the irrigation ditch the horse ran down the road to the vineyard of Dikran Halabian where it began to leap over vines. The horse leaped over seven vines before I fell. Then it continued running.

My cousin Mourad came running down the road.

I'm not worried about you, he shouted. We've got to get that horse. You go this way and I'll go this way. If you come upon him, be kindly. I'll be near.

I continued down the road and my cousin Mourad went across the field toward the irrigation ditch.

It took him half an hour to find the horse and bring him back.

All right, he said, jump on. The whole world is awake now.

What will we do? I said.

Well, he said, we'll either take him back or hide him until tomorrow morning.

He didn't sound worried and I knew he'd hide him and not take him back. Not for a while, at any rate.

Where will we hide him? I said.

I know a place, he said.

How long ago did you steal this horse? I said.

It suddenly dawned on me that he had been taking these early morning rides for some time and had come for me this morning only because he knew how much I longed to ride.

Who said anything about stealing a horse? he said.

Anyhow, I said, how long ago did you begin riding every morning?

Not until this morning, he said.

Are you telling the truth? I said.

Of course not, he said, but if we are found out, that's what you're to say. I don't want both of us to be liars. All you know is that we started riding this morning.

All right, I said.

He walked the horse quietly to the barn of a deserted vineyard which at one time had been the pride of a farmer named Fetvajian. There were some oats and dry alfalfa in the barn.

We began walking home.

It wasn't easy, he said, to get the horse to behave so nicely. At first it wanted to run wild, but, as I've told you, I have a way with a horse. I can get it to want to do anything *I* want it to do. Horses understand me.

How do you do it? I said.

I have an understanding with a horse, he said.

Yes, but what sort of an understanding? I said.

A simple and honest one, he said.

Well, I said, I wish I knew how to reach an understanding like that with a horse.

You're still a small boy, he said. When you get to be thirteen you'll know how to do it.

I went home and ate a hearty breakfast.

That afternoon my uncle Khosrove came to our house for coffee and cigarettes. He sat in the parlor, sipping and smoking and remembering the old country. Then another visitor arrived, a farmer named John Byro, an Assyrian who, out of loneliness, had learned to speak Armenian. My mother brought the lonely visitor coffee and tobacco and he rolled a cigarette and sipped and smoked, and then at last, sighing sadly, he said, My white horse which was stolen last month is still gone. I cannot understand it.

My uncle Khosrove became very irritated and shouted, It's no harm. What is the loss of a horse? Haven't we all lost the homeland? What is this crying over a horse?

That may be all right for you, a city dweller, to say, John Byro said, but what of my surrey? What good is a surrey without a horse?

Pay no attention to it, my uncle Khosrove roared.

I walked ten miles to get here, John Byro said.

You have legs, my uncle Khosrove shouted.

My left leg pains me, the farmer said.

Pay no attention to it, my uncle Khosrove roared.

That horse cost me sixty dollars, the farmer said.

I spit on money, my uncle Khosrove said.

He got up and stalked out of the house, slamming the screen door.

My mother explained.

He has a gentle heart, she said. It is simply that he is homesick and such a large man.

The farmer went away and I ran over to my cousin Mourad's house.

He was sitting under a peach tree, trying to repair the hurt wing of a young robin which could not fly. He was talking to the bird.

What is it? he said.

The farmer, John Byro, I said. He visited our house. He wants his horse. You've had it a month. I want you to promise not to take it back until I learn to ride.

It will take you *a year* to learn to ride, my cousin Mourad said.

We could keep the horse a year, I said.

My cousin Mourad leaped to his feet.

What? he roared. Are you inviting a member of the Garoghlanian family to steal? The horse must go back to its true owner.

When? I said.

In six months at the latest, he said.

He threw the bird into the air. The bird tried hard, almost fell twice, but at last flew away, high and straight.

Early every morning for two weeks my cousin Mourad and I took the horse out of the barn of the deserted vineyard where we were hiding it and rode it, and every morning the horse, when it was my turn to ride alone, leaped over grape vines and small trees and threw me and ran away. Nevertheless, I hoped in time to learn to ride the way my cousin Mourad rode.

One morning on the way to Fetvajian's deserted vineyard we ran into the farmer John Byro who was on his way to town.

Let me do the talking, my cousin Mourad said. I have a way with farmers.

Good morning, John Byro, my cousin Mourad said to the farmer.

The farmer studied the horse eagerly.

Good morning, sons of my friends, he said. What is the name of your horse?

My Heart, my cousin Mourad said in Armenian.

A lovely name, John Byro said, for a lovely horse. I could swear it is the horse that was stolen from me many weeks ago. May I look into its mouth?

Of course, Mourad said.

The farmer looked into the mouth of the horse.

Tooth for tooth, he said. I would swear it *is* my horse if I didn't know your parents. The fame of your family for honesty is well known to me. Yet the horse is the twin of my horse. A suspicious man would believe his eyes instead of his heart. Good day, my young friends.

Good day, John Byro, my cousin Mourad said.

Early the following morning we took the horse to John Byro's vineyard and put it in the barn. The dogs followed us around without making a sound.

The dogs, I whispered to my cousin Mourad. I thought they would bark.

They would at somebody else, he said. I have a way with dogs.

My cousin Mourad put his arms around the horse, pressed his nose into the horse's nose, patted it, and then we went away.

That afternoon John Byro came to our house in his surrey and showed my mother the horse that had been stolen and returned.

I do not know what to think, he said. The horse is stronger than ever. Better-tempered, too. I thank God.

My uncle Khosrove, who was in the parlor, became irritated and shouted, Quiet, man, quiet. Your horse has been returned. Pay no attention to it.

Yoimut

I AM THE LAST

I AM THE LAST full-blood Chunut left. My children are part Spanish. I am the only one who knows the whole Chunut or Wowol language. When I am gone no one will have it. I have to be the last one. All my life I want back our good old home on Tulare Lake. But I guess I can never have it. I am a very old Chunut now and I guess I can never see the old days again.

Now my daughter and her Mexican husband work in the cotton fields around Tulare and Waukena. Cotton, cotton, cotton, that is all that is left. Chunuts cannot live on cotton. They cannot sing their old songs and tell their old stories where there is nothing but cotton. My children feel foolish when I sing my songs. But I sing anyway:

Toke-uh lih-nuh Wa-tin-hin nah yo
Hiyo-umne ahe oonook miuh-wah.

That is all.

From *The Way We Lived: California Indian Reminiscences, Stories, and Songs* copyright 1981, 1993 by Malcolm Margolin, published by Heyday Books. First appeared in *Handbook of the Yokuts Indians*, revised and enlarged edition copyright 1977 by Bear State Books. This lament was written down in 1933.

John Steinbeck

from THE HARVEST GYPSIES

THE SQUATTERS' CAMPS are located all over California. Let us see what a typical one is like. It is located on the banks of a river, near an irrigation ditch or on a side road where a spring of water is available. From a distance it looks like a city dump, and well it may, for the city dumps are the sources for the material of which it is built. You can see a litter of dirty rags and scrap iron, of houses built of weeds, of flattened cans or of paper. It is only on close approach that it can be seen that these are homes.

Here is a house built by a family who have tried to maintain a neatness. The house is about 10 feet by 10 feet, and it is built completely of corrugated paper. The roof is peaked, the walls are tacked to a wooden frame. The dirt floor is swept clean, and along the irrigation ditch or in the muddy river the wife of the family scrubs clothes without soap and tries to rinse out the mud in muddy water. The spirit of this family is not quite broken, for the children, three of them, still have clothes, and the family possesses three old quilts and a soggy, lumpy mattress. But the money so needed for food cannot be used for soap nor for clothes.

With the first rain the carefully built house will slop down into a brown, pulpy mush; in a few months the clothes will fray off the children's bodies while the lack of nourishing food will subject the whole family to pneumonia when the first cold comes.

Five years ago this family had fifty acres of land and a thousand dollars in the bank. The wife belonged to a sewing circle and the man was a member of the grange. They raised chickens, pigs, pigeons and vegetables and fruit for their own use; and their land produced the tall corn of the middle west. Now they have nothing.

If the husband hits every harvest without delay and works the maximum time, he may make four hundred dollars this year. But if anything happens, if his old car breaks down, if he is late and misses a harvest or two, he will have to feed his whole family on as little as one hundred and fifty.

But there is still pride in this family. Wherever they stop they try to put the children in school. It may be that the children will be in a school for as much as a month before they are moved to another locality.

Here, in the faces of the husband and his wife, you begin to see an expression you will notice on every face; not worry, but absolute terror of the starvation that crowds in against the borders of the camp. This man has tried to make a toilet by digging a hole in the ground near his paper house and surrounding it with an old piece of burlap. But he will only do things like that this year. He is a newcomer and his spirit and decency and his sense of his own dignity have not been quite wiped out. Next year he will be like his next door neighbor.

This is a family of six; a man, his wife and four children. They live in a tent the color of the ground. Rot has set in on the canvas so that the flaps and the sides hang in tatters and are held together with bits of rusty baling wire. There is one bed in the family and that is a big tick lying on the ground inside the tent.

They have one quilt and a piece of canvas for bedding. The sleeping arrangement is clever. Mother and father lie down together and two children lie between them. Then, heading the other way, the other two children lie, the littler ones. If the mother and father sleep with their legs spread wide, there is room for the legs of the children.

There is more filth here. The tent is full of flies clinging to the apple box that is the dinner table, buzzing about the foul clothes of the children, particularly the baby, who has not been bathed nor cleaned for several days. This family has been on the road longer than the builder of the paper house. There is no toilet here, but there is a clump of willows nearby where human feces lie exposed to the flies—the same flies that are in the tent.

Two weeks ago there was another child, a four year old boy. For a few weeks they had noticed that he was kind of lackadaisical, that his eyes had been feverish. They had given him the best place in the bed, between father and mother. But one night he went into convulsions and died, and the next morning the coroner's wagon took him away. It was one step down.

They know pretty well that it was a diet of fresh fruit, beans and little else that caused his death. He had no milk for months. With this death there came a change of mind in his family. The father and mother now feel that paralyzed dullness with which the mind protects itself against too much sorrow and too much pain.

And this father will not be able to make a maximum of four hundred dollars a year any more because he is no longer alert; he isn't quick at piece-work, and he is not able to fight clear of the dullness that has settled on him. His spirit is losing caste rapidly.

The dullness shows in the faces of this family, and in addition there is a sullenness that makes them taciturn. Sometimes they still start the older children off to school, but the ragged little things will not go; they hide in ditches or wander off by themselves until it is time to go back to the tent, because they are scorned in the school.

The better-dressed children shout and jeer, the teachers are quite often impatient with these additions to their duties, and the parents of the "nice" children do not want to have disease carriers in the schools.

The father of this family once had a little grocery store and his family lived in back of it so that even the children could wait on the counter. When the drought set in there was no trade for the store any more.

This is the middle class of the squatters' camp. In a few months this family will slip down to the lower class. Dignity is all gone, and spirit has turned to sullen anger before it dies.

The next door neighbor family of man, wife and three children of from three to nine years of age, have built a house by driving willow branches into the ground and wattling weeds, tin, old paper and strips of carpet against them. A few branches are placed over the top to keep out the noonday sun. It would not turn water at all. There is no bed. Somewhere the family has found a big piece of old carpet. It is on the ground. To go to bed the members of the family lie on the ground and fold the carpet up over them.

The three year old child has a gunny sack tied about his middle for clothing. He has the swollen belly caused by malnutrition.

He sits on the ground in the sun in front of the house, and the little black fruit flies buzz in circles and land on his closed eyes and crawl up his nose until he weakly brushes them away.

They try to get at the mucous in the eye-corners. This child seems to have the reactions of a baby much younger. The first year he had a little milk, but he has had none since.

He will die in a very short time. The older children may survive. Four nights ago the mother had a baby in the tent, on the dirty carpet. It was born dead, which was just as well because she could not have fed it at the breast; her own diet will not produce milk.

After it was born and she had seen that it was dead, the mother rolled over and lay still for two days. She is up today, tottering around. The last baby, born less than a year ago, lived a week. This woman's eyes have the glazed, far-away look of a sleep walker's eyes. She does not wash clothes any more. The drive that makes for cleanliness has been drained out of her and she hasn't the energy. The husband was a share-cropper once, but he couldn't make it go. Now he has lost even the desire to talk. He will not look directly at you for that requires will, and will needs strength. He is a bad field worker for the same reason. It takes him a long time to make up his mind, so he is always late in moving and late in arriving in the fields. His top wage, when he can find work now, which isn't often, is a dollar a day.

The children do not even go to the willow clump any more. They squat where they are and kick a little dirt. The father is vaguely aware that there is a culture of hookworm in the mud along the river bank. He knows the children will get it on their bare feet. But he hasn't the will nor the energy to resist. Too many things have happened to him. This is the lower class of the camp.

This is what the man in the tent will be in six months; what the man in the paper house with its peaked roof will be in a year, after his house has washed down and his children have sickened or died, after the loss of dignity and spirit have cut him down to a kind of subhumanity.

Helpful strangers are not well-received in this camp. The local sheriff makes a raid now and then for a wanted man, and if there is labor trouble the vigilantes may burn the poor houses. Social workers, survey workers have taken case histories. They are filed and open for inspection. These families have been questioned over and over about their origins, number of children living and dead. The information is taken down and filed. That is that. It has been done so often and so little has come of it.

And there is another way for them to get attention. Let an epidemic break out, say typhoid or scarlet fever, and the country doctor will come to the camp and hurry the infected cases to the pest house. But malnutrition is not infectious, nor is dysentery, which is almost the rule among the children.

The county hospital has no room for measles, mumps, whooping cough; and yet these are often deadly to hunger-weakened children. And although we hear much about the free clinics for the poor, these people do not know how to get the aid and they do not get it. Also, since most of their dealings with authority are painful to them, they prefer not to take the chance.

This is the squatters' camp. Some are a little better, some much worse. I have described three typical families. In some of the camps there are as many as three hundred families like these. Some are so far from water that it must be bought at five cents a bucket.

And if these men steal, if there is developing among them a suspicion and hatred of well-dressed, satisfied people, the reason is not to be sought in their origin nor in any tendency to weakness in their character.

Wilma Elizabeth McDaniel

PICKING GRAPES 1937

Magic seventeen
and new in California

working in bursting
sweet vineyards

hot sand on soul
one strap held by a
safety pin

a girl could be whatever
she desired

the first breath of
Eve in Paradise

the last gasp of Jean Harlow
in Hollywood

Gerald Haslam

THE DOLL

MRS. HOLLIS SAW THEM walking up her driveway toward the front porch, two ragged boys holding hands, the larger pulling the smaller. Okies, she thought, from the camp in Riverview; I wonder what they want, walking so boldly in a respectable neighborhood. Something really ought to be done about them. A person didn't know what to expect next. Just yesterday a skinny, one-legged man had knocked at her door and tried to sell her a can of salve; before that it had been another scrawny man trying to sell garden seeds. Lord, they were spreading into nice neighborhoods like vermin. They'd be wanting to move in next.

Before the boys reached her front steps Mrs. Hollis walked heavily into the doorway and talked through the screen: "Yes?" she said with no warmth.

The day was oppressively hot, even in the shade of the porch, but the Okies stopped in full sunlight on the front lawn. The larger boy, lean with dirty-looking yellow hair that contrasted with his deeply sun-browned skin, answered in a flat nasal voice: "Lookin' fer work, lady. Kin we mow yer lawn er anythang?"

Although her lawn was indeed shaggy, she didn't want this drippy-nosed Okie near her any longer than necessary. "Well I haven't got anything for you boys to do, and I doubt if anyone in this neighborhood does," she replied curtly, feeling an immediate rush of satisfaction at having put them in their place. Turning, she reentered the relative coolness of her house.

"Lady," she heard the boy whine.

"What is it?" She returned to the doorway. They don't even know when they're not wanted, she thought; they're like animals.

"Could we have us a drank from yonder hose?" The boy's voice was inflectionless, almost exhausted; he pointed toward a rubber hose curled like a sunning serpent near a metal spigot.

Well that's typical, thought Mrs. Hollis, next thing you know they'll be

From *Okies* copyright 1973 by Gerald Haslam. This story of Oklahoma migrants is reprinted by permission of the author.

asking for a meal. "Be quick about it. I've got my bridge club coming, you know!" She turned away from the door again, but abruptly changed her mind, deciding to watch them: they'll steal anything if a person isn't careful.

The yellow-haired boy ambled toward the hose, saying over his shoulder to the smaller boy, "Come own Henery." Mrs. Hollis glanced at the smaller boy, gazed away, then shot her eyes back at him in a huge, swallowing look. He wasn't a boy at all! He was a little man! His flat, pasty face was covered by a filigree of fine wrinkles, and he grinned vacantly at her, his teeth pointed like a shark's, his eyes like two tiny ball bearings set in uncooked dough. She felt something like a soft flower opening in her middle, warming her stomach, filling her with an imprecise sense of foreboding and fascination. The little lined face continued to grin at her, its features immobile. Mrs. Hollis fidgeted, shifting her weight from one thick leg to the other. "Uh, hello," she finally mumbled, her voice less harsh and certain than before.

The face, moon-like with metallic eyes, its contour broken by a thin nose as sharp and curved as an owl's beak, remained unchanged, sharp teeth grinning, reminding her vaguely of a wetting doll she'd owned as a little girl.

Before the spigot holding the hose, the yellow-haired boy knelt and slurped heavily the gushing water. He stood after a moment and called to the little man: "Come own Henery. Take you a drank." Henry obeyed, shuffling with the undisciplined jerks of a string-tangled marionette.

"What's wrong with him?" Mrs. Hollis asked the boy.

"He's a idyet," the boy answered absently, "born in sin."

"Born in *what?*"

"In sin. His ma warn't mur'd."

My Lord, thought Mrs. Hollis, what's wrong with those people, letting a boy know about such things. Yet her fascination compelled her to look once more at the little man who was drinking now, then ask the boy: "Is he your brother?"

"No'm. He's m' uncle."

The idiot stood up and said to the boy, "Na nwa goo."

"Shore is," the boy replied.

"What did he say?" asked Mrs. Hollis.

"Said the water's good."

After turning the water off and grasping one of Henry's hands, the boy said to his uncle, "What do ya say?"

"Nan nyou," Mrs. Hollis heard the idiot mumble; she noticed how small his hands were as he wiped his glistening mouth, and how perfect, like an exquisite doll's. Her own plump hands were swollen and splotched with glandular middle age, and her attempts to fade them with creme had only bleached them so that she appeared to be wearing pink gloves, with splotches.

The boy feigned a departure, hesitated, then turned to Mrs. Hollis. "You ain't got some little thang we could do fer a samich do you lady. We're mighty hongry."

Still transfixed by the idiot's guileless grin, Mrs. Hollis vacillated. "We'll do most anythang," said the boy hopefully. Glancing at him, Mrs. Hollis noticed he too had an owl's beak nose and that his gray eyes were rimmed with red and streaked. His face was, save for its grainy tan, colorless and void. Well, she thought, I can't turn away the hungry, but I've got to hurry because my bridge club's coming. "You wait on the porch," she ordered with a hint of her previous curtness, "I'll bring you something."

As she walked back through the house past the card table laden with cookies and little liverwurst (she called it "paté") sandwiches without crusts, she heard the boy's whining voice talking to his uncle. She supposed the boy to be about twelve-years-old, large for his age though thin, with a viperene face too wide at the forehead, too narrow at the chin; his eyes were not aligned. He wore faded blue jeans that exposed bony ankles, and a torn tee-shirt that was on backwards so that it hugged his throat in front (Mrs. Hollis's own plump throat had grown a little uncomfortable looking at it) and drooped in back. His neck was dirty. The little man, whose hair was mousy gray, wore short bib overalls with no shirt. Neither wore shoes.

She brought each of them a peanut butter sandwich wrapped in wax paper. "You eat these after you've gone, hear," she told them, "I've got my bridge club today."

"You ain't told us what work to do, lady."

"Never mind that. Take your sandwiches. I'm expecting my bridge club any time."

"We can come back after if you want, lady," said the boy, making no effort to take the sandwiches.

"Oh, all right. You come back tomorrow and I'll have some job for you."

"Thank you, lady," he said, taking the food. "What do you say, Henery?"

"Nan nyou."

Later, at the bridge table, Frances Bryant had brought it up. "There was an Okie came by my house," she reported between nibbles at a frosted cookie, nodding slowly and raising one eyebrow, "selling some kind of remedy. He was the awfulest thing I ever saw, all skinny and with one leg cut off."

"Well," replied Mrs. Tatum, the newest club member, knowledgeably.

"And," Frances continued, "have you ever noticed their skin, how it's so dark and blotchy. They've all got colored blood, you know."

"My Ev says they're part Gypsie, kid," interjected Mary Cannon.

Frances nodded. "Could be. Lord knows they act like Gypsies, what with never settling down or taking steady jobs. All the same, Gypsie or not, they're still part colored."

"They're certainly not much better than niggers," Hope Cuen added. "They're certainly not."

They sat with corpulent dignity around a small card table, their soft bodies scented against sweat with strong colognes, their plump little fingers poised away from their hands—like fat white worms tempting unwary fish—while the ladies nibbled cookies.

"Well, Winnie, You're certainly quiet today," observed Frances. "You and Claude have a fight?" The other ladies giggled.

"Oh, no," Mrs. Hollis hid her welling desire to put Frances in her place, "I've just got a lot on my mind."

"You're always day-dreaming Winnie," Frances said with a laugh, exchanging a knowing look with Mrs. Tatum. "You'd better be careful or you'll have those Okies moving in on you. You have to be alert in this world."

"Amen," added Hope Cuen. "Amen."

"Not much chance of that—them moving in, I mean," Mrs. Hollis sputtered impotently. She was always helpless before Frances and wasn't even sure she liked her; she could only think of things to say to Frances after their one-sided verbal exchanges were long over.

Frances continued, ignoring Mrs. Hollis: "Well you girls heard about what happened at the First Baptist Church, I guess." She looked about her; they hadn't heard. "Well, last week a whole family of Okies came to the services just as big as you please."

"Oh no, kid."

"What next? What next?"

"What did the congregation do?" asked Mrs. Hollis.

"What could they do, being Christians and all? They left them alone, of course. But right after services Reverend Willis and some men went over to them and, nice as you please, told them they weren't wanted."

"Of course."

"That was a nice way to handle things."

"Well, that's just what I thought when I heard," hissed Frances, "but listen to this: that Okie man got mad and started a fight!"

"Oh no."

"The Sheriff had to come and break it up and take him to jail."

"It's too good for them," Mrs. Tatum observed with finality.

"That's just what I think," Frances said. "Now listen to *this*"—she paused as though to savour their attention—you know that crazy Durant woman, the one who drinks and whose daughter is so loose, well she got angry at Reverend Willis and caused a scene right in front of everybody."

"What in the world did she do, kid?" drooled Mary Cannon.

"Well, I don't know. *Everything.* Screamed. Shouted. Quoted the Bible incorrectly as usual. She even claimed Jesus said everyone was supposed to love their neighbors, so that means everyone's supposed to love Okies! You know how crazy she is."

"What a shame," said Mrs. Hollis.

"They certainly aren't my neighbors," Mrs. Tatum gasped, "not those Okies."

"Well I'll tell you one thing"—Frances nodded with certainty—"Jesus didn't mean *them.*" She glared triumphantly at the other women for a moment. "He didn't mean them," she said once more.

Later, just as the ladies were readying to leave, Mrs. Hollis thought to ask Frances about her brother who hadn't been right. What had ever happened to him?

Frances looked nervous. "Why in the world do you ask?"

"Oh, I just got to thinking that the last time I saw him was when you had bridge club two years ago and your parents were visiting. Remember? And they'd brought your brother with them. He was such a quiet little man."

"Well, when Mother and Father died so close to one another, Brother didn't have anywhere to stay, so Charles and I arranged for him to live up north." She hurried with her final cookie, but Mrs. Hollis wasn't letting go so easily.

"Where up north, dear?"

"Well, at Sonoma. It's lovely there, not hot," she added glancing round the table.

"At the *state* hospital?" Mary Cannon asked. "Oh, kid."

"Well. Well, he's happier with his own kind."

Mrs. Hollis almost inquired if his sister wasn't his own kind, but she wasn't certain Frances would take too kindly to that in her present agitated state, so she let it pass and the afternoon ended for her on an unusual note of triumph.

Mrs. Hollis hummed in her too-warm kitchen; only if she could convince Claude to accept a salad for dinner in the summer, only if she could. Still, no mere physical heat could disturb the deep pleasure that softened and sweetened her: Frances put in her place for once. And almost an accident at that, for she hadn't dreamed Frances had done—or even *could* do—such a thing. Imagine, her own brother sent to a lunatic asylum. Her own brother. And all Mrs. Hollis had intended was to ask about the little man; those Okies had reminded her of him.

Dinner was unusually gay and afterwards she and Claude walked slowly through the Oildale heat to the ice cream shop on Woodrow Street and each slurped a double cone. Then she sat on the front porch as usual while he watered the front lawn in thick, dimming twilight, Claude shouting pleasantries across the street into other yards where people whose houses they'd never entered shot sleek, silvery sprays over yellow grass. A clatter of children swept up and down the block on or after a roller-skate scooter. When Claude finished, they went into the house and Mrs. Hollis impulsively suggested a game of Monopoly. It was an unusually happy evening for them, and later, in spite of the heat, they made it even happier.

Mrs. Hollis always slept late—Claude had for years arisen early and taken breakfast at a diner—so she hadn't finished her morning coffee when the door bell rang. Stealing a clandestine glance at the front porch through a barely-drawn side curtain, she was stunned to see a flat, grey eye pressed against the glass opposite her's, staring back; it was the idiot. Mrs. Hollis smiled her embarrassment, then joggled to the door and opened it, keeping the screen locked.

"Mornin' lady," said the straw-haired boy.

"Oh. Why, good morning."

They stood facing one another through the fine mesh for several moments, then the boy spoke again. "We come about that job of work."

"Oh. Why, yes. Just let me finish my coffee and I'll tell you what to do." Out the corner of her eye she noticed that the little man still stared into the side window.

"Awright lady."

She turned back toward the kitchen only to hear: "Lady?"

"Yes."

"You ain't got no extra coffee do you. Me and Henery ain't had us no breakfast to eat." The boy's voice was flat, with no hint of either demand or plea or even expectation. An automatic "No!" flashed into Mrs. Hollis, and she turned quickly, her jaws tightening; she ought to put that boy in his place. She hesitated, looking from the boy to the man who still stood next to the window, then back to the boy again. "Well, I'll get you something. You can pick up the dog dirt from the lawn—there's a shovel in the garage—and then start mowing. Make sure you rake up the clippings; I don't want them turning brown on the grass. The lawn mower's in the garage too."

Mrs. Hollis put a fresh pot of coffee on to perk and made sandwiches; something they could eat outside. And she watched them working on the lawn through the window over the kitchen sink. The large thin boy pushing the lawn mower, the delicate little man raking industriously, if sloppily, behind him. The boy stopped every now and then and said something to the man, who nodded vigorously, then went on raking.

Why they really weren't half-bad she mused, not when you knew how to handle them. They worked. And there was that idiot man living right with them, she guessed, right with his own family, working, being taken care of, really happy. Wait until Frances and the bridge club hear about this! It's a shame, though, them coming to California where they're not wanted. They should have stayed in Oklahoma or wherever it was they really belonged. But still, they weren't half bad if you knew how to treat them.

The boy wolfed down his sandwich in an instant and drank the hot, unblown coffee with grimacing gulps. The man ate slowly, nibbling and chewing with rapid grins for a long time before swallowing, his tiny hands squeezing the bread a bit too tight so that catsup and mayonnaise dripped—a pink blot—unnoticed by him onto his overall bib. The boy followed Mrs. Hollis's eyes to the gooey stain and told his uncle to wipe himself, handing him a kerchief that was clearly a ragged square torn from an old sheet.

The boy drank a second cup of coffee, then they returned to work, mowing and raking steadily over the nearly half-acre of lawn, resting occasionally in the hot morning and drinking from Mrs. Hollis's garden hose. She watched them from the cool cavern of her house, sitting in front of the blower much of the time. They're hard working people, she thought, and not half bad. At least they take care of their own; not like some I know.

By the time they finished the lawn and trimmed the edges it was nearly lunch, so she brought them sandwiches and lemonade. She tried to prompt the boy into telling her about his family; much to her dissatisfaction, he said little, answering her with apparent candor but no enthusiasm. She began to feel there was something *basic* about these people, though she didn't know exactly what. The doll like little man grinned and ate, ate and grinned. A slightly triumphant, exhilarated mood crept into Mrs. Hollis as she watched them eat. She had tamed them; she had them doing her bidding and acting properly. Mrs. Hollis gave each of them one of her good chocolate-covered cherries after lunch, the ones wrapped in red foil.

She set them to digging up dandelions and crab grass in the afternoon, telling them it was perfectly all right if they did the patches in shade under trees first in the glaring heat, then she retired to the couch under the blower for her afternoon nap.

She had just fallen asleep—surely she hadn't slept long—when a persistent tapping at the front door roused her roughly from the freedom of empty slumber; she drowsed awake but unaware for several moments before her mind focused on the sound. Arising heavily, Mrs. Hollis's legs felt thick, and her dress pressed to her with perspiration; her face remained creased where the pillow had been creased.

It was the boy. "What do you want?" she groggily asked.

"Kin we use yer bathroom, lady? We gotta piss."

His final word hit her with a dull jolt of revulsion. "You *what?*"

"We gotta piss," he repeated.

The word even sounded filthy. Mrs. Hollis felt her neck and face swelling with anger. "There are limits, young man," she sputtered, "there are limits. You can not use my bathroom. You can finish your work and leave!" She slammed the door, her stomach churning, near tears, and hurried into the kitchen to fix a cup of coffee. "Piss" echoed within her, and deep inside her mouth, just where her throat started, she felt the warm discomfort that preceded vomiting, but she caught herself, grasping the sink with both hands until nausea passed; she finally fully awakened. That nasty word. She went into her bathroom, her pride, with its pink rug and curtains, and washed her puffed face, brushed her hair, then bathed herself lightly with cooling toilet water.

Back in the kitchen her coffee was ready. She looked out the window and saw the Okies under her pepper tree talking. They're not children she reassured herself, they can hold it. I did the right thing. You don't just invite people you don't know to use your bathroom; after all, there are limits. Besides, there was that nasty word. They can hold it.

In spite of herself, Mrs. Hollis mellowed as she sipped her coffee and watched them work. They do have to go just like anyone else, she thought, but where? Not in her bathroom surely. And not outside where the neighbors could see something so nasty. No solution came to her, so she again decided they could hold it until they got home, or at least to the Golden Bear gas station up on Chester Avenue. She hated that word, but was a little sorry she'd told them to finish work and leave. Okies probably couldn't help it, talking like that. Why her own brother-in-law sometimes slipped and said it. With all the little jobs around the house that wanted doing, she decided to forgive them and tell them to come back tomorrow to fix the roof.

After taking a dollar from her purse, she walked onto the porch and called them from their tasks; they could finish weeding tomorrow in the morning when it wasn't so hot. They walked to the porch, the little man bent forward a bit as though he'd stiffened-up while weeding, and for an instant she wondered how old he really was. The Okies stood below her on the walk side-by-side.

"You can quit for today. Here's a dollar for your work"—the boy quickly snatched it—"and I'll want you back tomorrow morning to finish the weeds and do another little job I have for you." She heard the boy cackle with a high, hollow sound. "What's so funny young man?"

The boy who had been looking at his uncle, turned smirking toward her and said: "Looky. He pissed hisself."

A growing wetness darked the little man's faded overalls, spreading shapelessly on his lap and his legs, urine running in thin yellow rivulets down his dusty ankles and feet, cutting faintly muddy paths, and puddling on Mrs. Hollis's cement walkway. His little doll face grinned stiffly.

A wave of nausea instantly overcame Mrs. Hollis; she slammed through the front door and lurched toward the bathroom, vomiting first in the hall, then in the bathroom on her pink rug. Again and again her body purged itself while she, like a suffering spectator, knelt before the toilet and her uncluttered brain vomited too, drawing deep from a reservoir within her: "Jesus didn't mean them," raced through her mind, "He didn't. He wouldn't. He didn't."

William Saroyan

FRESNO

A MAN COULD WALK four or five miles in any direction from the heart of our city and see our streets dwindle to land and weeds. In many places the land would be vineyard and orchard land, but in most places it would be desert land; the weeds would be the strong dry weeds of desert. In this land there would be the living things that had had their being in the quietness of deserts for centuries. There would be snakes and horned-toads, prairie-dogs and jack-rabbits. In the sky over this land would be buzzards and hawks, and the hot sun. And everywhere in the desert would be the marks of wagons that had made lonely roads.

Two miles from the heart of our city a man could come to the desert and feel the loneliness of a desolate area, a place lost in the earth, far from the solace of human thought. Standing at the edge of our city, a man could feel that we had made this place of streets and dwellings in the stillness and loneliness of the desert, and that we had done a brave thing. We had come to this dry area that was without history, and we had paused in it and built our houses and we were slowly creating the legend of our life. We were digging for water and we were leading streams through the dry land. We were planting and ploughing and standing in the midst of the garden we were making.

Our trees were not yet tall enough to make much shade, and we had planted a number of kinds of trees we ought not to have planted because they were of weak stuff and would never live a century, but we had made a pretty good beginning. Our cemeteries were few and the graves in them were few. We had buried no great men because we hadn't had time to produce any great men. We had been too busy trying to get water into the desert. The shadow of no great mind was over our city. But we had a play-

---

First appeared in *The Hairenik Daily* (1934) and later reprinted in *The Saroyan Special* (1948), published by Harcourt Brace & Co. Reprinted by permission of the William Saroyan Foundation.

ground called Cosmos Playground. We had public schools named after Emerson and Hawthorne and Lowell and Longfellow. Two great railways had their lines running through our city and trains were always coming to us from the great cities of America and somehow we could not feel that we were wholly lost. We had two newspapers and a Civic Auditorium and a public library one-third full of books. We had the Parlor Lecture Club. We had every sort of church except a Christian Science church. Every house in our city had a Bible in it, and a lot of houses had as many as four Bibles in them. A man could feel our city was beautiful.

Or a man could feel that our city was fake, that our lives were empty, and that we were the contemporaries of jack-rabbits. Or a man could have one viewpoint in the morning and another in the evening. The dome of our court-house was high, but it was ridiculous and had nothing to do with what we were trying to do in the desert. It was an imitation of something out of Rome. We had a mayor but he wasn't a great man and he didn't look like a mayor. He looked like a farmer. He *was* a farmer, but he was elected mayor. We had no great men, but the whole bunch of us put together amounted to something that was very nearly great. Our mayor was not above carrying on a conversation with an Armenian farmer from Fowler who could speak very little English. Our mayor was not a proud man and he sometimes got drunk with his friends. He liked to tell folks how to dig for water or how to prune muscat vines in order to get a good crop, and on the whole he was an admirable man. And of course we had to have a mayor, and of course *somebody* had to be mayor.

Our enterprise wasn't on a vast scale. It wasn't even on a medium-sized scale. There was nothing slick about anything we were doing. Our enterprise was neither scientific nor inhuman, as the enterprise of a growing city ought to be. Nobody knew the meaning of the word efficiency, and the most frightening word ever used by our mayor in public orations was *progress*, but by *progress* he meant, and our people understood him to mean, the paving of the walk in front of the City Hall, and the purchase by our city of a Ford automobile for the mayor. Our biggest merchant was a small man named Kimball, who liked to loaf around in his immense department store with a sharpened pencil on his right ear. He liked to wait on his customers personally, even though he had over two dozen alert clerks working for him. They were alert during the winter, at any rate, and if they sometimes dozed during the long summer afternoons, it was because our whole city slept during those afternoons. There was nothing else to do. This sort of thing gave our city an amateur appearance, as if we were only experimenting and weren't quite sure if we had more right to be in the desert than the jack-rabbits and the horned-toads, as if we didn't believe we had started something that was going to be very big, something that would eventually make a tremendous change in the history of the world.

But in time a genius appeared among us and he said that we would change the history of the world. He said that we would do it with raisins.

He said that we would change the eating habits of man.

Nobody thought he was crazy because he wore spectacles and looked important. He appeared to be what our people liked to call *an educated man,* and any man who had had an education, any man who had gone through a university and read books, must be an important man. He had statistics and the statistical method of proving a point. He proved mathematically that he would be able to do everything he said he was going to do. What our valley needed, he said, was a system whereby the raisin would be established as a necessary part of the national diet, and he said that he had evolved this system and that it was available for our valley. He made eloquent speeches in our Civic Auditorium and in the public halls of the small towns around our city. He said after we got America accustomed to eating raisins, we would begin to teach Europe and Asia and maybe Australia to eat raisins, our valley would become the richest valley in the whole world. China! he said. He shouted the exact number of Chinese in China. It was a stupendous figure, and the farmers in the Civic Auditorium didn't know whether to applaud or protest. He said that if he could get every living Chinaman to place one raisin, only *one* mind you, in every pot of rice he cooked, why, then, we could dispose of all our raisins at a good price and everybody in our valley would have money in the bank, and would be able to purchase all the indispensable conveniences of modern life, bathtubs, carpet-sweepers, house electricity, and automobiles.

Rice, he said. That's all they eat. But we can teach them to drop one raisin in every pot of rice they cook.

Raisins had a good taste, he said. People liked to eat raisins. People were so fond of eating raisins that they would be glad to pay money for them. The trouble was that people had gotten out of the habit of eating raisins. It was because grocers all over the country hadn't been carrying raisins for years, or if they had been carrying them, the raisins hadn't been packed in attractive packages.

All we needed, he said, was a raisin association with an executive department and a central packing and distributing plant. He would do the rest. He would have an attractive package designed, and he would create a patented trade-name for our raisins. He would place full-page advertisements in *The Saturday Evening Post* and other national periodicals. He would organize a great sales force. He would do everything. If our farmers would join this raisin association of his, he would do everything, and our city would grow to be one of the liveliest cities in California. Our valley would grow to be one of the richest agricultural centers of the world. He used big words like *co-operation, mass production, modern efficiency, modern psychology, modern advertising,* and *modern distribution,* and the farmers who couldn't understand what he was talking about felt that he was very wise and that they must join the raisin association and help make raisins famous.

He was an orator. He was a statistician. He was a genius. I forget his name. Our whole valley has forgotten his name, but in his day he made something of a stir.

The editor of the *Morning Republican* studied this man's proposal and found it sound, and the editor of the *Evening Herald* said that it was a good thing, and our mayor was in favor of it, and there was excitement all over our valley. Farmers from all over our valley came to town in surreys and buggies. They gathered in small and large groups in front of our public buildings, and they talked about this idea of making the raisins famous.

It *sounded* all right.

The basic purpose of the raisin association was to gather together all the raisins of our valley, and after creating a demand for them through national advertising, to offer them for sale at a price that would pay for all the operating expenses of the association and leave a small margin for the farmers themselves. Well, the association was established and it was called the Sun-Maid Raisin Association. A six-story Sun-Maid Raisin Building was erected, and an enormous packing and distributing plant was erected. It contained the finest of modern machinery. These machines cleaned the raisins and took the stems from them. The whole plant was a picture of order and efficiency.

Every Thursday in those days I went down to Knapp's on Broadway and got a dozen copies of *The Saturday Evening Post.* The magazine was very thick and heavy. I used to carry a dozen of them in a sack slung over my shoulder. By the time I had walked a block my shoulder would be sore. I do not know why I ever wanted to bother about selling *The Saturday Evening Post,* but I suppose it was partly because I knew Benjamin Franklin had founded it, and partly because I liked to take a copy of the magazine home and look at the advertisements. For a while I even got in the habit of reading the stories of George Agnew Chamberlain. One Thursday evening I had a copy of *The Saturday Evening Post* spread before me on our living-room table. I was turning the pages and looking at the things that were being advertised. On one page I read the words, *Have you had your iron*

Raisin Festival Arch, 1912. Courtesy of Fresno Historical Society Archives.

today? It was a full-page advertisement of our Raisin Association. The advertisement explained in impeccable English that raisins contained iron and that wise people were eating a five-cent package of raisins every afternoon. Raisins banished fatigue, the advertisement said. At the bottom of the page was the name of our Association, its street address, and the name of our city. We were no longer lost in the wilderness, because the name of our city was printed in *The Saturday Evening Post.*

These advertisements began to appear regularly in *The Saturday Evening Post.* It was marvelous. People were hearing about us. It was very expensive to have a full-page advertisement in the *Post,* but people were being taught to eat raisins, and that was the important thing.

For a while people actually *did* eat raisins. Instead of spending a nickel for a bottle of Coca-Cola or for a bar of candy, people were buying small packages of raisins. The price of raisins began to move upward, and after several years, when all of America was enjoying prosperity, the price of raisins became so high that a man with ten acres of vineyard was considered a man of considerable means, and as a matter of fact he was. Some farmers who had only ten acres were buying brand-new automobiles and driving them around.

Everybody in our city was proud of the Raisin Association. Everything looked fine, values were up, and a man had to pay a lot of money for a little bit of desert.

Then something happened.

It wasn't the fault of our Raisin Association. It just happened. People stopped eating raisins. Maybe it was because there was no longer as much prosperity as there had been, or maybe it was because people had simply become tired of eating raisins. There are other things people can buy for a nickel and eat. At any rate, people stopped eating raisins. Our advertisements kept appearing in *The Saturday Evening Post* and we kept asking the people of America if they had had their iron, but it wasn't doing any good. We had more raisins in our Sun-Maid warehouse than we could ever sell, even to the Chinese, even if they were to drop *three* raisins in every pot of rice they cooked. The price of raisins began to drop. The great executives of the Association began to worry. They began to think up new ways to use raisins. They hired chemists who invented a raisin syrup. It was supposed to be at least as good as maple syrup, but it wasn't. Not by a long shot. It didn't taste like syrup at all. It simply had a syrupy texture. That's all. But the executives of our Association were desperate and they wanted to dispose of our surplus raisins. They were ready to fool themselves, if necessary, into believing that our valley would grow prosperous through the manufacture and distribution of raisin syrup, and for a while they believed this. But people who were buying the syrup didn't believe it. The price of raisins kept on going down. It got so low it looked as if we had made a mistake by pausing in the desert and building our city in the first place.

Then we found out that it was the same all over the country. Prices were low everywhere. No matter how efficient we were, or how cleverly we wrote our advertisements, or how attractive we made our packages of raisins, we couldn't hope for anything higher than the price we were getting. The six-story building looked sad, the excitement died away, the packing house became a useless ornament in the landscape. Its machinery became junk, and we knew a great American idea had failed. We hadn't changed the taste of man. Bread was still preferable to raisins. We hadn't taught the Chinese to drop a raisin in their pots of cooking rice. They were satisfied to have the rice without the raisins.

And so we began to eat the raisins ourselves. It was amazing how we learned to eat raisins. We had talked so much about them we had forgotten that they were good to eat. We learned to cook raisins. They were good stewed, they had a fine taste with bread. All over the valley people were eating raisins. People couldn't buy raisins because they were a luxury, and so we had to eat them ourselves although they were no luxury to us.

David Mas Masumoto

FIREDANCE

THE WHIRL OF COLORS, the crack of *kachi-kachi,* the familiar voice of a Japanese singer, Kenjiro felt as if he had returned to Kumamoto and the summer *obon* had begun.

But Kumamoto lay three thousand miles away. This *obon* was celebrated in Del Rey, a small town located in the rich farmlands of the San Joaquin Valley in California.

Del Rey was settled by many Japanese who had immigrated to the area two decades before. Some lived and worked on the surrounding farms, others opened businesses and built a thriving "J-town." Initially, the *Issei* were attracted to Del Rey because of the miles and miles of vineyards and plentiful work in the fields, where grapes were grown and dried into raisins. Over the years, the Japanese continued to work but dreams of wealth and fortune began to fade. Now they worried about simply surviving, for the year was 1938, the middle of the Great Depression.

Kenjiro watched the *obon,* his thoughts were lost in the scratchy sounds of an old recording of a Japanese folk song. He was hypnotized by the flashing *kimono* colors and the rough movements of the dancers.

"Ah, the *obon,*" he said to himself, "to honor ancestors. We dance for your enjoyment, for our joy."

Once a year the Japanese community of Del Rey gathered at the *obon* to celebrate. Families left their shack homes and dressed in their best clothes, forgetting their depression, ignoring for a moment the summer heat, the dirty farmwork and the poverty in their lives. Almost a thousand *Issei* and their *Nisei* children crowded the streets of Del Rey for the festival. Like Kenjiro, these Japanese shared a common life: all worked hard, endured

---

From *Silent Strength* copyright 1985 by David Mas Masumoto, published by New Currents International. This Depression-era story is reprinted by permission of the author.

their shortcomings and continued to hope and dream of one day owning farmland.

"Ancestors," Kenjiro spoke softly, "I wonder if you have a better life than we here?"

Across the railroad tracks a group of *hakujin* men sat drinking and yelling at the Japanese.

"Hey you Japan-boys, what crazy dancing you doing?"

"Look at them, them folks are dancing in the streets."

"Awwww, that noise, you call that music?"

"Reminds me of a Western movie. They all look like crazy Injuns dancing around."

They drank cheap whiskey from a bottle, each taking a swig, wiping their lips with their shirt sleeve, and then passing it on to another. All the while they yelled and laughed at the Japanese and the *obon*.

Usually *hakujins* spoke very little to the Japanese because only a few of the *Issei* knew English. Most everyone communicated via waving hands and a few English words, or through *Nisei* sons and daughters who had American schooling. But this Saturday night the gang of men drank, escaping their own poverty with each gulp, finding enjoyment in yelling at the "Japan-boys" and believing themselves superior.

The railroad tracks divided Del Rey into two halves, one side contained the Japan-town with the *Issei* businesses: a small grocery store, a hardware store, some boarding houses and a few gambling halls. On the other side sat the *hakujin* world, the older homes of wealthy landlords and businessmen, a bank, a post office, general store and a huge raisin processing and packing plant.

The raisin plant was built of red brick, it rose above the rest of the town, its windows shimmered in the afternoon sun, glistening over the surrounding vineyards, a luster visible for miles. Most everyone in the surrounding countryside was connected to the plant. Some worked in the plant while others were farmers who delivered their raisins to the plant for packaging and marketing. The raisin company intertwined itself within the lives of both the *hakujin* and Japanese communities.

Two Armenian brothers named Setz and Ara Minasian had built the plant 25 years ago. The Armenians had immigrated to the area about the turn of the century, the same time Japanese arrived. Both ethnic groups struggled for acceptance and respect. The Armenians carved a community by gradually buying farms and building an empire. The Japanese became known for their hard work and silent tenaciousness, a quiet people with quick hands and feet and strong backs suited for field work.

"Thump, thump. Whack, whack, whack. Thump, thump." The *taiko* drum beat out a rhythm. Kenjiro's family passed in front of him, his wife with her frail build that disguised her strength and his three children, two daughters and Masao, his son. The children, all under ten years old, mimicked their mother, a beat or two late with each step that evoked laughter from the audience. Kenjiro nodded his head in approvement. He was content this evening, his family was strong and healthy and he was slowly nearing his goal: with each passing of the *obon* he had saved more money and could soon purchase a farm.

Kenjiro began to think of the years of work, leasing land, saving money by doing all the work themselves, handling each raisin bunch with care and respect. Despite the depression, they had accumulated some wealth and in two years Kenjiro would buy land through his son's name. California laws stated that Kenjiro and his wife, both Japanese citizens, could not buy land.

"Ah, two years more," Kenjiro thought as his mind wandered from the *obon* dancers. His eyes turned to stare at the lurking raisin plant in the distance. "You Minasians. Only two years more. Then I will have land like you people."

Bitter memories danced in Kenjiro's mind. Every year he had to deliver his raisins to the Armenian's plant and every year Kenjiro grew to hate the Minasians more and more. He recalled last year's harvest and delivery to the plant. At the entrance gate Kenjiro stopped the truck he had borrowed from a friend, a truck loaded with Kenjiro's raisin crop. Setz Minasian climbed onto the truck and squatted atop the boxes filled with raisins. As he inspected and priced the load of raisins, a soured look grew on his face.

"These raisins," said Setz, "I can't use this crap. No good. No good. And you expect good money for this junk?" He continued shaking his head.

Kenjiro stood silently behind his raisins. He stared at the Armenian then looked downward. Kenjiro knew his raisins were of high quality, as good as any others. But he had no choice, he had to accept the Minasian's price because he had to save money for his own farm. Kenjiro answered in silence.

"Here," said Setz, "I'll give you 145 dollars a ton. Take it or leave it. If I were you, I'd take it and be happy. Better take it while your raisins are still good." Setz had a slip of paper and bent down to hand it to Kenjiro. Kenjiro nodded, accepted the paper and crumpled it in his hand as he walked to the cashier's window.

Along the way Kenjiro's neighbor, Carl Sapazian, stopped him and said: "Say Ken, I saw what happened. It happens to a lot of us little farmers. Don't be angry, it's just Setz's way of doing business." Carl patted Kenjiro on the back. "I don't know if you understand but we Americans have an expression for it: we just 'grin and bear it.'"

Kenjiro nodded and continued walking forward. "Grin and bear it. Grin and bear it," Kenjiro said over and over to himself.

The *obon* wound its way through Del Rey's J-town. Beginning at the Buddhist Hall, the dancers traveled down the town's dirt roads, past Miura's General Store and the Shoji and Harada boarding houses where many of the bachelors lived. The course then rounded the corner where the Tokyo Club sat, the building that housed a gambling hall and a brothel that the Japanese men spoke little of but often visited. A series of stores, Fujimoto Sakanaya and Mochida's Barber Shop and many other small businesses and a few more boarding houses lined the rest of the route. The dancers then turned and circled back to the Buddhist Church, the railroad tracks next to them, opposite the J-town storefronts.

Along the roadside, the Japanese positioned themselves, watching for family and friends, talking and exchanging stories of their life and work. For many, the *obon* offered one of the few opportunities for the community to gather, distances overcome as families journeyed from the country dwellings into town for this annual event. For many of the young folk, the *obon* represented the only chance they had to see each other and renew friendships. They said that many romances began at the *obon*, a vital occasion for the young adults nearing marrying age.

At first Kenjiro didn't see it, he only heard voices yelling and saw a few men racing toward the Tokyo Club building.

"Yaaa, fire!" shouted one man racing by. "Fire! Hurry!"

The sound of the word "fire" sent a cold panic through Kenjiro, like an electric shock jolting him awake. Running toward the center of J-town, he saw the flames dancing atop the wooden building and flashing out of the windows. Some men already had buckets, tossing water onto the roaring flames, then running back to the town well. Ignoring these splashes, the fire swept through the Tokyo Club and quickly jumped to the next set of buildings. The Fujimoto and Mochida families were running in and out of their stores, trying to rescue valuables and some supplies as the fire leaped around them.

"Where the fire department?" yelled Mr. Fujimoto. "When they come?"

Del Rey's volunteer fire department was located on the other side of the tracks, the fire truck and water and hoses and crew should have been racing toward the fire. The Japanese listened for the clanging bell of the fire engine, hoping and helplessly waiting for assistance. But only the crackling flames and yells of frantic men and families could be heard.

Kenjiro helped with a bucket brigade. Everyone lined up in a path from the fire to the town well, passing buckets full of water to douse the fire then handing back empty buckets to be filled. Even the women dressed in their fine *kimono* rolled up their sleeves and tucked them back out of the way with cords wrapped about their bodies. The fire spread quickly, rolling over wooden structures, advancing to the next. The bucket line gradually slowed as everyone watched the fire consume the town. Tanaka's grocery store was engulfed with flames, Shoji's and Harada's boarding houses crackled in a brilliant fire. The *obon* decorations, colorful banners and signs, were all burnt beyond recognition.

The fire engine never arrived, the J-town vaporized before everyone's eyes. Kenjiro glanced across the tracks. The *hakujins* lined the rails, watching the fire like a Fourth of July parade and spectacle. A few had crossed to the Japanese side and helped with water, but most stood and watched, pointing to the flames, chattering when a structure collapsed and sent sparks exploding into the night sky.

As quickly as the fire spread, stories circulated about the Tokyo Club.

"How did it begin?"

"A group of *hakujins."*

"Ah?"

"Hakujins. Behind the Tokyo Club. We saw them with torches."

When Kenjiro heard this he turned and scanned the line of faces across the tracks. His eyes stopped at the Armenian, Setz Minasian. An anger rose within. Kenjiro clenched his fists, his arm muscles tensed. Setz and his friends were pointing at the leaping flames, two of the men were laughing at something Setz had said.

An explosion rocked the crowd, people ducked, some threw themselves onto the ground. From Sasahara's Garage, flaming rockets shot into the air, balls of fire soared into the night sky.

"It's cans of gas," someone yelled.

Everyone stood frozen for a moment, hypnotized by the streaking comets. Most of the cans shot straight up, landing back in the flames or atop the charred ruins of J-town.

"Sasahara-san, was there much gas in your place?" Kenjiro asked.

"Saa." He slowly answered. Black streaks ran down his face where sweat had trickled and blended with soot that covered his skin. "Saa. Not too much," he said shaking his head, "not too much I think." He stared at his garage, his flesh had grown hoary, shock settled into his soul.

Boom!

Everyone whipped their heads back to the garage. A series of flaming streaks shot out of the fire and across the tracks. They arched above the earth, a rainbow of sparks in the night. For a brief moment, the entire town seemed to be entranced, an uneasy silence captured everyone's attention. The fire ball crashed in an open field of dried weeds and grass. Flames immediately burst alive, igniting the parched field and heading toward the lone structure silhouetted against the moon-lit sky: the Minasian Raisin Plant.

The *hakujin* crowd lunged toward the fire, men raced to prevent it from spreading.

"Get the damn fire engine," someone yelled.

"The fire, hurry."

"Get water, water!"

The yells and screams blended into a panic, a confusion of voices, commands and directions.

At first Kenjiro ran toward the fire, but then he began to slow down near the tracks. Many of the other Japanese slowed with Kenjiro, a hesitation gripped them, a few stopped at the tracks.

A hakujin woman turned to them and pleaded, "Hurry, please hurry. Help us. The raisin plant, it's our town that's burning!"

The Japanese slowed, stumbling to a stop. Behind them, their J-town had burned, an entire block blackened.

Kenjiro bolted from the pack and ran across the tracks. The other Japanese surged behind him and joined the *hakujins* with their buckets of water and wet rags. But like an attacking army, the flames marched toward the plant a few hundred yards away, undaunted by the brigade of townspeople.

"Faster, faster, it's spreading."

"Help us, over here."

"More water."

"The fire engine, where in the hell is it?"

Voices, screams, desperate cries.

The clanging bell of the fire engine broke the night air. The old, horse-drawn water tank raced out from behind the raisin plant toward the flames. *Hakujin* men with heavy coats and red helmets ran alongside, accompanying the engine. Smoke blanketed the entire area, in the still night air visibility grew worse and worse. The fire engine's driver tried to head directly toward the fire, guiding the horses through the smoky chaos.

Kenjiro looked up and as if in slow motion witnessed the entire accident. The racing horses bolted, jerking the engine forward. A front wheel dipped into a hole, a gully in the field, then it leaped out of the depression. For a second the entire front end of the wagon hung in the air, suspended with the wheel spokes whirling in a blur. Then it crashed against the earth, shattering the wheel axle and sending pieces of wood flying through the air. The right side of the engine tilted as the wheel-less frame grabbed the dirt, jerking the horses to the right. The weight of the water tank and water sent the entire vehicle in a spin, the end whipped around and flipped, the wagon was sent rolling onto its side. The tank burst open, a wave of water gushed out into the dry field. Meanwhile the flames seemed to lunge forward as if intensified by the accident.

Kenjiro was startled by the scream of the man next to him.

"Nooooo. Nooooo," shrieked the man, his voice a biting, unearthly wail. Veins protruded from his temples, his body was shiny from sweat as he continued to beat the fire with a wet rag. The man looked up at Kenjiro and both stared at each other. For the first time, Kenjiro's eyes had met with the Armenian's.

Setz turned toward Kenjiro, his face was contorted and twisted; his head was shaking and his mouth locked open; his eyes darted from Kenjiro to the dancing flames and back to Kenjiro. Setz had stopped whipping the flames, his arms hung limp, his head shook back and forth as if saying, "no, no, no."

Kenjiro broke off his stare when he saw the dry rag in Setz's hand ignite. But Setz seemed oblivious to the flames, he clung to his rag and stumbled backward, away from the fire toward his plant. He fell to the ground, tried to rise but his legs refused to respond and he rolled in the dry weeds. Kenjiro lunged to catch him. Grabbing his arm, Kenjiro propped the broken man against his body and they both retreated toward the raisin plant. They staggered to the red brick wall and tumbled to the ground. Beneath the "Minasian Brothers Raisin Company" sign, both men lay, panting from exhaustion, sweat pouring from their foreheads, black smudges painted on their cheeks.

The fire advanced, the dry grass crackled in the heat, the cries of the town's people lost in the deafening popping and snapping of the fire. Kenjiro looked up and saw that the Armenian's rag had left a trail of small fires behind them. When Kenjiro sprang up, his head spun and pain shot through his stiff body. He staggered to the first fire and began to stamp it out. The smoke blinded and burned his eyes, he tried to cough but a choking tightness clenched his throat.

Kenjiro continued stamping at the flames and he succeeded in stopping its advance, but he could not control its expanding power. The fire consumed the weeds and raced away from him. All Kenjiro could do was to continue his vigil along one front. Dizzy and weak from exhaustion and pain, he raced back and forth on one side of the fire, the Minasian Raisin Plant directly behind him.

The flames danced into the field, dashing away from Kenjiro, continually fed by dry weeds. The fire marched away and left behind a blackened strip. Kenjiro quickened his pace, racing back and forth, not allowing the small fire to outflank him. He glanced up and realized that the fire was indeed racing away, and leaving behind a natural firebreak.

Sumida-san and Mochizuki-san had noticed the new smoke and saw Kenjiro. They ran to help him and all three together contained one flank, keeping the fire in check. A group of *hakujins* also looked up at the running men. They yelled and scrambled toward the raisin plant to investigate. Both parties then worked along one edge of the fire, passing instructions to each other, their feet combining to stamp out the wandering flames. They yelled, one in English and the other in Japanese and worked as a team.

"*Oi*," said a Japanese, pointing to flames that were swinging around the flank. A group of men ran to check the advance.

The flames grew and marched in an opposite direction, away from the men and plant. It moved faster and faster towards the parent fire. The raging heat scorched the earth, everyone held handkerchiefs over their faces to subdue the smoke. They raced back and forth, stamping and guiding the fury. Crouched low to avoid smoke, waving their arms and pointing at new flames, they looked like dancers enacting an ancient ritual.

The fire crackled in a deafening wave as the two fronts met and consumed each other. Some stopped to watch the two forces clash, others kept working, only when the crackling began to die down did they look up. Like a thunderstorm that abruptly stops with no warning, the fire lost its power and ceased to grow. It soon became tame, some flames struggled to keep alive, and along one edge a small fire remained but was quickly contained by stamping feet.

Smoke lingered in some places and a few embers still glowed in the night, but the fire had been conquered. The firefighters wandered through the blackened field, staggering from exhaustion, their vision impaired from the smoke.

A few of the *hakujins* were talking, smiling.

"We did it Jim, by God, we did it."

"Saved the damn town. And the plant."

"Yea, the plant. Only burned up a field, that's about all."

The men patted each other on the back and walked back to their town.

"How about a drink buddy. After all that work, I could sure use one."

"A celebration, that's what's called for. We saved the damn town."

Some people seemed to be dancing as they strolled back; they had been spared disaster.

The Japanese trudged to the remains of their town, the chared ruins of a J-town. Most everyone had forgotten that it was the *obon*. Instead they silently walked, weary and beaten; the silence mirrored their spirits.

Kenjiro joined his people, they were crossing the tracks, gathering family together amid the town's remains. Some were packing what they could find, preparing to return to their country shacks. Others were sifting in the chared buildings, salvaging what they could.

"Tomorrow we'll begin the clean up," said Fujimoto-san.

"Ah, tomorrow," answered Mochida-san.

"Let's get some water, wash our faces and go to bed," said Miura-san. "At least my bed didn't burn," said Fujimoto-san.

"Ya, my bed. Our beds are safe, no? That's all I want now," answered Mochida-san.

Someone tapped Kenjiro on the shoulder. Slowly he turned back, glancing to his right. Setz Minasian slouched behind Kenjiro, his head drooped down, his eyes blinking and squinting from the smoke. He coughed and slowly turned his eyes upward toward Kenjiro. Kenjiro stopped and peered at Setz, both men had soot creases on their faces and necks, lines that matched the wrinkles in their skin.

Setz began to say something, his dry mouth opened but no sound emerged. Instead he tightened his lips then gradually bent at the waist: dipping his head slightly, slouching his shoulders, a jerking motion like an uneasy bow. He straightened himself and once again faced Kenjiro. Kenjiro hesitated, then slowly leaned forwards, bending at the waist, returning a bow.

For a few moments both men surveyed the other, then Setz turned and hobbled back toward his plant. Kenjiro watched the Armenian limp, then he too turned to his home and community.

William Everson

SAN JOAQUIN

This valley after the storms can be beautiful beyond the telling,
Though our city-folk scorn it, cursing heat in the summer and drabness
in winter,
And flee it—Yosemite and the sea.
They seek splendor, who would touch them must stun them;
The nerve that is dying needs thunder to rouse it.

I in the vineyard, in green-time and dead-time, come to it dearly,
And take nature neither freaked nor amazing,
But the secret shining, the soft indeterminate wonder.
I watch it morning and noon, the unutterable sundowns;
And love as the leaf does the bough.

THE RAIN ON THAT MORNING

We on that morning, working, faced south and east where the sun was
in winter at rising;
And looking up from the earth perceived the sky moving,
The sky that slid from behind without wind, and sank to the sun,
And drew on it darkly: an eye that was closing.
The rain on that morning came like a woman with love,
And touched us gently, and the earth gently, and closed down delicately
in the morning,
So that all around were the subtle and intricate touchings.
The earth took them, the vines and the winter weeds,
But we fled them; and gaining the roof looked back a time,
Where the rain without wind came slowly, and love in her touches.

THE RUIN

The year through September and the veils of light
Broke equinox under; south darkened;
The-wind-of-no-rain, northwest and steady,
All day running the valley,
Swung with the dusk, strengthened;
And the cloud gathered, raiding the open sky.

Under the whisper we watched it come over,
The raisins heavy yet in the fields,
Half-dried, and rain a ruin; and we watched it,
Perceiving outside the borders of pain
Disaster draw over:
The mark of the pinch of the coming months.
There was above us the sheet of darkness,
Deadly, and being deadly, beautiful;
Destruction wide for the dreading eyes;
What was hardly of notice another month
Now burned on our sight;
And it rode us, blown in on the wind,
Above and beyond and the east closed under;
It let down the ruin of rain.

FOG

The gray mask of the fog, the pale plate of the sun,
The dark nudeness of the stripped trees
And no motion, no wave of the branch:
The sun stuck in the thick of the sky and no wind to move it.
The sagged fence and the field
Do not remember the lark or her mate or the black lift of the rising crows.
The eye sees and absorbs; the mind sees and absorbs;
The heart does not see and knows no quickening.
There has been fog for a month and nothing has moved;
The eyes and the brain drink it, but nothing has moved for a number of days;
And the heart will not quicken.

Franz Weinschenk

MADERA

WE PASS AROUND the thin Rand McNally gazeteer Dad bought at the bus depot. Madera, California—just about in the center of the state, both from top to bottom and side to side. The county its in produces cattle, alfalfa, cotton, fruits, grapes, lumber. There is a picture of an old car driving through the base of a giant redwood. I follow our route on the map. Right now we're in Pennsylvania, green and lush. He lights up a cigar even though the sign says no smoking. I sit next to him blowing the smoke to one side so that the driver won't notice.

Mother sits across the aisle from us knitting, glum and silent. I sense her mind: for the second time, the wandering Jews are forced to roam the earth to God knows where—among God knows whom. All because that gutter snipe Hitler, that uncultured streetbrawler, that latrine painter, that brutal, murdering fanatic had willed it. She clicks off her hearing aid and retreats into her all-encompassing rage—resigned to sit alone in hollow deafness—not much interested in where the bus is heading—stolid and sour—doggedly working her misery into the dress she's knitting.

America is flying by—so vast it takes a powerful, rumbling diesel with its throttle wide open three and a half days to get from sea to shining sea. City after city, town after town, state after state, mile after mile, the bus drones on over a seemingly inexhaustible ribbon of asphalt. The United States of America—an unbelievable, limitless abundance of just about everything.

On the morning of the fourth day, as we leave L.A. going north, we know it will take only a few more hours, and the looking becomes intense. We follow Highway 99 north over the Tehachapis. Once past Bakersfield, I keep hoping for some mountains or at least maybe a hill or two, but there are only desolate farms way off into the hazy distance on both sides of the highway. Contrary to what I had expected, there is no golden glow, no orange hue, no verdant opulence. The San Joaquin Valley is flat, denuded, arid and monotonous, and the bus is getting hot. When the driver leans forward, I notice his shirt is totally soaked. Dad had changed into his only

---

suit in the bathroom in L.A.—a shop-worn blue gaberdine—because he wants to make an impression. He takes his coat off; still the sweat pours off him.

We arrive at 1:30. The bus stops in an alley next to the depot, and the driver helps us off. The sign reads Madera, Pop. 7,520. It is hot, sweltering. We stand on the soft blacktop next to the bus which is idling, gurgling out smelly exhaust, as he unloads our luggage. The sun is so bright you have to squint to see. It beats down from above and radiates up off the ground. We walk over and into the merciful shade of the tiny bus depot. There is a ticket window, a small freight counter and a truncated soda fountain. The ceiling fan rotates lazily around a strip of yellow fly paper heavy with dead flies. The thermometer on the wall registers 102. Mother glowers at me in pain, "My God, so boiling I am! Such a heat!" she scowls—an exclamation I would hear many times over in the years to come. She gives my father a pitiful look, "*Ach, so eine Glut,*" she says. But he straightens up, shakes his glistening bald head, "*Nah, ja* . . . of course, a little hot it has to be, *nicht wahr?* Otherwise, how can there be sugar in the grapes for the wine, no?" She doesn't answer. "Look," he says, "It's not that bad now, eh? I mean, I don't feel nothing so terrible." To which the tall thin man behind the counter replies laconically, "Oh, this ain't nothin', folks. It's only May—I mean June." He picks up his cigarette from the edge of the counter and takes a drag. "Wait till you get to July, August and September, whew! Last year, we had two weeks in a row all over 105—some even over 110."

Nobody from the ranch has come to pick us up. So we sit at the counter and he orders apple pie and cokes. I notice how old and strained they look—he in that ill-fitting, baggy suit, with a limp spotted tie pinned to his shirt by a tie-clip made from a real Roman coin in the likeness of the Emperor Hadrian and she with her red hair turning white, her lipstick uneven and smeared on her teeth, forever fussing with the volume control of her hearing aid.

After waiting for a half hour, he calls the ranch and it takes another hour and a half before a faded green flat-bed Chevy truck arrives. The driver smilingly introduces himself as Manuel. He is a big, burly, friendly Mexican guy with a pug nose and thick black, curly hair. He helps us hoist our luggage up onto the truckbed.

"Whazat?" he says pointing to the cello, "a baby bass? Hey, we could use one of those in the band."

"It's a cello," I offer. " Belongs to my father."

"Can he play *rancheras* or *paso dobles* or *huapangos?*"

"I don't know, but I don't think so."

I help them into the dusty, tattered cab, slam the door and swing myself onto the back of the truck where I sit right on the bed amid our luggage and some unfamiliar tools and machinery, cradleing the cello. In just minutes we are beyond the city limits and out in the country. We pass orchards, vineyards, cotton and alfalfa fields, pastures with horses and cattle grazing. In about twenty minutes we come to a long dirt road lined with tall palm trees on either side—the entrance to the ranch. The truck plows through powdery silt raising a cream-colored plume of dust behind it. Beyond the palms, on both sides of the road, there is a never ending sea of deep green grapevines clear to the horizon. And suddenly, the 'home ranch' looms up ahead of us—a massive barn dwarfs the structures around it. Soon we're driving among the buildings and I begin to realize how old and run-down this place is. Each structure is made of rough faded red boards. The roofs are shingled in green and sag with age.

The driver makes a sharp turn just beyond a tall water tower. He shuts off his motor, and we coast to a stop with the faint squeal of his brakes. So this is it—the 'free roof' about which Julius had been so euphoric. Though small and ramshackle, it sits beneath several tall eucalyptus trees which, I speculate, will grant us valuable shade. Before unloading, Manuel gives us a tour. There is a large kitchen with a smoke-smudged wood stove up against the interior wall, two bedrooms, and an unspeakably filthy bathroom that can only be accessed through the front porch. If you need hot water, the driver explains, you have to make a fire in the wood stove which, he shows us, has pipes—"coils" he calls them—circling the inside of the firebox. But who would want to make a fire when the temperature is already over a hundred?

Since it's too late to prepare supper, the driver invites us to eat in the cookhouse, family style, with about 20 ranchhands. Their silent bronzed faces are several shades darker than their balding scalps. They look tired and eye us politely. Underdone shortribs in tepid tomato sauce with potatoes, bread, margarine and coffee.

We sleep in California for the first time on three soiled mattresses on top of metal cots on loan from the bunkhouse. Early the next morning, I take a look around. The huge barn and its attached corral, which I had noticed the day before as we were driving up, still dominate. I count 34 mules. Big reddish-brown beasts with long shifting ears and greying nostrils. They crowd together in front of a long trough, swishing their tails nervously, and pull straw through the tines, anxious to get a last mouthful before a long day in the field hitched to plows, vineyard wagons, sulfur dusters, hay mowers and manure spreaders. Just west of the barn is the bunkhouse where the ranch hands stay. In contains a long row of rooms, each with two cots in them, that face a long extended screened-in porch.

Next to it, the cookhouse. On the opposite side of the barn is the tank house, the bottom floor of which would turn out to be my father's office. There is also a blacksmith shop, several storage sheds, and a long shelter to park tractors. Toward the periphery of the compound, almost swallowed up by the vineyards, are several other ramshackle homes with red sides and green roofs like ours for employees who qualify for company housing. About a quarter of a mile down a dirt road, I see more industrial-type buildings. These, I learn later, are a dehydrator and a fruit drying shed.

They call it the 'home ranch,' and it turns out to be in excess of 2,000 acres—a substantial holding—90 per cent grapes—pungent sweet dark varieties like Malagas, Zinfandels, and Muscatels that ferment into thick, sugary dessert wines and apple green Thompson seedless that sun-dry into sweet, sticky raisins. Almost as an afterthought, there are small acreages of apricots, peaches, nectarines, and alfalfa. About a mile west of the main compound is what they call "Camp Two," a long low building with simple rooms on each side and an acrid-smelling, leaky, common bathroom at its end. Camp Two is used only in the fall at picking time as migrant housing.

Constant use of caterpillar tractors has ground the soil into such fine silt that any vehicle traveling anywhere lifts up a cloud of dust. Fine particles of earth hover in the atmosphere and darken the sky most every day. Nothing moves without stirring up dust. It invades everything—the house, its rooms, closets, drawers, cupboards. Even things carefully wrapped and hidden away can not escape the gritty, powdery film.

The owner and management, all the way down to the straw bosses, are either first or second generation Armenians—with one notable exception—Ira Cutler, the ranch foreman newly arrived from Texas. Behind his back, everybody calls him "Beech Nut" because that's the brand of tobacco he chews. In all the years I knew him, he was never without a telltale bulge between cheek and jaw and a tinge of brown oozing down the crease from the corner of his mouth to his chin. But the people who actually do the work are almost all Mexicans.

At a quarter to seven most every morning, Beech Nut ambles out to the barn and climbs up and sits on the top rung of the corral fence. He carefully loads up a wad of chew from the rectangular blue packet it comes in. Slowly and deliberately he folds the package back to its original shape and slides it back into his shirt pocket. Then he spits a couple of loose splashers in the direction of the mules who take little notice and proceeds to assign jobs for the day. At seven, the stable boy pulls the rope next to the tankhouse to activate the bell in its rafters, and the workmen begin to disappear into the ocean of grapevines and orchards—some to irrigate, some to drive tractor, some to "French" or "single" plow, some to harvest, prune, tie, dig up weeds, load, unload, spray, dust. The bell rings again at twelve and one, punctuating the lunch hour, and finally at five—quitting time.

Dad pays $7.50 for a bicycle from a ranch hand so that I can pedal three miles down the narrow country road to a tiny country store run by some Japanese people and buy as many supplies as I can fit into my rucksack—cleanser and Brillo pads, a pound of ground meat, a cube of butter, a small bag of potatoes, a quart of milk, a loaf of bread carefully balanced on top. They use an abacus to add up the bill.

About a month later, without telling anyone, he gets a ride to Madera and buys a 1931 Chevrolet coupe with a rumble seat in the back for $85. Being as short as he is, he looks under the steering wheel to see where he is going. When making a turn, he never pulls the wheel around in a continuous arc, but pushes it up a small distance with one hand, holds it momentarily with the other, and repeats the process quickly over and over. On his way home, right at the entrance of the ranch by the palm trees, he misses the turn completely and drives point blank into a half-filled irrigation ditch. Both front wheels and most of the motor compartment end up under water. The first person to come by is the blacksmith, Mr. Alvira, who laughs broadly but promises help.

Alvira and his son Gilbert start up the small John Deere tractor, the one they call 'the Jitterbug,' and amid a lot of chuckling from a small crowd of ranchhands and children, we all head down to pull him out. They get a big kick out of "helpin' the little guy out," as Alvira puts it. It confirms their suspicions about city people—lots of book learning but little know how. After it dries out, the car starts up without hesitation—only a couple of nicks to remind us of the embarrassment.

Dad tells me Cutler says it's ok for me to get a job. So at a quarter to seven the next morning, dressed in Levi's and a blue workshirt, with my father's dollar watch in my pocket, I go out to the corral with the rest of them and wait my turn. Beech Nut puts me to picking apricots. Right away, Mike, my strawboss, a real sourpuss from the old country, starts yelling.

"Hey, boy! Hey, boy . . ." he strides over in a huff, hands flailing the air, "You pickee too damn green, too damn green, boy. If you no *see* fruit is ripe—feel, feel. Take your finger like dis and *feel,* God dammeee!"

They use long, 10-foot, three-legged ladders to get the fruit down from the trees. One time the back leg of my ladder sinks unexpectedly into soft, wet dirt and the whole thing tips over with me on it. I wrench my wrist and scuff the palm of my hand trying to stop my fall and spill and bruise just about all the fruit in my bag. Thank God, Mike wasn't there to see it.

As the afternoon heat begins to soar, the time drags and drags interminably. I keep looking at Dad's watch every five minutes disappointed that so little time has elapsed. What a relief to see five o'clock finally come around. I wash up, hurry through dinner and crash into bed.

The next day Gilbert, the blacksmith's son, who is 15 like me, is also assigned to pick and things are a little easier. At least there is someone to talk to. Gilbert is tall, lanky, with a handsome Indian profile, jet black hair and glossy black eyes. He tells me about high school, how he is on the baseball team, about the ranch, about how to pick fruit, about the need to wear a broad-brimmed hat to ward off the sun, where to find cool water, and how to get along with Mike by always saying "Sure, Mike, sure."

Gilbert and his family turn out to be our neighbors since his father, as the blacksmith, qualifies for company housing. That night I walk over to their house and meet everybody, his mom, his dad and his younger brother Tony. Their place reminds me of a petting zoo—chickens, ducks, pigeons, a milk cow, two noisy peacocks, a huge garden, even a beehive. Gilbert's mother loads me down with two big bags of onions, tomatoes, cucumbers, squash and carrots.

When the apricots peter out, Beech Nut assigns me to help Gilbert's father in the blacksmith shop.

I love it. Mr. Alvira is a solid, thick-chested guy with steel-gray hair. When he's working, he always wears a greasy black skullcap and an equally spotted leather apron.

"Pancho," he calls me, "I can see your daddy is a pretty honest fella. At least now we know that if we work, we'll get what the hell is coming to us and no *cabrón* is gonna deduct a bunch of bullshit like extra meals we never ate. It's good to have an honest guy in that office for a change. You lemme know if anybody gives him a hard time."

When I work in the blacksmith shop, the days pass like nothing ever. Mr. Alvira built his own hearth out of firebrick. It's my job to turn the rotary bellows and get the fire white hot. Then he positions the plowshares right into the hottest part of the coals and in just a few minutes, the blades glow cherry red. He plucks them out deftly with metal tongs, lays them on the anvil and pounds them to the desired shape and thickness with obvious strength and an artist's touch. While they are still blazing hot, he plunges them into a large barrel of black oil and they explode with a loud, steamy hiss. "That's the way you temper steel, *hijo,*" he explains. After they cool, I get to sharpen them on a rotary power grinder wearing the goggles he gives me. Every once in a while, I notice him going into the storage shed and down something hidden in a paper bag—booze I figure. No question about it, by the end of the day, he has a buzz on. "They hate us, Pancho," he tells me, "the *gabachos,* the whities, the red necks like Beech Nut. They hate our guts because we're Mexicans. First they took our land and then our women. They use us. And all we do is bow and scrape. Shit, man, it's a fuckin' shame, Pancho."

My stint in the blacksmith shop lasts only a couple of days, and because the apricots are all finished, there is nothing for Gilbert and me to do but cut weeds—the worst job on the ranch—digging up Bermuda and Johnson grass. They call it "workin on the chain gang" especially when Mike is your strawboss. Both Bermuda and Johnson have long segmented roots botanists call rhizomes. Each little segment, if left in the soil, will start a whole new plant. Bermuda is low and thick with matted roots; Johnson gets to be 5 to 6 feet tall with long tough, intertwined roots.

"Get root! Get root, boy, God dammee!" Mike yells as he stands there watching me struggle. By lunchtime, I'm exhausted. We look for some shade under a vine, and the Mexicans start a small fire from dead pieces of grapevine and roast tortillas on makeshift grills fashioned from loose vineyard wire. A bunch of them play craps until Mike makes us get back to work. With the temperature somewhere around 110, covered with dust and sweat, and pesky little thrips flitting into our faces, the going gets tough and tougher especially after three in the afternoon. What a relief to hear the truck plowing through dust to pick us up at five.

Mike tells us that he has to go to town the following Friday afternoon to answer some questions about his income tax. That leaves ten of us all alone in the middle of this vast vineyard miles from anywhere—on our honor. Even before we see his black Plymouth disappear into the dust, someone has already taken up a collection and two of our guys hurry down to where one of the irrigators is working and get him to drive them to the store for beer. They come back with two cases of Regal Pale in quart bottles. We cart the beer into the middle of a row so nobody can spot us, sit under some vines, and pass around bottle after bottle. I get the idea they all want Gilbert and me to get smashed—the elders initiating the young. And it isn't long before I become woozy, then dizzy and finally nauseous. They are singing the most beautiful songs I think I ever heard—all in harmony—and every once in a while, one of them lets out a great cry—"*un grito,*" they call it—a long piercing "Aaa-aa-yy, Aaa-aa-yy," like "Here we are world! In this boiling, blistering, fucking vineyard, but still full of manhood, full of dignity . . . full of *orgullo*—pride—AAAAYYYYYEEEE!"

"Hey, Pancho, you don't look so good," Enrique teases. General laughter.

"If I look as bad as I feel, I must look awful," I confess.

"Well," he says, "how would you like a special kinda cigarette?" he gives me a knowing wink. "Make you feel real good again, just like that—just 25 cents."

I give him the money but almost choke smoking it. It makes me so ill I crawl beneath a couple of vines into another row and retch until my sides ache. They help me both on and off the truck.

I expected my folks to really give me hell, but they think my being drunk is hilarious. I hadn't seen my mother laugh like that in years.

"Get the boy something to eat," my father shouts, "and some hot coffee."

"*Der ist besoffen*," she gloats, giggling.

I just head for bed.

The next morning, my father explains that Mike has claimed two exemptions on his income tax which of course is a lie. Everybody knows he's single. That's probably why the IRS wanted to talk to him he tells us. Not only that, but he marked himself down for 9 hours on the time sheet for the previous day, even though we all know he was in Madera all afternoon. "Oh, but keep shush," he whispers sarcastically. "About the annointed we don't say a word—nothing, right?" he says. "Sshhhh," and puts his finger to his lips.

My mother gets a job cutting fruit in the drying shed. They set a 25 lb. box of nectarines in front of her. She has to cut each fruit in half, pit it, and place the two halves face up on a wooden tray. After the box is finished, the "fruit boy" brings her another box, while the checker, who just happens to be Beech Nut's wife, a matronly lady with purple hair, comes over like the queen and punches her ticket with aloof reserve. Each punch is worth 12 cents.

Many a night Julius puts the front leg of his chair through the hole of the short piece of wood I fashioned for him. The metal peg of his cello bites into the board so that the instrument won't slip. Boccherini, Haydn, and Bach—his jaw jutts back and forth as he bows away with raw fury, the sweat pouring down his face onto his blue workshirt. And on Saturdays, my mother places her hearing aid right in front of the speaker of our radio and listens to the Metropolitan Opera with Milton Cross all the way from New York. I can see her lips move with the text especially when they do German operas—even the Wagner.

"Why do you listen to that stupid Nazi crap?" I taunt.

"Babelee, babelee bab," she answers making a deprecating hand motion. "What do you know? Just a baby, you are—still wet behind the ears. *Ja*. Babelee, babelee, bab—you chust shut up," she laughs.

Art Coelho

PAPA'S NATURALIZATION

PAPA HURRIED THE HOLSTEIN COWS into the milk barn. We all got out of his way quickly. All week long Papa showed little patience and we could tell something was eating at him. Since we'd gotten out of school he had been dishing out orders right and left. He made Ida so nervous she almost spilled her pail of milk. We were all milking by hand back then. You had to wash down the udders first, and strip 'em to clean 'em, and squirt the warm milk into the bucket. You had to get into a rhythm to do it right. But Papa had gotten us all outta whack 'cause we were watching him instead of paying attention to what we were supposed to be doing, which was milking cows like I said. Most of the time we squirted each other and cut up like crazy to break up the old routine, and Papa would smile at us, even chase us around the wooden stanchions. He hadn't gone after us in a long spell. His lips had been firm lately too. Firm as all getout. There was a gruffness in his voice that really put us on edge. He really got on our youngest sister's case something fierce. "You still milk Bertha," he said. Well, *said* ain't exactly right 'cause he *shouted* it. Laura told me later that she heard a couple of pigeons fly outta the barn when Papa raised his voice. "Belle, quit damn daydream too much."

"I ain't daydreaming, Papa."

"No. Why you milk udders? Nothing come out!"

"Oh, I guess—"

"You guess with the lollygag," he bleared, and marched up to the other end of the barn to check on how the rest of his brood was keeping to the grindstone.

I moved closer to Belle to find out the scoop. "What's Papa on his highhorse about?"

"Mama gave him the devil this morning."

"What for?"

"Naturalization."

"That again."

"Yeah."

"Papa says he's gonna go in later today to become a naturalized citizen. He's gonna leave Mama home and she threw one of those conniption fits. The hairy kind as Johnney puts it. She said he wasn't going without her, and that was that. You know how Mama swings her little head around when she means business. Here Papa is six-feet-two. She's five-foot-three. And he's listening close."

"Why doesn't Mama go too?"

"She's not ready yet. She's had no time to prepare. Papa done studied them papers and memorized all the answers by heart 'cause he don't go nowheres without his lessons. He even takes them in the old outhouse that nobody uses now but him, since we got a modern toilet that flushes right inside the house. You know how dark and dingy it is in there. Papa says he gets enough light from the cracks in the boards, but I don't see how he can. Papa says he's not no foreigner and folks better not call him that bad *word* no more."

"You mean *greenhorn*?"

"Yeah."

"Papa says he owns his own dairy farm here in Kings County and that he's gonna be a citizen of the United States of America and nothing can stop him under God's Heaven."

"He's really determined."

"I know."

"What else did Papa say?"

"Said he was going to take the test before supper. You know how Papa always messes up our American slang. He told Mama, 'Come this Hell in high water, I full Americano before this sundown comes.'"

"I bet Mama was angry after that."

"Like Papa always says her Terceira Island blood was boiling. He tried to kid her about her running the bulls over him."

"What do ya mean?"

"You know the Terceira story of Mama's island where at the Bay of Salga the farmwife turned the bulls loose from their pens and drove them down onto the invading Spanish soldiers—pushed them right back into the sea. Defeated them right in their tracks. Papa always teases her about that farmwife, saying it was a cousin of hers. Papa says you can't win an argument 'cause Terceira women are so stubborn and defiant."

"What'll you think Mama will do?"

"I don't know. I'm afraid to think about it."

"You know Mama when she makes her mind up. Papa always gave into her before. But this time he says she could be getting her citizenship papers too if she had wanted to bad enough like him."

"Did Mama come unglued?"

"It really burned her up. She said that raising eleven kids, looking after Papa, working side by side shocking hay, plus splitting apricots in Armona at the open shed cannery never left her no time for no book studying to become naturalized. Papa's lips shut tighter than a Pismo clam then. So I think Mama thought she'd won out."

"Does Mama know he's still planning to go after the cows are all milked."

"No. He never mentioned it after their fight at breakfast this morning. They were so mad at each other they never spoke as much as a word since. Their eyes would meet and it made me shiver to see them glaring like that. Papa cut into his beef steak at lunch and he had so much force in it that I could hear the blade cutting into the plate."

"It ain't gonna be long and something will hit the fan."

"You got that right."

We finished milking half an hour early at Papa's hectic pace and some of the older kids went out to the haystack to do the other chores. Papa told Tony to wash up the milk buckets and he headed straight for the house. I went to the side of our home and got myself scratched by the rosebush. We had us a good spot through the window, kept it cracked a little so we could listen in on what was going on. Mama was peeling potatoes for supper and at first they didn't speak. I guess Mama thought it was kinda funny that Papa was in early from milking and her curiosity made her break her vow of silence. I put my ear up close to the window and I could hear everything like I was in the next room. I realize it ain't polite to eavesdrop, but we never had no TV back then. And we'd only get to go to town once every three or four months in the buggy, so when something juicy around the house took shape we'd go whole hog and get it while it was hot.

"You done milking already," I heard Mama say.

"Finish up early."

"Cows no produce as much this evening?"

"Oh, produce lots."

"That's good. But supper won't be for awhile yet."

"For me—sandwich and soup. That's all."

Mama looked at Papa hard then, measuring him up and down with her piercing brown eyes, which were wide as they could go now 'cause of the shock in 'em. Papa poured himself some fresh milk from the pitcher in the icebox. I was listening so close and watching so intently that I almost fell off the little stool we'd placed under the window outside the house.

"Sandwich and soup!" Mama's voice cut like a scythe through ripe wheat stubble. "What you mean sandwich and soupa? I'm preparing this big meal."

"The test. This evening is test. It's what I been wait for so long. These many years been come to something now."

"I thought we settled it over linguiça and eggs this morning."

"I no give you idea like that to believe settle."

"You no mention at noon when we eat. I think you have change of heart."

"That's a good one, but not for me."

"You no go without me. I see to that."

"How you go to get those facts in your head so quick? It take time. Much time. When do you take time? No, you don't. I do. I take time. I study. I study crazy. You no study crazy. You no take papers to barn with pigeons."

"Next month. You'll see. I ready then. Next month I promise you."

"You been say 'Next month' every since we come from The Açores. Next month, Papa. Next month is next month. Years have go by. Now is now. I am now to go. This evening. Where's your facts? The naturalization. You got to have facts. Papers say you know this. You know that. Then you know what to say on test. You pass. You no longer *greenhorn*. Americano all the way. That's what I want now."

"I tell you Papa we should go together. It's only right. Man and wife do things side by side. No go behind the bush like you see everybody in this country do. Kill two birds with one stone. *Matar duma cajadada dois coelhos;* that's what I mean. The way you want to do it—feathers fly all over the place."

"Sorry Mama. I have to dress now. If I don't go, I don't know. I go crazy maybe."

"You no ignore my wish."

"It's no sense in talking argument like wild dogs." Papa walked into the bathroom and Mama grabbed her rosary and started praying. I was just glad she didn't light a candle for Papa. He'd been a goner then. She prayed for what seemed like a long time. Like she was summoning all the saints to her side at once.

While Papa was washing up Mama snuck in the bedroom and took out Papa's Sunday church coat and carried it out to the back porch. Papa came out of the bathroom whistling his favorite São Jorge Island tune. It sounded real happy and free. He started to get dressed. He put on his pants and a clean white shirt and a tie Mama had ironed earlier in the week for mass that Sunday. Papa had been shining on his good shoes every night, and even now, he took a brush to them; then he stood up, looked in the full-length mirror on the door and picked off a piece or two of lint on his slacks. He reached in the closet for his coat. He shoved and separated the clothes on the rack to one side and even looked through Mama's dresses. Finally in desperation he walked to the hallway and called out.

"Mama, you see my dress coat? I no can find."

"I'm looking right now at it."

"You what?"

"I see it."

"No, it's gone. No in closet here."

"Here."

"Here where?"

"Right here."

"You take out?"

"Yes."

"It no need cleaning. Good the way it is."

"No, but it will soon. Very soon."

"Mama, don't play game. No time for that now. I be late. I'm important business I tell you in town. In courthouse you be on time. I got to go *naturalize* myself."

Papa walked into the kitchen with a bewildered look on his face. Mama got up from the kitchen table and walked into the front porch that led to the farm driveway outside. I jumped off my stool from under the window and ran out into the yard where I could see. Mama took Papa's dress coat and removed it from the clothes hanger and walked down the steps. There had come a big shower the day before and rain puddles were visible and still deep out in the yard. Mama walked up to the biggest one she could find and furled out Papa's Sunday coat like a flag and picked up a stick from the ground and pushed the clean coat down into the muddy water till it was completely submerged in the puddle. Papa watched from the top of the porch steps. He seemed frozen there. He saw the defiant look on Mama's face. She looked up at him without fear and Papa couldn't say anything at first. He was really flabbergasted. Finally he blurted out.

"Holy Mother of God, what do you think you do to me?"

Before Mama spoke she turned to all us kids and smiled. We had not seen a smile on Mama's face for several weeks. She turned to Papa and said, "Well, you know they Baptize you first, then they give you Holy Communion, after that Confirmation. This is your NATURALIZATION!"

I thought Papa was going to get real mad. You couldn't tell what kind of expression he had on his face. He looked numb. But he finally started to laugh. He laughed so hard he had to sit down on the porch steps. I watched the tears rolling down his cheeks and Mama came over and sat down beside him and put her arm around his neck and kissed him. "Next month," she said. "We go next month." "It's a good month," Papa said. "The best," Mama agreed.

Lee Nicholson

A NEW SONNET FROM THE PORTUGUESE: A *FESTO DO ESPIRITO SANTO*—CORONATION DAY

Will this July Sunday turn out too hot?
Will we sweat? Will we fry like eggs in this
Noon sun? Will our mouths dry up? We are caught
In questions of the old until one Miss

In white appears and we forget to fear.
She is so young, but already she shines
In this light. Her dress is hand-made, as near
To regal with beads in rows, pearls in lines

As two grandmothers, calloused, faded brown,
Could sew all year. She, the newest of their
Line, has been placed here too, girl in a gown,
A walking pearl, their Beauty, latest heir.

Clearly this is a crowning moment for
All—girl, light, pearl, lines, even us—and more.

Amado Muro

SOMETHING ABOUT FIELDHANDS

IN BAKERSFIELD, I lined up at Nineteenth and M with dozens of other fieldhands waiting for day hauler's buses to carry them out to the pea fields.

A full-bearded fieldhand, called Raggedy Man on account of his torn clothes, waited with me. When I told him I'd just come in from Stockton, he smiled and allowed I'd arrived just in time for the harvest. But when I asked how the peas were, he frowned.

"It's been too cold—peas need warm weather and they're not filling up like they should," he said. "They aren't going to be swollen like they were last year. Buddy, we made it good then. Stiffs were earning their six or seven dollars a day. You could see jungle fires all along that spur track behind the San Joaquin Cotton Oil Company at night. Field tramps were cooking up, and it was nice to see everyone getting enough to eat. But it's not like that this year on account of the rains. There's a lot of hungry blanket stiffs running up and down the railroad tracks, and if I'm lying my elbow's named Dennis."

It was cold and frosty, and the wind blew steadily from the north. We sat down on the curb near Uncle Dave's Pawnshop, and the Raggedy Man said there hadn't been much field work in Kern County since the carrots.

"I been sleeping in a condemned boxcar on a spur track just off the Edison Highway," he said. "And I been living mostly on potatoes that Pacific Fruit Express icemen sweep out of reefers when they clean them. That's pride-breaking eating, but a stiff's not himself when he's hungry."

Another fieldhand, small, pale, and nicknamed Shades because he wore pinch-nose glasses, sat down beside us on the curb. He had a pinched, bony face, and there were tired circles under his eyes. He drew his breath in convulsive jerks, making a sharp whistling sound against his teeth, and each

---

exhalation was like the sudden deflation of a balloon. He said he'd hustled soda bottles all winter.

"Groceries pay four cents for every empty bottle," he said. "One day I walked clear to Greenfield looking for bottles on the road shoulders. I found eighteen and that meant 72 cents so I bought four pounds of pinto beans, and still had twelve cents left over for snuff. But it's hard to scuffle up something—the weather's cold for bottle-hauling."

The dim-lighted thoroughfare of small cafes, walkup hotels, and dingy scuffle joints, hummed like a running belt with the voices of many fieldhands. A cheery man, wearing a gray rope sweater, a faded brown derby and mud-caked gabardine trousers patched with Durham sacks, shook hands with the Raggedy Man. His good-natured face was fiery red from the wind, and he said he'd topped carrots and then sold whole blood and plasma on L.A.'s Nickel Street skid row after Kern County's field work gave out.

"Field tramping hasn't hurt me any—I don't look a day over 200," he said.

A day hauler's bus with WILLIAM BLEVINS stenciled on its blue sides nosed into the curb, and a lanky, red-haired bullhorn bawled: "Let's get those peas, men—let's make that farmer go to the bank at ten o'clock in the morning."

The street livened up. Fieldhands milled around the parked bus with heads lowered to keep the whistling wind from cutting their faces.

"I'll be glad when winter's over," the Raggedy Man mumbled. "I'd rather hunt a shade than a stove."

It was growing lighter. The sky was gray in the east and the moon had set, but the half-dark city was still only faintly streaked with daylight. The farm bus began to fill up.

Shades blinked sleepily when he got on the bus, and his breathing was like a sigh.

"I'm sleepier than Rip Van Winkle, and I wouldn't get on this bus at all if it wasn't for Mr. Need More," he averred. "You know him don't you, buddy? He's the guy that loads these farm buses. It don't matter though. This'll be country gravy compared to what I've went through this winter."

The stocky, barrel-chested fieldhand picking peas in the next row filled his hamper about the same time I did. He straightened up, grimacing from the strain on his back, calves, and thighs.

"Easy money ain't it, Pancho?" he remarked pleasantly.

Away ahead, in the distance, a long range of sharp hills embroidered the horizon. Wind-driven clouds raced before a smoky blob that seemed to pursue them. The wind's chill crawled under my trouser cuffs and down the neck of my collar and I turned up my lapels.

"Wind's about to blow away the mortgage on the widow's old cow," the stocky man said.

He had a handkerchief tied like a mask over his nose and mouth to keep out the dust. He closed his teeth down over his lower lip, and kept his head bent while he untied it. After that we hefted the hampers to our shoulders, and plodded toward the grading bins. There was a long line ahead of us, and we waited our turn while graders culled out short, flat, and spotted pea pods.

The stocky man rolled a cigarette down between his knees, where the wind couldn't get at it, and then passed the makings to me. After that, with his mouth and nostrils dense with smoke, he rewrapped his sack of tobacco and matches in waxed bread paper and put it back in his shirt pocket. He had a strong candid face with mild eyes, and his hair was like pulled taffy in streaks and gray for the rest of it. He wore a peanut straw haying hat, and it was pushed back on his brows. Occasionally he would show his teeth in a slow smile. After a while he told me he'd been a coal miner in Hazard, Kentucky, before his wife died. He talked without gestures, and without opening his lips very far.

"After she died I went to field tramping," he said. "That was two years ago, and I been on a ramble ever since."

The long line thinned out. We moved on to the grading bins, and the stocky man said he'd stayed with his married son in Houston part of the winter.

"My son's a longshoreman, and a fine boy," he said. "He works mostly trimming grain and sulphur, some cotton and washdown jobs too. He treated me about as nice as a father could ask. He threw up his hand and said goodbye to me when he went to work every morning, and he bought me instant coffee and Prince Albert. Beside that I had two changeovers of clothes. That boy didn't want to see me leave. He said: 'Don't go, Dad—stay here with me and Constance.' I'd have liked to have stayed, but I knew it wouldn't be right. So I said to him: 'Son, if I was disabled I'd stay but I'm able to make it. I got to go—I can't set down on you no longer.'"

The graders culled out our bad peas, and we moved on to the scales and waited in line for a weigh-up. The sky had begun to spot with rain, and the wind had come up stronger. We could hear far off in the distance the occasional whistle of a freight train which, dying down, left the field quiet again.

"After I left my boy's house I went out to the SP yards and got on a through Man, and hung it to Yuma," the stocky man said. "Then I got on another fast fellow, and I doubled back to Phoenix, I made the day hauls in the carrots and stayed in a double-decker flophouse on Jefferson Street. But the Sweet Lucy's went to juggin' and fussin' all night so I moved out and slept in that jungle stiffs call the Icehouse. Before long farm work slacked up though, and I couldn't catch a day haul. They was tying onions by then, and the day haulers left a whole lot of hands behind. So I went to the soupline at St. Vincent de Paul's on Ninth Avenue. I counted 356 in the soupline that day, not including women and children. So many lined up the meat was out of the stew when my turn come. After that I heard the peas was about ready so I come here. That was a beautiful ride from L.A. That empty boxcar didn't bounce—just waved a little—it was a hydra-cushion car. Going over the hump it got cold though. I had three quilts and my overcoat, and when I woke up there was snow on me. But I snuggled down deeper, and kept warm."

The line thinned out again, and we carried our hampers to the scales. The weigh-hand was middle-aged with chapped, scaling lips, and a body that bulged in the middle like a top. He said we each had 25 pounds of peas, and we showed the paymaster our work cards. The paymaster had on clean khaki work clothing, and his clean-shaven face had a rich, healthy coloring. He wore a black Sam Browne belt, and his pistol was pearl-handled. He marked our work cards, and paid us. After that we went back to our rows.

Rain was falling lightly now, like mist. The stocky man set down his hamper and then pulled up his foot, and examined the loosened sole of his field shoe.

"Ain't so had as the short-handled hoe is it, Pancho," he said.

He smiled and fixed his eyes on me.

"I'd rather tough it out in this pea patch than live off my son, and just rock on his front porch," he said.

After that he bent almost double over the peavines again.

Chitra Banerjee Divakaruni

THE BRIDES COME TO YUBA CITY

The sky is hot and yellow, filled
with blue screaming birds. The train
heaved us from its belly
and vanished in shrill smoke.
Now only the tracks
gleam dull in the heavy air,
a ladder to eternity, each receding rung
cleaved from our husbands' ribs.
Mica-flecked, the platform
dazzles, burns up through thin
chappal soles, lurches
like the ship's dark hold,
blurred month of nights, smell of vomit,
a porthole like the bleached iris
of a giant unseeing eye.

Red-veiled, we lean into each other,
press damp palms, try
broken smiles. The man
who met us at the ship whistles
a restless *Angrezi* tune
and scans the fields. Behind us,
the black wedding trunks, sharp-edged,
shiny, stenciled with strange men-names
our bodies do not fit into:
Mrs. Baldev Fohl, Mrs. Kanwal Bains.
Inside, folded like wings,
bright *salwar kameezes* scented
with sandalwood. For the men,
kurtas and thin white gauze
to wrap their uncut hair.

Laddus from Jullundhar, sugar-crusted,
six kinds of lentils, a small bag
of *bajra* flour. Labeled in our mothers'
hesitant hands, pickled mango and lime,
packets of seeds—*methi, karela, saag*—
to burst from this new soil
like green stars.

He gives a shout, waves
at the men, their slow
uneven approach. We crease our eyes
through the veils' red film,
cannot breathe. Thirty years
since we saw them. Or never,
like Harvinder, married last year
at Hoshiarpur to her husband's photo,
which she clutches tight to her
to stop the shaking. He is fifty-two,
she sixteen. Tonight—like us all—
she will open her legs to him.

The platform is endless-wide.
The men walk and walk
without advancing. Their lined,
wavering mouths, their
eyes like drowning lights.
We cannot recognize a single face.

YUBA CITY SCHOOL

From the black trunk I shake out
my one American skirt, blue serge
that smells of mothballs. Again today
Neeraj came crying from school. All week
the teacher has made him sit
in the last row, next to the fat boy
who drools and mumbles,

picks at the spotted milk-blue
skin of his face, but knows
to pinch, sudden-sharp,
when she is not looking.

The books are full of black curves,
dots like the eggs the boll-weevil lays
each monsoon in furniture-cracks
in Ludhiana. Far up in front
the teacher makes word-sounds
Neeraj does not know. They float
from her mouth-cave, he says,
in discs, each a different color.

Candy-pink for the girls
in their lace dresses, marching
shiny shoes. Silk-yellow
for the boys beside them,
crisp blond hair, hands raised
in all the right answers. Behind them
the Mexicans, whose older brothers,
he tells me, carry knives,
whose catcalls and whizzing rubber bands
clash, mid-air, with the teacher's
voice, its sharp purple edge.
For him, the words are
a muddy red, flying low and heavy,
and always the one he has learned to understand:
idiot, idiot, idiot.

I heat the iron over the stove. Outside
evening blurs the shivering
in the eucalyptus. Neeraj's shadow
disappears into the hole
he is hollowing all afternoon.
The earth, he knows, is round, and if
one can tunnel all the way through,
he will end up in Punjab,
in his grandfather's mango orchard,
his grandmother's songs lighting

on his head, the old words
glowing like summer fireflies.

In the playground, Neeraj says,
invisible hands snatch at his uncut hair,
unseen feet trip him from behind,
and when he turns, ghost laughter
all around his bleeding knees.
He bites down on his lip
to keep in the crying. They are
waiting for him to open his mouth,
so they can steal his voice.

I test the iron with little drops of water
that sizzle and die. Press down
on the wrinkled cloth. The room fills
with a smell like singed flesh.
Tomorrow in my blue skirt I will go
to see the teacher, my tongue
stiff and swollen
in my unwilling mouth, my few
English phrases. She will pluck them
from me, nail shut my lips. My son
will keep sitting in the last row
among the red words that drink his voice.

Carlos Bulosan

from AMERICA IS IN THE HEART

I COUNTED THIRTEEN short tunnels before we came out to the border of California, rolling across a wide land of luxuriant vegetation and busy towns. Then there was a river, and not far off the town of Marysville loomed above a valley of grapes and sugar beets, all green and ready for the summer harvest.

I wanted to stop and walk around town, but some of the hoboes told me that there were thousands of Filipinos in Stockton. I remained on the same train until it got to Sacramento, where I boarded another that took me to Stockton. It was twilight when the train pulled into the yards. I asked some of the hoboes where I could find Chinatown, for there I would be sure to find my countrymen.

"El Dorado Street," they said.

It was like a song, for the words actually mean "the land of gold." I did not know that I wanted gold in the new land, but the name was like a song. I walked slowly in the streets, avoiding the business district and the lights. Then familiar signs glowed in the coming night, and I began to walk faster. I saw many Filipinos in magnificent suits standing in front of poolrooms and gambling houses. There must have been hundreds in the street somewhere, waiting for the night.

I walked eagerly among them, looking into every face and hoping to see a familiar one. The asparagus season was over and most of the Filipino farmhands were in town, bent on spending their earnings because they had no other place to go. They were sitting in the bars and poolrooms, in the dance halls and gambling dens; and when they had lost or spent all their money, they went to the whorehouses and pawed at the prostitutes.

I entered a big gambling house on El Dorado and Lafayette Streets, where ten prostitutes circulated, obscenely clutching at some of the gamblers. I

Reprinted by permission of Harcourt Brace & Company. These excerpts describe events in the 1930s.

went to a stove in the middle of the room where a pot of tea was boiling. I filled a cup and then another, and the liquid warmed my empty stomach. This was to save me in harsher times, in the hungry years of my life in America. Drinking tea in Chinese gambling houses was something tangible, and gratifying, and perhaps it was because of this that most of the Filipino unemployed frequented these places.

I was still drinking when a Chinese came out of a back room with a gun and shot a Filipino who was standing by a table. When the bullet hit the Filipino, he turned toward the Chinese with a stupid look of surprise. I saw his eyes and I knew that the philosophers lied when they said death was easy and beautiful. I knew that there was nothing better than life, even a hard life, even a frustrated life. Yes, even a broken-down gambler's life. And I wanted to live.

I ran to the door without looking back. I ran furiously down the street. A block away, I stopped in a doorway and stood, shivering, afraid, and wanting to spit out the tea that I had drunk in the gambling house. When my heart ceased pounding, I walked blindly up a side street. I had not gone far when I saw a building ablaze.

"What is it?" I asked a Filipino near me.

"It is the Filipino Federation Building," he said. "I don't agree with this organization, but I know why the building is burning. I know the Chinese gambling lords control this town."

I did not know what he meant. I looked at him with eager eyes.

"I don't know what you mean," I said. "I've just arrived in the United States."

"My name is Claro," he said, extending a long, thin hand, and coughing behind the other. "I came from Luna, in the province of La Union. Let us go to my restaurant and I will explain everything to you. Are you hungry, boy?"

He was not much older than I, and he spoke my dialect.

"I have not eaten for two days," I said. "You see, I took the freight train in Sunnyside, Washington."

Claro hugged me. When he entered the restaurant, he locked the door and put down the shades.

"I don't want the swine in the street to see us," he said, going to the stove. "They disgust me with their filthy interest in money. That is why I am always behind in my bills. I like good people, so I am keeping this restaurant for them."

I watched him prepare vegetable soup and fry a piece of chicken. When the pot started to boil, Claro put a record on a portable phonograph at the other end of the counter; then there was a sudden softness in his face, and his eyes shone. He had put on a Strauss waltz. Going back to the stove, Claro raised his hands expertly above his head in the manner of boleros and started to dance, swaying gracefully in the narrow space between the stove and the counter. He was smiling blissfully, and when someone knocked on the door he stopped suddenly and shouted:

"Go away! The place is closed for tonight!"

When he had placed everything on the counter, Claro took a chair and sat near me.

"Listen, my friend," he said. "The Chinese syndicates, the gambling lords, are sucking the blood of our people. The Pinoys work every day in the fields but when the season is over their money is in the Chinese vaults! And what do the Chinese do? Nothing! I see them only at night in their filthy gambling dens waiting for the Filipinos to throw their hard-earned money on the tables. Why, the Chinese control this town! The local banks can't do business without them, and the farmers, who badly need the health and interest of their Filipino workers, don't want to do anything because they borrow money from the banks. See!"

I was too hungry to listen. But I was also beginning to understand what he was trying to say.

"Perhaps in another year I will be able to understand what you are saying, Claro," I said.

"Stay away from Stockton," he warned me. "Stay away from the Chinese gambling houses, and the dance halls and the whorehouses operated by Americans. Don't come back to this corrupt town until you are ready to fight for our people!"

I thanked him and walked hastily to the door. I hurried to the freight yards. I was fortunate enough to find an empty boxcar. I sat in a corner and tried to sleep, but Claro's words kept coming back to me. He wanted me to go back to fight for our people when I was ready. I knew I would go back, but how soon I did not know. I would go back to Claro and his town. His food had warmed me and I felt good.

• • •

I arrived in Bakersfield and walked from poolrooms to gambling houses. The season for picking grapes was still far off. The vines were just pruned. There was no work for the cold months of winter. From the gambling houses I went to the whorehouses, hoping to find someone I knew. There were no other places where Filipinos could go. I sat in the living room and watched lonely Filipinos paw at the semi-nude girls. I felt angry and lost. Where in this wide country could I go? I felt the way other Filipinos felt. I rushed out and cursed the cold night.

I discovered that three Filipino farm labor contractors controlled the grape industry. Nearly three thousand Filipino workers depended on them. They lived in crowded bunkhouses operated by these men. It was exploitation everywhere, even among ourselves. It was the same thing I had known years before.

I wanted to see one of the contractors. I was introduced to Cabao, who had nearly eight hundred Filipinos under him. He was younger than most contractors, but I was skeptical. He drove me to his ranch. His house was large and gaudy. I saw a college diploma on the wall above his writing desk.

I was looking at it when a car drove into the yard. Cabao rushed to the window and looked out.

"My wife," he said.

She burst into the house and came to the study with a bottle of whisky. I was shocked when I saw her face. Where had I seen her before? I stumbled to my feet when Cabao introduced me to her. But she did not stay long. I heard her drive out of the yard. Then I remembered where I had seen her! I looked at Cabao sadly.

"I'm sure you have seen my wife before," he said apologetically. "Everybody knows what she was before I married her. She worked in every important town in California. That is why everybody knows her. She followed the seasons, the way Filipinos follow the crops."

I wanted to find out why he married her. He had almost everything he wanted. He had had a good education.

"I saw your wife once some years ago," I said carefully. "But I didn't mean to ask you about her life history."

"It's all right," he said. "When I talk about it I feel free. Do you think it's money she wants? I give her enough. But she still is eager for the attention of men. I guess they are all the same."

"Why did you marry her?" I asked.

"She was young when I saw her in Watsonville," he said. "I was young, too. I had gone there to work for the summer, because I wanted to earn enough money to pay my college fees that year. I was taking Sociology at the University of California. I took her with me and worked for her. There were years of desperation. But when I came here and made a little money, I bought this house for her. I thought she would settle down. I was wrong. Do you know where she is going tonight?"

I did not want to know. But I could guess. I got up and started moving to the door.

"I'll drive you to town," Cabao said.

At the station, when Cabao had left, I discovered that he had put some money in my coat pocket. I took the bus and sat silently in a corner.

I was on my way north again. Familiar towns. But I could not erase Cabao from my mind. I recalled his gentle, educated voice, his delicate hands. There was something lost and futile, something utterly defeated in him.

"It's all right for me to suffer," I said to myself. "I'm stronger than he is. He has no right to suffer. . . ."

I arrived in Stockton during a strike. Filipino asparagus workers were in the midst of a general walkout. A long parade was moving down El Dorado Street, but the strikers were orderly and quiet. I stood on a corner reading the pennants and placards carried by some of the men. I noticed that all the stores and other buildings were closed on either side of the street. Even the gambling houses and liquor stores were closed.

I saw Claro leading a section of strikers. He was boldly carrying a sign which said:

PAISANOS! DON'T PATRONIZE JAP STORES!
IT MEANS HUNGER!

His chin was up, his face animated. There was a grin on his mouth. Suddenly I felt an urge in me to run to him. When was it that I first saw him? It seemed so long ago! I shouted to him and pointed to the sign. He looked in my direction but did not recognize me. I ran to him and shouted into his ear.

"Don't you remember me?"

For a moment he stopped, his eyes wandering wildly into the past, and then he flung his arms about me. There was genuine affection in his voice. The gesture of Claro, similar to the moving salutation of the French, was to spread to the members of our circle—to the Filipinos in the labor and progressive movements. It was to become a sign of affinity and affection.

"You have changed, Carlos!" he said.

I ignored him. But I said, "I don't understand some of your placards. I thought this was a general walkout of asparagus workers."

"Yes, it is!" he shouted with anger. "But a Japanese woman is breaking it. She is supplying laborers." He walked on, looking from side to side, shouting greetings to friends watching from doorways. When he saw a Japanese face he became furious.

Where had I met a similar character? Was it in Gorki's *Decadence*? In this novel, in one of the crowded streets, a revolutionist was walking with a surging crowd, anonymously. There was a powerful secret in his heart, and as he moved with the crowd remembering comrades who had fallen and thinking of the promise of the future, his eyes glowed with happiness and his whole face became animated with sudden joy.

I tugged at his sleeve. Could I tell him that I had come back to fight? Would he remember that he had sent me away long ago but advised me to return when I was ready to fight for our people? Would he remember that autumn day when I ate hungrily in his restaurant?

"This strike means more than dollars and cents in the asparagus fields," Claro said. "This very day the trade union movement and other progressive groups in Manila are demanding that the government boycott Japanese products. But it is deeper than you think. Tons and tons of scrap iron are going to Japan from the United States. These are made into bombs that are being dropped upon the peaceful Chinese people."

"I thought you didn't like the Chinese people," I said.

"There are good and bad men in every people," he said. "For instance, I didn't like the Chinese vice lords in Stockton. I still don't like them, but they are co-operating in this strike."

"What do you mean?" I asked.

"They have closed all their gambling houses. Do you see what this means? The Pinoys will keep their money and spend it only on food. The strike will last longer, and the farmers will lose two million dollars this season. Of course, the Chinese will also lose—but they figure that they will win in the end."

"There is no sense to it," I said. "If you win from one side, you lose to the other. Is there no way of winning from both sides? Isn't there, Claro?"

The parade moved eastward to Main Street and into the huge auditorium where local leaders were assembled to address the strikers.

"The UCAPAWA is now in power in the agricultural areas of the coast," Claro said. "But we have a strong independent union here."

I went to the back room and sent a dispatch to a labor paper in San Francisco about the strike. A representative from the Philippine government in Washington went to the rostrum and offered his support. This man was a spectacular figure in Filipino life. A labor commissioner in Hawaii as a young man, he was also a writer and an editor. He was multilingual. He was a leader for the common man, and he tried, in his brief career among Filipinos in California, to bring their predicament to the attention of the home government. Unfortunately he died before he could accomplish his mission.

Alan Chong Lau

"SUN YAT SEN COMES TO LODI"

for my great grandfather, ou ch'ü-chia

1

SUN YAT SEN COMES TO LODI
grandfather in pinstripes
mouth sporting a toothpick
tells friends, "no sin, no sin,
no sir, no sin to get excited"

mr. yee's four-year-old beaming in a pink meenop
hair's done up in pinktails
sam wo has closed his laundry
only day of the year he would do this
excepting new year's

the good doctor smiles
from a sedan's back seat
cheers resound
delta dust flies

there is the speech
"china will be china again"
this brings tears

not losing a minute
to sip
he tells us all that money buys arms
money drives out manchus

most people understand, there is little hesitation
the new york yankees have not yet won the pennant

---

it is too early to predict weather or the lucky number
but money is dug from pockets
pulled from cloth bags

when the time comes
he says thank you
a cry of genuine sadness
a rush to take seats for a last picture

photographer tong yee
fumbles underneath a black shroud like a soul leaving body
poses change   legs shift position
nobody seems to mind too much
only local banker wong hesitates
meeting the public often, he declines
offering a bigger contribution instead

grandfather sits by the doctor's side
pausing only to doff his hat
remove a coin from the ear
and drop a wet toothpick in a spittoon

2

he is proud of that picture
brown and bent in one corner
the only photo left in the family album
since big sister's marriage

there is also a newspaper clipping
with the headline
"SUN YAT SEN COMES TO LODI"
spread out all in characters
that could be relatives telling a story
or scales of a black bass dripping evidence of water

never having learnt the language
i just have to go by hearsay

Maxine Hong Kingston

from THE AMERICAN FATHER

THE BEST OF THE FATHER PLACES I did not have to win by cunning; he showed me it himself. I had been young enough to hold his hand, which felt splintery with calluses "caused by physical labor," according to MaMa. As we walked, he pointed out sights; he named the plants, told time on the clocks, explained a neon sign in the shape of an owl, which shut one eye in daylight. "It will wink at night," he said. He read signs, and I learned the recurring words: *Company, Association, Hui, Tong.* He greeted the old men with a finger to the hat. At the candy-and-tobacco store, BaBa bought Lucky Strikes and beef jerky, and the old men gave me plum wafers. The tobacconist gave me a cigar box and a candy box. The secret place was not on the busiest Chinatown street but the street across from the park. A pedestrian would look into the barrels and cans in front of the store next door, then walk on to the herbalist's with the school supplies and saucers of herbs in the window, examine the dead flies and larvae, and overlook the secret place completely. (The herbs inside the hundred drawers did not have flies.) BaBa stepped between the grocery store and the herb shop into the kind of sheltered doorway where skid-row men pee and sleep and leave liquor bottles. The place seemed out of business; no one would rent it because it was not eyecatching. It might have been a family association office. On the window were dull gold Chinese words and the number the same as our house number. And delightful, delightful, a big old orange cat sat dozing in the window; it had pushed the shut venetian blinds aside, and its fur was flat against the glass. An iron grillwork with many hinges protected the glass. I tapped on it to see whether the cat was asleep or dead; it blinked.

BaBa found the keys on his chain and unlocked the grating, then the door. Inside was an immense room like a bank or a post office. Suddenly no city street, no noise, no people, no sun. Here was horizontal and vertical

---

 This excerpt describes events in the 1940s.

order, counters and tables in cool gray twilight. It was safe in here. The cat ran across the cement floor. The place smelled like cat piss or eucalyptus berries. Brass and porcelain spittoons squatted in corners. Another cat, a gray one, walked into the open, and I tried following it, but it ran off. I walked under the tables, which had thick legs.

BaBa filled a bucket with sawdust and water. He and I scattered handfuls of the mixture on the floors, and the place smelled like a carnival. With our pushbrooms leaving wet streaks, we swept the sawdust together, which turned gray as it picked up the dirt. BaBa threw his cigarette butts in it. The cat shit got picked up too. He scooped everything into the dustpan he had made out of an oil can.

We put away our brooms, and I followed him to the wall where sheaves of paper hung by their corners, diamond shaped. "Pigeon lottery," he called them. "Pigeon lottery tickets." Yes, in the wind of the paddle fan the soft thick sheaves ruffled like feathers and wings. He gave me some used sheets. Gamblers had circled green and blue words in pink ink. They had bet on those words. You had to be a poet to win, finding lucky ways words go together. My father showed me the winning words from last night's games: "white jade that grows in water," "red jade that grows in earth," or—not so many words in Chinese—"white waterjade," "redearthjade," "firedragon," "waterdragon." He gave me pen and ink, and I linked words of my own: "rivercloud," "riverfire," the many combinations with *horse, cloud,* and *bird.* The lines and loops connecting the words, which were in squares, a word to a square, made designs too. So this was where my father worked and what he did for a living, keeping track of the gamblers' schemes of words.

We were getting the gambling house ready. Tonight the gamblers would come here from the towns and the fields; they would sail from San Francisco all the way up the river through the Delta to Stockton, which had more gambling than any city on the coast. It would be a party tonight. The gamblers would eat free food and drink free whiskey, and if times were bad, only tea. They'd laugh and exclaim over the poems they made, which were plain and very beautiful: "Shiny water, bright moon." They'd cheer when they won. BaBa let me crank the drum that spun words. It had a little door on top to reach in for the winning words and looked like the cradle that the Forty-niner ancestors had used to sift for gold, and like the drum for the lottery at the Stockton Chinese Community Fourth of July Picnic.

He also let me play with the hole puncher, which was a heavy instrument with a wrought-iron handle that took some strength to raise. I played gambler punching words to win—"cloudswallow," "riverswallow," "river forking," "swallow forking." I also punched perfect round holes in the corners so that I could hang the papers like diamonds and like pigeons. I collected round and crescent confetti in my cigar box.

While I worked on the floor under the tables, BaBa sat behind a counter on his tall stool. With black elastic armbands around his shirtsleeves and an eyeshade on his forehead, he clicked the abacus fast and steadily, stopping to write the numbers down in ledgers. He melted red wax in candle flame and made seals. He checked the pigeon papers, and set out fresh stacks of them. Then we twirled the dials of the safe, wound the grandfather clock, which had a long brass pendulum, meowed at the cats, and locked up. We bought crackly pork on the way home.

According to MaMa, the gambling house belonged to the most powerful Chinese American in Stockton. He paid my father to manage it and to pretend to be the owner. BaBa took the blame for the real owner. When the cop on the beat walked in, Baba gave him a plate of food, a carton of cigarettes, and a bottle of whiskey. Once a month, the police raided with a paddy wagon, and it was also part of my father's job to be arrested. He never got a record, however, because he thought up a new name for himself every time. Sometimes it came to him while the city sped past the barred windows; sometimes just when the white demon at the desk asked him for it, a name came to him, a new name befitting the situation. They never found out his real names or that he had an American name at all. "I got away with aliases," he said, "because the white demons can't tell one Chinese name from another or one face from another." He had the power of naming. He had a hundred dollars ready in an envelope with which he bribed the demon in charge. It may have been a fine, not a bribe, but BaBa saw him pocket the hundred dollars. After that, the police let him walk out the door. He either walked home or back to the empty gambling house to straighten out the books.

Two of the first white people we children met were customers at the gambling house, one small and skinny man, one fat and jolly. They lived in a little house on the edge of the slough across the street from our house. Their arms were covered with orange and yellow hair. The round one's name was Johnson, but what everyone called him was Water Shining, and his partner was White Cloud. They had once won big on those words. Also *Johnson* resembles *Water Shining,* which also has *o, s,* and *n* sounds. Like two old China Men, they lived together lonely with no families. They sat in front of stores; they sat on their porch. They fenced a part of the slough for their vegetable patch, which had a wooden sign declaring the names of the vegetables and who they belonged to. They also had a wooden sign over their front door: TRANQUILITY, a wish or blessing or the name of their house.

They gave us nickels and quarters; they made dimes come out of noses, ears, and elbows and waved coins in and out between their knuckles. They were white men, but they lived like China Men.

When we came home from school and a wino or hobo was trying the doors and windows, Water Shining came out of his little house. "There's a wino breaking into our house," we told him. It did occur to me that he might be offended at our calling his fellow white man a wino. "It's not just a poor man taking a drink from the hose or picking some fruit and going," I explained.

"What? What? Where? Let's take a look-see," he said, and walked with us to our house, saving our house without a fight.

The old men disappeared one by one before I noticed their going. White Cloud told the gamblers that Water Shining was killed in a farming accident, run over by a tractor. His body had been churned and plowed. White Cloud lived alone until the railroad tracks were leveled, the slough drained, the blackbirds flown, and his house torn down.

My father found a name for me too at the gambling house. "He named you," said MaMa, "after a blonde gambler who always won. He gave you her lucky American name." My blonde namesake must have talked with a cigarette out of the side of her mouth and left red lip prints. She wore a low-cut red or green gambling dress, or she dressed cowgirl in white boots with baton-twirler tassels and spurs; a stetson hung at her back. When she threw down her aces, the leather fringe danced along her arm. And there was applause and buying of presents when she won. "Your father likes blondes," MaMa said. "Look how beautiful," they both exclaimed when a blonde walked by.

But my mother keeps saying those were dismal years. "He worked twelve hours a day, no holidays," she said. "Even on New Year's, no day off. He couldn't come home until two in the morning. He stood on his feet gambling twelve hours straight."

"I saw a tall stool," I said.

"He only got to sit when there were no customers," she said. "He got paid almost nothing. He was a slave; I was a slave." She is angry recalling those days.

After my father's partners stole his New York laundry, the owner of the gambling house, a fellow ex-villager, paid my parents' fares to Stockton, where the brick buildings reminded them of New York. The way my mother repaid him—only the money is repayable—was to be a servant to his, the owner's, family. She ironed for twelve people and bathed ten children. Bitterly, she kept their house. When my father came home from work at two in the morning, she told him how badly the owner's family had treated her, but he told her to stop exaggerating. "He's a generous man," he said.

The owner also had a black servant, whose name was Harry. The rumor was that Harry was a half-man/half-woman, a half-and-half. Two servants could not keep that house clean, where children drew on the wallpaper and dug holes in the plaster. I listened to Harry sing "Sioux City Sue." "Lay down my rag with a hoo hoo hoo," he sang. He squeezed his rag out in the bucket and led the children singing the chorus. Though my father was also as foolishly happy over his job, my mother was not deceived.

When my mother was pregnant, the owner's wife bought her a dozen baby chicks, not a gift; my mother would owe her the money. MaMa would be allowed to raise the chicks in the owner's yard if she also tended his chickens. When the baby was born, she would have chicken to give for birth announcements. Upon his coming home from work one night, the owner's wife lied to him, "The aunt forgot to feed her chickens. Will you do it?" Grumbling about my lazy mother, the owner went out in the rain and slipped in the mud, which was mixed with chicken shit. He hurt his legs and lay there yelling that my mother had almost killed him. "And she makes our whole yard stink with chicken shit," he accused. When the baby was born, the owner's wife picked out the scrawny old roosters and said they were my mother's twelve.

Ironing for the children, who changed clothes several times a day, MaMa had been standing for hours while pregnant when the veins in her legs rippled and burst. After that she had to wear support stockings and to wrap her legs in bandages.

The owner gave BaBa a hundred-and-twenty-dollar bonus when the baby was born. His wife found out and scolded him for "giving charity."

"You deserve that money," MaMa said to BaBa. "He takes all your time. You're never home. The babies could die, and you wouldn't know it."

When their free time coincided, my parents sat with us on apple and orange crates at the tiny table, our knees touching under it. We ate rice and salted fish, which is what peasants in China eat. Everything was nice except what MaMa was saying, "We've turned into slaves. We're the slaves of these villagers who were nothing when they were in China. I've turned into the servant of a woman who can't read. Maybe we should go back to China. I'm tired of being Wah Q," that is, a Sojourner from Wah.

My father said, "No." Angry. He did not like her female intrigues about the chickens and the ironing and the half-man/half-woman.

They saved his pay and the bonuses, and decided to buy a house, the very house they were renting. This was the two-story house round the corner from the owner's house, convenient for my mother to walk to her servant job and my father to the gambling house. We could rent out the bottom floor and make a profit. BaBa had five thousand dollars. Would the owner, who spoke English, negotiate the cash sale? Days and weeks passed, and when he asked the owner what was happening, the owner said, "I decided to buy it myself. I'll let you rent from me. It'll save you money, especially since you're saving to go back to China. You're going back to China anyway." But BaBa had indeed decided to buy a house on the Gold Mountain. And this was before Pearl Harbor and before the Chinese Revolution.

He found another house farther away, not as new or big. He again asked the owner to buy it for him. You would think we could trust him, our fellow villager with the same surname, almost a relative, but the owner bought up this house too—the one with the well in the cellar—and became our landlord again.

My parents secretly looked for another house. They told everyone, "We're saving our money to go back to China when the war is over." But what they really did was to buy the house across from the railroad tracks. It was exactly like the owner's house, the same size, the same floor plan and gingerbread. BaBa paid six thousand dollars cash for it, not a check but dollar bills, and he signed the papers himself. It was the biggest but most rundown of the houses; it had been a boarding house for old China Men. Rose bushes with thorns grew around it, wooden lace hung broken from the porch eaves, the top step was missing like a moat. The rooms echoed. This was the house with the attic and basement. The owner's wife accused her husband of giving us the money, but she was lying. We made our escape from them. "You don't have to be afraid of the owner any more," MaMa keeps telling us.

Sometimes we waited up until BaBa came home from work. In addition to a table and crates, we had for furniture an ironing board and an army cot, which MaMa unfolded next to the gas stove in the wintertime. While she ironed our clothes, she sang and talked story, and I sat on the cot holding one or two of the babies. When BaBa came home, he and MaMa got into the cot and pretended they were refugees under a blanket tent. He brought out his hardbound brown book with the gray and white photographs of white men standing before a flag, sitting in rows of chairs, shaking hands in the street, hand-signaling from car windows. A teacher with a suit stood at a blackboard and pointed out things with a stick. There were no children or women or animals in this book. "Before you came to New York," he told my mother, "I went to school to study English. The classroom looked like this, and every student came from another country." He read words to my mother and told her what they meant. He also wrote them on the blackboard, it and the daruma, the doll which always rights itself when knocked down, the only toys we owned at that time. The little *h*'s looked like chairs, the *e*'s like lidded eyes, but those words were not *chair* and *eye.* "'Do you speak English?'" He read and translated. "'Yes, I am learning to speak English better.' 'I speak English a little.'" "'How are you?' 'I am fine, and you?'" My mother forgot what she learned from one reading to the next. The words had no crags, windows, or hooks to grasp. No pictures. The same *a, b, c*'s for everything. She couldn't make out ducks, cats, and mice in American cartoons either.

During World War II, a gang of police demons charged into the gambling house with drawn guns. They handcuffed the gamblers and assigned them to paddy wagons and patrol cars, which lined the street. The wagons were so full, people had to stand with their hands up on the ceiling to keep their balance. My father was not jailed or deported, but neither he nor the owner worked in gambling again. They went straight. Stockton became a clean town. From the outside the gambling house looks the same closed down as when it flourished.

Wilma Elizabeth McDaniel

THEM CHINAMEN HAS GOT A LOTTERY

Only eight years old
Bobby Gene was a tattletale
told everything he heard
like the time
he rushed in out of breath
and gasped

Harvey and Lowell played the
lottery
and won seven dollars
I ain't lyin' to you
them Chinamen has got a lottery
in Merced

right back of where we eat
the noodles
at Sing Lum's Rice Bowl

Oncet I peeked behind the
curtains
and seen 'em wearin' funny
robes

Lawson Fusao Inada

TROMBPOEM

"'. . . and a Japanese guy we had, Paul Hagaki,
from San Francisco, who could hit high notes
on the trombone that I have not heard anybody else do to this day.'" —trombonist
Al Grey in WAITING FOR DIZZY, Gene Lees

Right—1948, and I was a kid
old enough to make morning deliveries
for my grandfather's fish store,

which consisted of, if you can dig this—
since everything's boarded-up or torn-down now—
just being a happy kid out of the camps
with the privilege of going into front-doors
of restaurants of Fresno's swinging
colored section, strolling all
purposeful, proud, and casual
into the kitchen, and handing over a package
in exchange for a sip and a snack—

because all the black, brown, yellow cooks
knew me, and would feed me,
just for making my delivery!

And on this Saturday morning
ol' "hepcat" Ben Tagami comes bopping into the store
to make this announcement:

"Man, it was wild last night
at the Palomar—there was this
Buddhahead cat on trombone with
Hamp! And *he* could *blow!*"

Uh oh. I remember that moment
exactly, those exact words
registering in my mind
like notes of a solo: a Japanese guy with Lionel Hampton!

If you can believe that. If you can dig that.
A Japanese guy with Lionel Hampton . . .

And about that same time, as I was
establishing my tradition
of hanging around record stores and digging jazz,
ol' Bob Mar behind the counter delivered yet another
valuable fact:

> "Satchmo got this Indian cat, Big Chief Russell Moore.
> I saw him in the movie called 'New Orleans.'"

And even though I never saw the movie
(I have the soundtrack; Billie Holiday,
if you can dig it, played a maid),

I can still see the scene plain as day:

> a Chinese merchant
> telling a Japanese boy
> about an Indian man
> playing in a black band—

and playing *trombone* at that!

Trombone? Hmmm. Oh, oh . . .

The rest is history:

> The kid goes and drags his grandfather
> across the tracks over to that Okie pawnshop
> and comes out blowing, baby!—
>
> just a natural on the 'bone!—

which is why you got all them albums
by the fabled
"Sansei Slider
& His
Rainbones
of
Rhythm!"

Well, not quite . . .

But I'll tell you what:

There's more than one way
of making
a mouthpiece;

there's more than one way
of playing
brass;

and I've got to believe,
"to this day,"
there's ways of delivering
highs and lows, choosing
suitable tempos, tunes,
and folks to play with

so's the whole happy community can take note of its beautiful music!

ELEMENTARY SPANISH

When Teatro Azteca opened up
right there on "F" Street
in the heart of "Chinatown,"

all us kids—"Hispanic"
and otherwise—got excited—

because with a few little coins
you could go in there
with the Wongs and Washingtons

to enjoy some serious cinema.
An "alternative," so to speak,
to what was already going on
in the West Side's, count em,
movie houses: Cal, Ryan's, Lyceum . . .

And, of course, since we were
all geographically versed
in advanced or at least
elementary Spanish,

"Hoy Cantinflas" on the marquee
meant just what it said: Laughs!

Well, "El Fin," as you may know,
means "The End," and though
we didn't know it then, los
estados unidos or somebody
was to invade and destroy

our community, leaving a lot of
nada and Teatro Azteca standing
with "Hoy" still on the sign

but without the need
to announce, all over
the "ash tree" which is Fresno—

"cerrado," which means *closed*.

Sherley Anne Williams

THE ICONOGRAPHY OF CHILDHOOD

i

A town less
than ten stories tall

Spring rains wash the wind
light annihilates
distance
snow flecked Sierras
loom at land's end

Land flat as hoecake
Summers hot enough
to fry one Crops fanned

out in fields far as
eyes can see
every
Time we work a row
another appears
on the horizon

ii

These are tales told in darkness
in the quiet at the ends
of the day's heat, surprised in
the shadowed rooms of houses
drowsing in the evening sun.

In this one there is music
and three women; some child is
messing with the Victrola.
Before Miss Irma can speak
Ray Charles does of "The Nightime"
and *Awww* it Is the fabled

music *yo'alls* seldom given
air play in those Valley towns
heard mostly in the juke-joints
we'd been told About; and so
longed for in those first years in
the Valley it had come to

seem almost illicit to
us. But the women pay us
no mind. We settle in the
wonder of the music and
their softly lit faces listening
at the songs of our grown.

iii

Summer mornings we
rose early to go
and rob the trees
bringing home the
blossoms we were told
were like a white girl's
skin And we believed
this as though we'd
never seen a white
girl except in
movies and magazines.

We handled the
flowers roughly
sticking them in oily
braids or behind
dirty ears laughing

as we preened ourselves;
savoring the brown
of the magnolias'
aging as though our color
had rubbed off
on the petals' creamy
flesh transforming some
white girl's face into
ornaments for our
rough unruly heads.

iv

We never knew the
woman her brothers
later told us was
Lena though we could
see that mamma was
a shadow of some
former self, her down
home ways worn down to
nubs.

They laid this on
"yo'daddy," the way
he'd changed her name
because Lena didn't
suit him. Well, "Lelia"
never seemed right to
them bowing and
scraping (as they phrased
it) before "yo'-
daddy"—after the
way she'd fought *mens* as
a girl.

Daddy blamed
it on her own
foolishness from which
even he couldn't save

her—that Siler blood
he told us; he had
rescued her, brought her
out of Texas. And
she'd have already
what her brothers wrote
home about; if it
wasn't for her own
ol country ways.

v

The buildings of the
Projects were arrayed
like barracks in
uniform rows we
called regulation
ugly, the World in
less than one square block.
What dreams our people
had dreamed there seemed to
us just like the Valley
so much heat and dust.

Home training was
measured by the day's
light in scolds and
ironing cords; we
slipped away from chores
and errands from
orders to stay in
call to tarry in
the streets: gon learn what
downhome didn't teach.

And
Sundown didn't hold us
long. Yet even then
some grown-up sat still

and shadowed waiting
for us as the sky
above the Valley
dimmed.

vi

Showfare cost a lot
but we ran the
movies every chance
we got, mostly grade
B musicals that
became the language
of our dreams. Baby
Lois sang in the
rain for the hell of
it; Helen was a
vamp. Ruise was the
blood-red rose of
Texas, her skin as
smooth and dark as a
bud with just a hint
of red.

Sweating and
slightly shamefaced, we
danced our own routines
seeing our futures
in gestures from some
half remembered films.
We danced crystal
sidewalks thrilled in the
arms of neighborhood
boys and beheld our
selves as we could be
beyond the Projects:
the nine and ten year
old stars of stage and
screen and black men's hearts.

vii

My mother knew what
figure she cut in
the world and carried
that hurt in silence,
once in great whiles roused
by some taunt or threat
to rage mutely then
settling back to
mutter angrily
and to sleep. By the
time I come to my
first memory of
her face she was
already mamma
as I knew her for
the rest of her life.
I saw a ghost in
flashes in lumbering
fury and shaking
laughter glowing
pretty. In these
remembered glimpses
I know the woman
Lena who was sister
to my uncles John
and Jimmy who
married Jesse Winson
and died on the
Texas Panhandle
years before my birth:
taciturn, quick
tempered, *hell-thay,*
my uncles said.

THE WISHON LINE

i

The end of a line
is movement the
process of getting
on, getting off, of
moving right along

The dank corridors
of the hospital
swallowed him up
(moving right along
now—from distant
sanatorium
to local health care
unit—the end of
that line is song:
T.B. is killing
me We traveled some
to see Daddy on
that old Wishon route
but the dusty grave
swallowed him up.

ii

These
are the buses of
the century running
through the old wealth of
the town, Huntington
Park, Van Ness Extension
the way stops of
servants; rest after
miles of walking and
working: cotton, working
grapes, working hay. The
end of this line is

the County: County
Hospital, County
Welfare. County Home—
(moving right on—No
one died of T.B.
in the 50's; no one
rides that Line for free.

THE GREEN-EYED MONSTERS OF THE VALLEY DUSK

sunset knocks the edge from the
day's heat, filling the Valley
with shadows: Time for coming
in getting on; lapping fields
lapping orchards like greyhounds
racing darkness to mountain
rims, land's last meeting with still
lighted sky.

This is a car
I watched in childhood, streaking
the straightaway through the dusk
I look for the ghost of that
girl in the mid-summer fields
whipping past but what ghosts lurk
in this silence are feelings
not spirit not sounds.

Bulbous
lights approach in the gloom
hovering briefly between
memory and fear, dissolve
into fog lamps mounted high
on the ungainly bodies
of reaping machines: Time
coming in. Time getting on.

Private road. Photo by Roman Loranc, courtesy of the photographer.

Delta-Mendota canal. Photo by Roman Loranc, courtesy of the photographer.

Panoche Hills looking east with fog over the Fresno area, Diablo Range. Photo by Roman Loranc, courtesy of the photographer.

View of flooded fields from a levee road in the Delta. Photo by Roman Lorance, courtesy of the photographer.

Orestimba Hills, Westley, Diablo Range. Photo by Roman Loranc, courtesy of the photographer.

Oak on the Cosumnes River near Galt. Photo by Roman Loranc, courtesy of the photographer.

Los Banos Reservoir, site of the oldest Yokuts Indian settlement. Photo by Roman Loranc, courtesy of the photographer.

Tuolumne River in fog. Photo by Roman Loranc, courtesy of the photographer.

Ernest J. Finney

PEACOCKS

When he ran he kept his eyes closed as long as he could. The connection between himself and the dirt path that paralleled the orchard was what kept him from running into a tree. The loamy soil was just soft enough to carry his heelprint. Enough to keep the connection. On the other side of the path was the levee that kept the water in the slough away from the orchard. Whenever he felt himself veering up the bank, he forced himself to keep his eyes shut until he was sure he was back on the path by the plum trees his grandfather had planted. He knew them as well as anyone could. He'd been up in every single tree. He was well connected with the orchard and his grandfather.

It was easy keeping his eyes closed because he knew where he was going. He could run forever this way: his feet kept sending his imagination the right direction. But the pain was getting worse, his kidneys were burning red hot. It was hard to breathe, but he made himself keep going. He wanted to make it all the way back to the house for the first time. Grandpa would be sitting on the porch waiting, watching his pets. He tried to outrun it, went faster. He must be almost past the last section, Queen Rosas. When the bile started coming up he remembered what Jesse, his coach, said: it doesn't really hurt until after you stop, so keep going. Then he started vomiting and opened his eyes so he wouldn't get any on his shoes. He leaned forward, his arms stiff against his knees, spitting, trying to get it all out. Gasping, feeling his whole body hurt. When he could breathe, he stood up and started for the house. He still had another half hour left of his workout.

He could see the roof. The house was just right for the two of them. All on one floor, so his grandfather didn't have to climb steps, like in the main house. Each had his own door off his bedroom into the bathroom. Plenty of privacy for his grandfather, who never came out into the kitchen in the

---

long johns he wore summer and winter. He dressed for his first cup of coffee. The place was small enough to heat easily in the winter and cool with one fan in the summer. He had set up his telescope on a platform outside the attic window. And the porch was perfect for his grandfather, two steps from the ground and his peafowl.

Grandpa was watching him come up. The peafowl were pecking and scratching in the flower bed. The cocks were too young to have any of the big feathers. He got a kick out of how his grandfather treated them. Named the three males after the Three Stooges. The nine females had no names. Had trained Moe to hop up on the porch rail and beg for the cracked corn he kept in his shirt pocket.

He started out first. "Well, how's the stock, are they all fed and watered? Getting ready to roost for the night?" He untied the rope around his waist and started jumping, easy, just lifting the toes of his shoes enough to let the rope pass under.

"Can't say," Grandpa answered. "They're so slow. If they don't hurry up and grow some of those tail feathers I'm ready to give up on them. I won't be responsible. They're just like chickens.

"Now that I'm semiretired and only a consultant in the plum business, I thought I'd raise these birds. I'd seen them in the catalogue for years, but never felt I was ready to buy." Elmo kept jumping, the sound of his grandfather's voice caught up in the rhythm that the rope and his feet were moving to. They both knew they had heard this before. "I thought they were a sign of prosperity: when a person raised something he wasn't planning on eating and selling, it gave you a certain distinction."

Elmo looked over at his grandfather. He'd changed again. Not just thinner than before, but he seemed like he was shrinking too. He was shorter than the six and a half feet he used to be. Elmo himself still had a foot to go to catch up, but there wasn't the difference there used to be.

When he heard the car, it was too close for him to get away. He kept jumping, faster now, trying not to watch her ease out of the car. Her stomach was so big she had the seat all the way back. She reached inside for the hot dish with a blue towel over the top. Grandpa had stood up, stiff-legged, and he went to meet her. "Well, Greta, you didn't have to go to any trouble, but we appreciate it, of course." She stopped nearly every day with something.

"It's no trouble," she said.

He turned the rope slower, not to miss anything.

Grandpa took the bowl. "I got the others washed up," he said. "I'll get them."

"That would be nice. I forget where they are. I keep looking for them."

"I got a little behind," he said, opening the screen door with his foot, carrying the bowl with both hands.

She turned awkwardly to watch Elmo jump. It looked like she was leaning back so far, to compensate for her stomach, that she would go over backwards. "How have you been, Elmo?" she asked.

"I'm fine."

"We don't see you much any more."

He jumped slow now, barely bringing the slack loop around. "I've been busy," he said. Then he heard the pickup. He didn't turn around, just kept jumping. James stopped right next to him, so he had to bring his arm in to miss tangling the rope on the truck mirror.

"Like old home week," James said. Grandpa came out the screen door carrying a cardboard box rattling with Pyrex bowls and tin piepans. Elmo kept jumping.

"What I want to know is why don't you two come up and eat with us instead of making her bring the food down to you?"

The rope caught on Elmo's head and he let it hang on his shoulders, breathing hard. "We don't ask her to bring it," he said.

"Well, she can stop, then," James said.

"I don't mind," Greta said. "It's only a couple of miles."

"How did the pruning go?" Grandpa asked James, taking the box over to Greta's car.

"We're half done. Another three weeks, unless we can get the prima donna to help."

Elmo didn't answer; he wasn't going to fall into that trap again. "I do my share" sounded like an admission of guilt when he said it to James. And then James would snicker. The sound made him want to close his eyes.

James got out of the truck, hooked the heel of his boot on the bumper. He had a gut on him now. Almost as big as Greta's. His leather belt cut into the middle of it as if it were separating a couple of sausages. James looked bigger than any two heavyweights Elmo had ever seen. "What exactly have you been doing?" James asked. "Let's hear it; I'm interested."

Elmo saw the look on Grandpa's face, as if he was going to see something awful. Greta was holding her stomach from underneath with both hands. "Ever since we've come back from our honeymoon, you act like you're a guest here. When I can find you. You live down here—that's all right with me—but you're not doing your share. Those plum trees can't take care of themselves. Grandpa can't work any more. Greta can't either. That leaves you and me."

He didn't have to answer; he could just look at James. Infuriate him. His face would turn as red as the bulb on the porch thermometer. James was too easy. "I've been pretty busy," he said. "Going to school." Saying it out loud, he realized he'd been catching the school bus for eleven years now. It sounded like he needed more, so he added, "I was elected class president. I have to stay for council meetings. And I have sports."

James snickered. "What sports?" He'd been All League in his freshman year, playing football. All State in the shotput.

"I'm on the boxing team."

James laughed out loud. Elmo felt himself start to lose the connection he had with the ground.

"I had to sign a card," Grandpa said. "He's got a punching bag in the shed. I've got four clippings from the school paper too, if you want to see them."

"When are you going to fight next?" James asked.

There was no way out. "Next month at the Elks'."

"I've got to see that," James said, getting back in the truck. "Come on, let's go," he yelled over to Greta.

"That woman can cook," Grandpa said. He'd barely touched the Swiss steak Greta had brought over. Elmo was sopping up the last of the gravy with his fifth piece of bread. "Maybe we ought to go up there for dinner when she invites us. Say on a Sunday."

"You go ahead," Elmo said.

"He's your brother."

"Remember when we went up there?"

"It takes a lot of adjustment when you're first married," Grandpa said. "Two different people."

"Did you ever slap around our grandmother?" Elmo asked. He had to ask; it was worse not to. "I was just wondering," he added.

Grandpa took his time answering. "I was guilty of a lot of things, but not that."

They hadn't seen anything that time. Just heard the yelling, waiting for someone to open the door. Greta had been crying, face pink, eyes floating in tears. They sat down at the table and James started. "I never met anyone so stupid. I tell her to pick up my boots and get the case of shotgun shells I ordered. She forgets both of them. Drives all the way into town for a hot fudge sundae. Dumb."

"That's not necessary," Grandpa said.

"It is, believe me. She's stupid."

"If anyone is stupid it's you, James," Elmo said. "She graduated from high school; you didn't."

"Well, well," James said, smiling, "look who's speaking up."

"That's enough," Greta said.

"You're not big enough to take seriously," James said to Elmo.

No one said a word the rest of the dinner. He wouldn't go back. He didn't want to be connected with that business.

The next time, he came home from school and Greta was there, her lip cut, her eye swollen almost closed. Grandpa was putting hot cloths on the side of her face. "Do this for me, Elmo; I'm going to talk to James." He had no choice. He squeezed the cloth out, folded it into a rectangle, and put it over her eye. What was the connection between them? Earlier she had been like an older sister to him. No. That's what he made up for his grandfather, after she married James. But she should have married him—that's what he thought at the time. At fourteen. With seven years' difference between them. She should have waited for him. But now the connection between them was much less. He was almost a spectator. Someone who was watching a movie, then went home when it was over. "Oh, Elmo," she said, taking hold of his wrist. He didn't know what to do, how to act. He didn't understand this. How could anyone do something like that if he loved a person? Would he ever do that?

When Grandpa came home, she went back up to the house. "I told him," Grandpa said, "what kind of a man it took to hit a woman. I made him listen. Made it as plain as I could. I explained I wasn't going to have anyone on my place like that. It's not you boys' yet."

It stopped for a couple of months. Then Greta went to the hospital with her nose broken. She said she fell. James took off, was in town drinking with his friends. Grandpa went after him. Elmo drove, parked in front of the Rio cocktail lounge, and followed him inside. James was sitting with Claude at the bar, swigging beer out of a green bottle. "I hope you're satisfied," Grandpa told him. "She went back to her folks."

"Good," James said. "I was getting tired of her anyway," and he laughed.

Grandpa backhanded him across the face hard. It was like hitting a tree. The place quieted down. "You sober up," Grandpa said. "I want to see you at six out in that orchard. Or don't you come back." Then he walked out. Elmo stood there for a minute, staring at his brother. He didn't want to leave him. Then his grandfather called out at the door, "Come on, Elmo."

He was doing his math at the kitchen table when the phone rang. "What's wrong with him? Elmo, is that you?"

"It's me, Joyce." He didn't want to hear what was coming from Greta's stepmother. It was amazing, all these connections. It reminded him of TinkerToys; you just kept adding more and more, going up and up. Was it natural—all these connections people had? Greta, his sister-in-law, her real mother dead, had her father Eddie, and of course Joyce, three stepsisters, and James her husband, and she was going to have a baby. They all meant something. Where he had his grandfather. And James. Sometimes. But more, say, than the photos of his mother, whom he couldn't remember; she'd left when he was little. Or his father, whom he could remember, that he'd found dead. His parents were like someone you didn't know that you read about in the newspaper. Interesting. But nothing to do with him.

"I'm not letting her come back there, Elmo. You can tell him that for me. Not when he mistreats her like this. And I can tell you, Eddie is upset; he's ready to come down there. Somebody better start talking some sense into that boy before it's too late."

He wrote down *James* on the paper. Grandpa's name. Then his own name. Grandpa was fifty years older than he was. James was nine years. Was time important? He needed more information for an equation. For an answer. "Who was that?" Grandpa asked, coming out of his bedroom. Elmo wasn't going to tell him. He was having to go in to the doctor almost every week now. He was skin and bones.

"It was Joyce."

Grandpa shook his head. "What can we do?" He kept a shining new thirty-gallon galvanized can in the entryway full of feed for the peacocks. He lifted the lid and filled the pocket of his wool shirt with cracked corn. Went out. Elmo could hear him tilt his chair back and the sound of the first peacock hopping up on the porch.

He watched from the kitchen window as another peacock jumped up on the railing and displayed its train of feathers for Grandpa. It was like having your own rainbow handy to look at. They made him uneasy, those birds. The way they waited for you to admire them. As if there were more to it than that. They didn't belong here. They didn't have any connection to this place. He couldn't imagine anything else that came close to a peacock. He tried, too. Combining different kinds of ducks from the marsh, a kingfisher and a red-winged blackbird, spring plum blossoms with California poppies. It was no use. The harder he tried, the less anything would come. He'd seen something almost close enough once, but never got the chance to decide. It was a painting by someone with a name he couldn't pronounce. Two fat naked women in a field drying themselves after taking a swim in a pond. It wasn't only the colors or the figures; it was the feeling there was something else that was there that you couldn't see. But someone had cut the picture out of the magazine before he could bring it home from school to compare with the peafowl. One time when his grandfather went to town he tried to catch one. He was going to chase it down, get ahold of one, get a closer look. But it took off, train of feathers and all. It was the first time he'd seen one leave the ground. It flew better than a pheasant. It was like seeing a hydrangea bush take off. He kept away from them after that.

He wasn't nervous; it was just that he wasn't used to so many people watching. The whole Elks' club was filled, must be a thousand people. He had heard them from the storage room where they were dressing. He'd stayed in there after the bouts started, following the fights by the yelling, which was off and on like someone was opening and shutting a door. Then he moved over to the doorway.

He hadn't wanted Grandpa to come. But he'd insisted. James had to stick his big nose in. "What's wrong—you don't want him to see you get whipped?" It wasn't that; he was going to win. But he'd never watched him fight before, and it was going to make him feel exposed. Immodest. As if all the efforts they made to be clothed, even when they got up in the middle of the night to pee, were for nothing.

James brought him. They were sitting down in front, both of them so big the people around them were already trying to shift their folding chairs so they could see. Grandpa was still taller than James. By getting fatter, James seemed shorter at the same height. Elmo knew if he made six feet it would be a miracle.

After they awarded the trophies for the bouts before him and the cheering died down, he came down the aisle. This was all familiar. Jesse there with the stool. He didn't look at the other fighter coming into the ring. He wouldn't do that until the bell. There was no connection between them. He wasn't responsible.

They went out to the middle for instructions. The other fighter tried to stare him down. He decided it would be better to fight outside under the stars. The announcer gave their names and the other fighter's tournament win in a trip to Chicago. Not only wouldn't it be so smoky—there'd be the constellations he knew. Jesse was taking his robe off his shoulders, wetting his mouthpiece before putting it in.

He moved to the exact center, watching. The whole trick was not to allow his eyes to go to any part of the other boxer but his nose. He imagined he was a flying railroad spike aimed at the middle of his opponent's head. There was the usual feeling out. It was going to depend on who was just a little quicker and who was in better shape for three two-minute rounds. That's what Jesse always told them.

Jab, he had the reach. His opponent's face began to get red; his eyes seemed to shift around, trying to get out of his head. "Stick 'im," Jesse kept yelling from the corner. It was almost too easy. The other fighter was getting concerned because he wasn't hitting anything. Getting wild.

He breathed easy, sitting on the stool after the first round. He'd heard James yelling "Killum, killum." Now he was yelling "Don't let up on him, don't lose him." There was a certain satisfaction in doing everything right, in order, exactly as he'd planned. He didn't understand why, but he always had the feeling he was the only one in the ring. That would be another good reason to fight outside.

He went out the second time, moving toward the center, knowing what was going to happen. His opponent came out fast, charged, swinging furiously. Then he threw his right hand, the leather connecting, making a satisfactory sound as it struck. His opponent sat on the canvas with a bewildered look on his face, his coach waving a towel, the referee yelling, the spectators in an uproar. He leaned against the ropes, not breathing hard yet, wondering if he should order two or three cheeseburgers after. He hadn't eaten any dinner.

The other fighter got up. He didn't want to look surprised, didn't register anything. He went back out. It was the only time in any of his nine fights that anyone had got up. But the kid's eyes were fixed now, and Elmo came in hitting him with both hands at will.

The referee stepped in and pulled Elmo away, his arms locked around Elmo's chest. The crowd went crazy; then there was a new burst of yelling as the referee put Elmo's arm up. This had been the main event. He got the winner's trophy, which was the same as the one they gave the loser. He always thought that was funny. Then the judges selected their fight as the best of the evening and they both got another trophy. More yelling. James and Grandpa were at the ring-side when he climbed down.

"You looked good," James yelled above the noise. "I never thought you had it in you." Grandpa patted his shoulder. Jesse came over. "He's a natural," he told them. "He's got fast hands and can hit with either one. You saw that. He's easy to coach, I'll tell you that. He can go all the way to the national finals if he wants." Elmo nodded, deciding he'd get all three cheeseburgers.

There was something pleasant about people coming up and congratulating him. People in town. Kids at school. Betty, a girl he'd ridden the school bus with, eight years to grammar school and two and a half so far to high school, came up blushing. "You pack a wallop, my father told me to tell you." He never knew what to say, except thank you. For her he said something different: "That's a kind thought."

Since Greta left, Grandpa had been cooking. Liver. Heart, braised. Spent all day scrubbing tripe on a washboard and soaking it in salted water for stew, red stew with potatoes and carrots. Brain fritters. Kidney pie. It was as if he was trying to find something Elmo wouldn't eat. "This is what I was raised on because it was cheap. Rich people threw this away with the guts," and he'd hold up a piece of liver with his fork. He didn't eat much, but he made Elmo eat. And James, when he stopped in at dinnertime. He'd stopped drinking and going to town after work. He'd been up to see Greta once. But she wouldn't come back.

A couple of weeks after his fight at the Elks' club they were sitting around the table after a dinner of brains and scrambled eggs. "That was tasty," Elmo told his grandfather.

"You're just trying to get on my good side so I'll make them again tomorrow." They both laughed. They were cracking English walnuts for dessert. Elmo put his halves on the oil-cloth near his grandfather's plate. They looked like dried flowers there. It was one thing he would eat. Lately when he told his stories he'd been going back, Elmo noticed, to his wife or his sister, never staying in the present very long if he could help it.

"I want you to consider quitting that boxing business," Grandpa said. Surprised, Elmo waited. Cracked another nut. "You took that other boy apart at the Elks' club. Just like you were pruning one of our plum trees." He pointed out the window toward the orchard. "It was too easy for you. You were like an executioner up there."

"I don't pick my opponents," Elmo answered.

"Some people shouldn't do things without knowing the same thing could happen to them. You're one of them. If it were James, I wouldn't waste my breath. He'd do the opposite. So would I, probably, for that matter. But I'm going to leave it up to you."

"I don't understand what you mean."

He lowered his voice, as if someone could overhear them. "You have to be careful, Elmo. What you do. You're my grandson. In the end it's going to be you that gets hurt."

Elmo slowly shook his head.

"You're too smart, Elmo," he said louder. "And now your hands are too smart." He got up. "I cooked this lovely dinner; you wash up," he said, going for the feed can.

He kept running, working out after school. It was habit, mostly, but he couldn't decide whether to quit or not. He kept going over what his grandfather had said. What difference did it make if hitting someone were easy or not? Betty offered to give him a ride home after a student-council meeting. She had the family Plymouth for the day. Her father had the dealership in town, but they lived on thirty acres of pears down the road from the plum orchard. They stopped at the A and W, and kids in other cars came over to say hello. They passed James on the way to town. He honked when he saw Elmo. "Who's that?" Betty asked. "I don't recognize the pickup."

"Don't know," Elmo said.

Grandpa was sitting on the porch. His pet peacock Curly was on the rail, its plumage spread out. He got up out of his chair when Elmo introduced him, a few kernels of cracked corn dropping off his washed-out wool shirt. He took off his fedora to shake hands with Betty. "I know your father," he said. "You'll notice we have nothing but Dodge pickups."

"Best farm truck on the road," Betty said.

"She's slightly biased, Grandpa." Elmo went to get some cracked corn for her to feed to the peafowl. They ate right out of her hand from the first try.

"They must like you," Grandpa said. "They won't do that for everyone. Won't do that for Elmo."

"Because I run them off the porch with a broom. You don't hear them at four A.M. screeching."

After Betty had driven off, Grandpa said, "Nice girl."

"You're just saying that because your peacocks liked her," Elmo said.

"You don't see a flaming redhaired girl with two million freckles every day."

Teasing, Elmo asked, "You notice things like that still?"

"I'm not dead yet," Grandpa said, whacking Elmo on the shoulder with the fedora.

There was a boxing tournament in the city over the weekend. Elmo mentioned to Jesse that he wanted to move up a class to welterweight. "You're giving away your edge; they're bigger kids. They'll outweigh you." Elmo waited until he was through. Jesse was a gardener at the high school who had boxed in the navy. His team was the only one at school that was winning this year. Jesse almost strutted when he went into the teachers' lounge for coffee. "Don't worry, I know what I'm doing," Elmo said.

He won five matches and the division title. His picture was on local TV with Jesse, who was shorter than he, picking him up in a bear hug around the waist after the last fight. He was interviewed in the dressing room by a sports writer who covered high-school athletics for the city paper. When he got home, people he didn't know phoned to congratulate him. Grandpa didn't say anything, handed him the phone.

Betty phoned him. "I saw you on TV," she said.

"I guess I'm just a celebrity."

"My father said you could be in the Olympics."

"Me and Joe Palooka. I don't know if I want to go that far. It's kind of like astronomy. I used to think that people were almost like constellations. You can only know so much about them, as much as you can see, and then you have to rely on your imagination. I was wrong; because people are so much closer, it makes it harder to understand them."

"You're losing me, Elmo. But go on."

"I'm losing myself," he said. "Hitting someone in the ring is like making contact with a constellation. Draco, say. You get to know more. You see what I mean? There is a connection. But not as much as, say, between your sisters and you."

"Keep going."

"Maybe boxing is getting down to the essentials. Between yourself and who you think you are."

"I'm changing the subject," she said. "Did you get the assignment in math?"

He laughed. "You know better; I finished that on the bus."

"That was a silly question. A scholar like you. I have to go now. You phone me next. My mother thinks it's undignified."

"I promise," Elmo said.

Grandpa went into town for an appointment, and the doctor put him in the hospital. "They want to run a few tests," he told Elmo when he came into his room. "They must know I had some more money left." Elmo sat next to the bed to be able to hear him whisper: "Take care of my pets for me."

"I will, Grandpa."

"Don't be stingy with the feed."

"I won't." There was nothing to be worried about, Elmo kept telling himself.

The necessary three parking places were empty, and he maneuvered the truck just right, ten feet from his grandfather's window, then hurried inside. "I got a surprise for you," he said. His grandfather lay on his side, looking in the wrong direction. He started to turn toward him. Elmo knocked on the window with his knuckles for the bird. He had tethered the peacock up on top of the case he'd brought him in. He didn't think the stupid bird was going to cooperate. Just then he raised his train and the plumage spread out like a fan catching the summer light. The iridescent blues and greens moved like water.

"That's old Moe," Grandpa said, sitting up. He was looking better after this operation than he had after the first. He might be gaining weight. But he was still white as a sheet. He'd been in here the whole summer. The peacock collapsed its feathers. "Greta's just left; brought the baby in. He's going to be big like me and James," he said. "On the other hand, Greta's tall too, of course. Said she's pregnant again. That's rushing it, I'd say." Elmo stood still at the window, waiting for the peacock to answer his knocking again. "How's the picking going?"

"Right on time. We're in the California Blues."

"You'll be going back to school next week."

"If James gives me permission." Elmo laughed. "He's got Claude helping in the plums; he'll get by. Greta comes when she can."

"How are they getting along now? I didn't want to ask her."

"All right, I guess." He didn't want to tell him about the latest episode. How could they do that: call each other names, throw things, hit each other. Was he missing some connection between them?

"They wanted to name the baby after me," Grandpa said. "I said don't. Three Jameses are enough. Gary is a good name." Elmo looked down at his grandfather. Listening to him, he could feel his eyes tear up, his eyelids blinking them away. He bent down and hugged him. When Elmo let go, he saw his grandfather was embarrassed.

Greta had left him a pan of meatloaf and potatoes. He sat at the kitchen table eating, not bothering to get a plate. Between going in to the hospital and working in the plums he hadn't been doing much else. He was so tired when he came in he sometimes fell asleep eating. He decided to phone Betty. He had meant to a dozen times. He hadn't seen her since school let out in June. He knew she'd got a job in town as a lifeguard at the pool.

She answered the phone on the first ring.

"May I speak to Betty Jean Briscoe, please?"

"This is she."

"This is Elmo Clark. Would you be interested in attending a movie classic, Creature of the Lost Lagoon, this next Friday evening?"

"Is this a date we're discussing here, Mr. Clark?" She started to giggle. He could hear her whisper to someone in the background, "I have a date, I have a date."

"I think it would come under that category."

"Then I accept."

She could have said no. But then on the other hand he could be underestimating his own resources. He was looking at himself in the mirror over the sink. The pimples were gone except on his forehead, and his hair hid them. Five-o'clock shadow; he needed to shave once a week now. If he fought again he'd be a middleweight; he'd gained. But he was through with that. He looked closer, from a different angle, at his reflection. If he did look like his grandpa's sister Lorraine, she must have been a knockout.

Someone honked his horn out front. It could only be Mickey Conlin. He went out, the peacocks bobbing out of his way. "You want to go hunting tomorrow morning?" Mickey called out before he got near. Mickey was sitting in a new red sportscar convertible Elmo didn't know the name of until he got close. Mickey's father had leased the marsh from his grandfather for his duck club five years ago. He brought clients over from the city to shoot. Mickey was convinced he and Elmo were the same. Because they were both sixteen. He had been in the auditorium when Elmo had boxed in the city. He had his younger sister Virginia with him.

"Some of us have to work for a living," Elmo said, resting his hands on the door. It was what his grandfather would have said.

"Come on, Elmo, the place is loaded with doves."

It gave him some pleasure to refuse—probably, he thought, because Mickey wanted him to come so much. "Can't. If I don't help James, my name's mud. You've got Virginia, she's lucky. You going to hunt tomorrow?" he asked her. She didn't answer. "You got your .410 with you?"

She finally nodded, then spoke. "Yes. I remember how you showed me, Elmo. Keep the butt tight against the small of my shoulder." At nine she was already a better shot than Mickey.

"Don't shoot in the trees where there's still plums."

"I won't," she said.

"Next time," he said, backing away from the car.

Elmo brought his grandfather home Friday morning. Helped him out of the car. "I can walk," he said, holding onto Elmo's shoulder. All the peafowl came across the yard in mincing steps, the males expanding their feathers. "They're happy to see me," he said, hauling himself up the steps by the railing. He sat down in his porch chair, and Elmo went back to the truck for his things. The peafowl came up on the porch, wanting to be fed. Elmo got the juice can of cracked corn and handed it to him.

"No matter what happens, I'm not going back to the hospital," he said as if he were ending a long conversation on the subject. "I'm staying right here."

"The doctor says you're fine," Elmo said.

"That's right," Grandpa said. "He said that the last time too."

• • •

It was almost time for him to leave to pick up Betty. "You're sure you don't want to go along?" he asked his grandfather. They both laughed.

"You don't need a chaperone, and that movie you told me about would keep me up nights."

"You're going to be all right?"

"If I'm lonely I'll phone Greta. If I get hungry I'll cook me up one of my peacock friends out there. Did I tell you I met someone in physical therapy that told me he actually had one of those critters for dinner? I was telling him about mine and he told me the story. I didn't tell you this yet?"

"No you didn't," Elmo said, not looking at the clock.

"Thought it would taste like pheasant. Cooked two. With orange slices, stuffed them with carrots and onions, like we do with ducks. Low heat. Basted them with their own drippings and white wine. Put it out on the table in front of his guests. Had his carving knife all ready, sharpened it with the steel." Grandpa made the motions of sharpening. "Tried to take a slice off. Wouldn't cut. Kept trying. Everybody waiting. Put it back on the cutting board and got a heavier knife; didn't do the trick. Tried a cleaver next, brought it down, never dented it. He had to take the guests out to a restaurant." The last time he told it, Elmo remembered, the peacock had been put on a chopping block and hit with an ax.

Mr. Briscoe opened the door, looked past him. "How many miles you got on that pickup?" he asked.

Elmo was ready; he remembered coming to birthday parties here: Mr. Briscoe had once wrapped up a colt for Betty in tissue paper and led it into the dining room. "One hundred and seventy-two thousand and six tenths of a mile, give or take a few feet."

"You Clarks, you're running it into the ground; I can't make any money off you. I can't even remember when I sold that truck." He sounded indignant. He led Elmo into the front room and pointed to the couch for him to sit down. Elmo perched on the edge. He never had a chance to be nervous; Betty came right out and he stood back up. Being out in the sun must have made her freckles multiply, he decided. There were at least a million more, the color of red ants.

"Did he sell you a new pickup yet?" she asked him.

"Not yet." He wished he could seem as at ease as she was.

"How come I haven't seen your name in the paper lately? Give up boxing?" Mr. Briscoe asked.

He was going to say, "No time. I've had to work." But for some reason he heard himself going into a longer explanation. "I liked it at first because a person has to concentrate so much, with the conditioning and the training. The actual fighting was secondary." Mr. Briscoe was looking puzzled. "Then I got so I liked getting ready more than the actual boxing, so I've decided not to do it any more. I just work out now."

Betty's mother came out of another room. She'd put on lipstick, he noticed. She had the same red hair and freckles as Betty had. "How's your grandfather? I heard he was in the hospital."

"He's home now, feeling much better."

Betty took him by the arm and led him toward the front door and opened it. "Say hello for me," Mrs. Briscoe said at the door.

Mr. Briscoe followed them down the walk. "Tell him I can get him a good deal on this year's model."

"You promised," Betty said as she got into the pickup. Elmo shut the door for her and went around.

"I know," Mr. Briscoe said, "but it's not every day I see you hold your breath when I start to say something. I couldn't pass it up." He started laughing, took his glasses off to wipe his eyes.

Elmo drove down their road wondering how long it took an individual family like the Briscoes to develop the connections with each other and how long with an outsider. Would an outsider not having the same past ever be connected well enough? Greta wasn't a good example. Maybe the connection between a person and object was more important. His grandfather's peacocks. Or a person and what he thought. Just what he could think up inside his own head. The connection between himself and himself.

"What are you thinking?" Betty asked.

"Just when was the last time I changed the oil in this old wreck. What were you thinking?"

"If I hid my shampoo so my sister wouldn't use it."

They took back seats, right under the wall where the beam of light came out of the projectionist's booth. He must have sat next to her at least a hundred times on the school bus, in classes. But he felt uncomfortable, as if he had no control over his body. His stomach might gurgle, he might pass gas, drool, snort. He shifted himself again.

"You can hold my hand, Elmo," she said, "I'm done with the popcorn." He picked it up. It was warm, and he could feel the echo her pulse made from the flat of her thumb. After the cartoon was over and the movie started she whispered, "You can kiss me if you like, but only with a modicum of passion."

"Do you think I need all this direction?" he said. She closed her eyes, he noticed as he leaned toward her. He kissed her once and said, "You're going to have to wipe your mouth of the popcorn butter, Miss Briscoe, if I'm going to be able to gain purchase on your lips."

Walking back to the pickup, they passed the Rio. James must have seen them, because he and Claude came outside. James was yelling after them "Lookee here, what do I see?" and slurring his words. Elmo had to stop; there were people passing. James was so drunk his fly was half-open and he had to brace himself with one hand against the blue tiled front of the bar to stand. Claude, weaving slightly, stood alongside him. "Aren't you going to introduce me?" James asked.

He wasn't embarrassed, Elmo told himself. This was nothing. "This is my brother James," he told Betty, "and his friend Claude. Betty Briscoe." He had been holding her hand but now let it go.

"Pleased to make your acquaintance, Betty," James said. "Out on the town, huh. Why don't you come in and I'll buy you a drink." He tried to straighten up, and took a few steps toward them. Took Betty by the arm. "Come on, I know your old man." He tried to take a step backward and almost fell, pulling Betty with him.

"That's enough," Elmo said, getting between them and pushing James away.

"Who do you think you're pushing?" James said, and he tried to shove Elmo back, lost his balance, and fell over onto the sidewalk.

Claude punched Elmo from the side and knocked him down on all fours. "Son of a bitch," he said as he kicked Elmo in the stomach. Elmo noticed Betty had her hands over her face. He would have liked to reassure her. This was just funny. If he could get his breath to laugh, he would. There was no connection between her and what was happening. She shouldn't be bothered by this.

"Wait a minute," James was saying. "Just wait a minute." He got back up. "That's my little brother." Elmo got back up too, holding his stomach. His head was ringing like a bell, as if his tongue was the clapper. Claude had stepped back, both hands clenched and down low like an old-time boxer, ready for him. He wasn't as drunk as James. He blinked his eyes. Elmo aligned himself with the north star and moved in.

Claude swung too soon, and he stepped in and under, bringing his fist around. The skin over Claude's cheek bone split, and he felt his knuckle pop against the bone. Then, noting he was still aligned, he brought his right hand around. Claude crumpled like a piece of wastepaper.

James came toward him, weaving back and forth, his hands to his sides. Elmo hit him in the face. It was like hitting a sponge. "How do you like that?" Elmo told him, saying it every time he struck, until some customers came out from the bar and stopped it.

He remembered Betty then. She wasn't there. He walked as fast as he could to where the truck was parked, but she wasn't there either. He just started walking, not knowing where to look next. He looked in the couple of stores still open as he passed, going all the way back to the movie. He was worried now. She couldn't walk home; it was too far.

He phoned her house. Her mother answered. "Mrs. Briscoe, Betty and I seem to have got separated somehow."

"Her father went to pick her up. She phoned here. She was upset."

"Thank you, Mrs. Briscoe," he said, not knowing what else to say.

James's face looked like a plum that had been stepped on. He wouldn't come into the house so Grandpa could see him. Talked to Elmo from inside the cab of his pickup. "Whatever happened," James said, "it was probably my fault. I don't even remember what I did," he said. Elmo didn't believe him. "You don't know how bad I feel about this. I'm married and have a kid and another one on the way. I have to decide what comes first." Elmo was embarrassed. James was serious. "I was thinking this morning, what if I lost Greta. I'm going to straighten up." He started the pickup. "Claude said you got lucky last night. I told him we were lucky you didn't kill us both. If it makes any difference, I'm sorry. I'm sorry about the girl. I remember that much. I'm sorry, Elmo."

He was going to phone Betty but he put it off. Decided to wait until school started. They both had U.S. history first period. She sat at the front. After class he caught up with her outside in the hallway. He didn't know how to start out. He just walked along beside her while she ignored him. She finally stopped and sat down on a wooden bench. He sat beside her. "I don't know what you want me to say," he started out.

"We don't act like that in my family," she said. "We don't go in for excess. You were as much to blame as they were." She had placed her books on her knees. This was not important, he decided, starting to enjoy the situation. There was no connection between them. "You Clarks are always doing something. You're famous in the county for it. When I was a little girl, I heard about you. I used to watch you on the bus and think I might catch you changing into something else."

She's being dramatic, he thought, and he started to laugh. She jumped up and walked away, surprising him. He yelled after her, "You have too many freckles anyway. I can't see you without trying to count them." She kept going, never turned around. "It's a nuisance," he called after her.

• • •

"I feel like I've been pregnant all my life," Greta said, sitting in the kitchen with Elmo. He didn't answer; he was thinking of Betty, whom he'd taken to the drive-in movie last night. Greta had to repeat herself. Elmo managed to nod and grin. They heard Grandpa's feet hit the linoleum floor. Elmo got up fast. "I'm getting the hell out of here," his grandfather said. "You and me are going to take off, Elmo. I'm not staying here any more." Elmo took him by the shoulders and eased him back in bed. "Are my clothes handy, Elmo?"

"They're handy, Grandpa," he said, putting the covers up around his grandfather's chin. "You rest some more; then we'll make some plans. These trips take some time." He went back out into the kitchen.

"His mind wanders," Greta said. "What did the doctor say yesterday?"

"He wants to put him in a rest home. Again. He thinks he's had another stroke. He'll come back, get better; he's not that way all the time, you know that."

"Do you want me to come tomorrow morning?"

"If you would. I got them to let me go just eight to twelve. You don't mind?"

"No, it's no trouble. Gary sleeps all morning. I'm too round to do much but sit." All the peafowl were waiting by the screen door to see if anyone would feed them. The birds scattered as they went out to the car.

When Elmo came back inside his grandfather was sitting up again. There were veins that stayed out on his forehead now, surrounded by brown spots the size of quarters. His eyes stared out of his head, fixed at what he saw. He was looking around the room, lost. "Where were you, Elmo? I was looking all over for you."

"I was out shucking some early corn, for a surprise."

"Lorraine and I used to shuck corn. I was better than she was. I did twenty bushels in one day. Don't deny your family, Elmo."

"I won't, Grandpa."

"She was a good sister to me. Couple of years ago I tried to phone her, planning on making up."

"I didn't know that."

"She had no number." He lay back down and closed his eyes. "I was too late."

James came in while he was boiling the water for the corn. It was staying light longer now; it was almost 7:30. "How is he?" James whispered. "Maybe we should take him into town, like the doctor said."

"He wants to stay here. I can take care of him," Elmo said.

"Greta said you can't stay cooped up here all the time."

"I don't mind. She's going to come in the morning so I can go back to school. They got after me."

"Claude's got the packing shed ready to go. I'm going to give it a week before we start picking the Red Julys. I could come in the afternoon until then."

"No, half a day of school is enough," Elmo said.

James got up. "Call us, Elmo; I can get here in a minute."

When the corn was done, he took a knife and cut the kernels off a cob into a bowl and added butter and salt and pepper. He sat by his grandfather's bed. "Open up, Grandpa, I've got some corn for you. Open up." He put a few mashed kernels in between his lips and saw them slide back out when he tried to swallow. His grandfather began to ramble on with his eyes closed, how he'd caught three porcupines eating up his young trees. Put them in a gunnysack and hauled them out by boat to a small island. Couldn't knock them in the head, he said.

When he drifted off Elmo went back to the kitchen and ate an ear of corn. It was sweet. The peacocks were roosting; he heard them shifting their tail feathers. He ate the rest of the corn, then sopped up the butter with a piece of bread. "Elmo," his grandfather called out, "are you there?"

"I'm here, Grandpa. I'm here." He did his homework. Then he tried to remember the time difference between east and west. It would be eleven o'clock there. He picked up the phone anyway. He got the operator and told her what he was after. She said she'd phone back. He did the dishes, waiting.

The phone rang. "I found a Vivian, but no Lorraine."

"Would you try her?" He listened to the rings. They made him think he was going back into the past. Carrying all his grandpa's stories.

"Hello," someone said.

"Are you any relation to Lorraine Johnston?"

"May I ask who's inquiring?"

"My name is Elmo Clark. I live in California."

"My mother was named Lorraine. She died four years ago."

"Did she ever mention a brother? My grandfather's name is Jim Clark."

"I remember Uncle Jim. He came out here to see us. I'm her daughter. I must have been nine or ten when I saw him last."

"It's complicated," Elmo started out. "My grandpa is sick, and he's been mentioning your mother. His sister. I thought I might get them together, but I guess it's too late."

"I don't know what to tell you."

"There was a disagreement," Elmo went on, wanting to finish it. "Over a stickpin that belonged to your grandfather and was supposed to go to my grandfather."

"I don't remember anything like that."

"I'm sorry to bother you like this; it's just that he's sick. I wanted to see if I could tie up some loose ends, I guess."

"I've been to California. I have a niece in San Diego."

"Well, if you come out again, be sure and stop," Elmo said. "We'd be happy to see you."

Elmo was giving his grandfather a sponge bath the next morning, before Greta came. "I must be in sorry shape if I can't even wash my own neck," he said in his old voice, as if he hadn't had a stroke.

"You're just lazy," Elmo said. "You remember me shaving you last time?"

"You almost cut my nose off." He laughed. "You got coffee out there, or you going to keep it for yourself?"

"You finish and I'll get you a cup." He kept talking, wanting it to last. "Greta's coming over in a little while."

"What for? They having trouble?"

"No, no, she gets tired staying up there at the house by herself. She'll bring the baby and you can keep her company while I go straighten out that teacher on a few calculus problems."

"I like Greta," he said. Elmo came back with a cupful of hot coffee. He took a big gulp. "Now that's what I call coffee." He took a deep breath. "I never thought I'd get over your grandmother dying. We had been married less than fifteen years. But I had to, because of your father. When he killed himself, I had you and James to keep me getting up out of bed. You see what I'm getting at, Elmo."

"I see," Elmo said, "but you're not going anywhere."

He hadn't been in school in almost three weeks. But Betty took notes and got his assignments, then gave them to the bus driver, who put them in the Clark mailbox and picked up the homework. She sat next to him in math. "Truant," she said, when she walked into the classroom and saw him back. "I'm so far ahead of you now in the race for valedictorian you won't even be close enough to hear me give the speech."

"I'm the one that taught you how to invert in dividing fractions. In the fourth grade, remember that? How soon they forget their betters."

She reached across the aisle and gripped his wrist. "I missed you, Elmo." It surprised him for a minute; he didn't think she was going to let go. The teacher had come in and was writing on the chalkboard.

He wouldn't let himself think about her. He wasn't in control when he did. She wasn't in the fantasy that made him have to pull out his shirt in front to get up out of his seat, or jam both fists in his front pockets to hide the bulge. She made him forget things. Big things like why he had to go home after school. Or little things like how many trees did they have of Abundance. She made him forget. He didn't know if he had his eyes open or closed sometimes. If late at night, unable to sleep, he was seeing the peacocks in his grandpa's bedroom or if he was imagining it.

There was no math problem he couldn't find the answer for. He had to remember that. It was important. All he had to do was make himself think. Put his mind to work. Let his imagination pick up the answer and bring it back to him. He had even gone to church with the Briscoe family. When Betty had asked him, he'd said yes. He hadn't been in church since he was eight or nine—and then because one of the town churches sent a bus out for the kids. He waited every Sunday out by the mailbox because of the cookies. Each woman in church brought a different kind. There were peanut butter, date, oatmeal, brownies, lemon, raisin, and some with frosting and multicolored sprinkles. The bus had broken down after about a year and they never fixed it. He never went after that. He sat next to Betty, feeling her leg against his, the rest of the Briscoes lined up on either side. When they sang Betty held the book open and he could hear the sound of her voice next to his ear. He closed his eyes and imagined the springtime orchard, with him running. The singing was like what he saw then. The shape of the wind as it blew the spray of petals against him. The petals falling on his head and shoulders until he opened his eyes. His grandfather laughing at him when he came back to the house. "Your Easter hat," he said.

After class they walked together. "How's your grandfather?"

"He's getting better. I was talking to his niece last night in New Hampshire. I think we might go there for a visit. See that end of the family. As soon as he gets a little stronger."

"It's warmed up early this year. Time for our first swim." She waited; when he didn't comment, she went on. "I can hardly wait to dive off the barge. I could meet you up there today."

"I'll see," he said.

"I'm going to go anyway," she said.

When he got home everyone was asleep, Greta on the chair, the baby on the couch, and Grandpa in his bed. The oven was on and he could smell cake. She must have been baking all morning. He cut himself a slice of Greta's homemade bread, spread it with butter, then sprinkled a spoonful of sugar over the top. Went back out, taking big bites out of the bread. The peacocks all followed him, but he ignored them.

He started working on the twelve-foot wooden ladders for the coming season. The pickers busted hell out of them. He couldn't blame them, heavy as they were. He started replacing the broken steps. Taking his time. Examining each ladder, sorting out the ones that needed work.

James came by, stopped. "It's hot for May," he said. "It's going to hit eighty."

"They're all asleep," Elmo said.

"We're going to eat with you tonight, if that's all right. She's got enough grub cooked to feed a picking crew."

"I'll be back by then," Elmo said. "I'm going to take a ride soon as I finish these."

She was already there. He saw her father's car parked up on the levee. Then, coming down the bank, he saw her, already changed and sitting on the barge. He hurried. The wooden barge had drifted in here years before. It was tilted up so one edge was almost in the water. He'd been coming here since he was in the sixth grade; James had brought him. He would never have found it by himself. The tules closed in around it on three sides like tall grass. A raft of yellowed tules and driftwood that looked solid enough to walk on floated at the far end of the slough.

He could hear the lap of the water as he changed into his swimming suit in the cabin. He noticed her clothes halfway stuffed into a brown bag. Someone had put a quilt over the old mattress that had always been there. He folded his trousers on their crease and put his shirt over the back of one of the chairs. He made himself go back up the stairs and out into the sunshine.

He lowered himself next to her on the edge of the barge, put his feet in the water too. "It's too cold to go in," he said.

"You can't let your senses overcome your mind," she said back. He laughed; he had told her that once. "Have you ever seen a naked woman, Elmo?"

"Mamie Eisenhower," he said, and added when she started giggling, "in my religious dreams."

She started talking in an Okie drawl too. "Have you ever known a woman? In the biblical sense?" she asked.

"I think you're going beyond the bounds of good taste," he said, splashing her with his foot.

"Have you?" she asked in her normal voice.

"A gentleman never reveals a lady's name or past." He looked over and saw her face was flushed. She really wants to know, he thought. Not looking at her, he said, "Claude took me over to the city."

"How was it?"

He thought. "I don't know," he said. "On one level, interesting to do what I'd been hearing about since I was five and thinking about since I was eleven. On the other hand, I would say under those circumstances it was a little overrated, but I'm just starting out."

"What was her name?"

"I don't think she ever told me. But she was a Christian, which gave me some comfort—she was wearing a cross."

Embarrassed, Betty laughed. "I was wondering if we might try it one of these days," she said.

He could barely talk: the words wouldn't come out right. His mind felt like it was shutting down. "You sure you're ready for the big step into adulthood? This is the final pubic rite of passage." He didn't correct the word; he couldn't think of the right one.

She hadn't noticed. "I'm ready," she said. "Some days I feel like I'm ready to bust."

"Any particular day?" he asked. Neither was looking at the other.

"Maybe today," she said, getting up and walking back across the barge. He heard her go down the steps into the cabin. He sat there awhile before saying in a loud voice, "Well, I can't think of anything better to do on a sunny day like this."

"Where in the hell have you been?" James yelled, coming out of the house when Elmo drove up. "He's in there crying because he thinks you're not coming back." Elmo ran, scattering the peafowl. Greta was by the bed holding his hand, with the baby in her lap. Both the baby and his grandfather were sobbing as if their hearts would break. "I was out checking on the railroad fares," he said loud as he could, to be heard. "I think we can get tickets straight through, changing at Chicago. We can each take two suitcases," he said all in one breath.

His grandfather looked at him, stopped crying. James had taken the baby into the other room. "They still have sleepers?" he asked, his face all wet. Elmo nodded. "It's going to have to be the train," Grandpa said with a big sob; "you're not going to get me on an airplane."

He stayed home the next day, phoned Greta to say she didn't have to come. He did their wash in the old wringer washer. It was another warm day and he hung the clothes outside to dry. When the sun was directly overhead he brought out the rocker and set it on the porch. Then he picked his grandfather up, wrapped him in two blankets, and took him outside. He didn't seem to weigh anything.

"Well, sir, take a look at those peacocks of yours. They are nearly the perfect pet; all they do is eat and sleep. They must need all that rest and nourishment just to show off their feathers once or twice a month." He looked over at his grandfather, who was staring into space, eyes open, unblinking. The words made no connection. He remembered them verbatim from when his grandfather had spoken them.

He went in the house and changed the sheets on his grandfather's bed and then took him back inside. Tried to get him to eat something. "All right for you, Grandpa, you're going to have to eat twice as much for dinner." He went back in the kitchen with the bowl.

Betty phoned at 3:30. After he said hello she said, "Well, I'm glad you didn't die for love." He was able to laugh. "I've got your assignments; do you want me to bring them out? I didn't get to the bus on time."

"Put them on the bus tomorrow," he said. "I need a rest from all that brain work."

"I'd like to see you," she said in a lower voice. "Just stop and visit." He didn't answer. Nothing would come out. "How's your grandfather?" she finally asked.

"He's fine. I took him out today and he had a good time, sitting on the porch. I better get the wash in off the line," he said before she could say anything else, and he hung up.

He had all the clothes in the basket when Greta drove in and got out of her car carrying a Pyrex bowl and something wrapped in a towel. "I made some carrot cake, Elmo," she called out.

"He's going to like that," he said, following her inside. She put the things down and went into Grandpa's room. Elmo put the basket of clothes down, opened the towel, then the wax paper. It had white frosting.

"Elmo," Greta said. He licked his fingers, went toward the doorway.

"Grandpa," Greta said, leaning over him. She took a face mirror out of her purse and held it up to his nose.

"What are you doing, Greta?"

She felt around his wrist and put her ear to his chest, then shook him by the shoulders. The old head rolled. "Call the doctor, Elmo," she said. "Grandpa's dead."

"No, he isn't; he's fine, he's just asleep." Elmo went over and straightened the blankets. He could hear her call the doctor, then James. Elmo brought a kitchen chair in and sat beside the bed, holding Grandpa's hand. "You're not going to guess what we're going to have for dessert," he said. "Your favorite, carrot cake. She put white icing on, a thick layer. It's going to be so rich we're going to need a lot of hot coffee."

James came into the house, slamming the door shut. He was breathing hard and the sound filled the bedroom. He put his big fingers on Grandpa's neck. Left them a long time. "He's dead, Elmo."

He knew he was being calm and reasonable. "He's not dead, James. I'm talking to him."

"I know dead, Elmo. He's dead."

"He wouldn't leave me here, James, I know that. Not alone. He's not dead." James went out.

"I was thinking, Grandpa, about our trip out east. To the eastern seaboard, as you say. I think the fall is better. I'd like to see those leaves you're always talking about. We should go out there before it gets too cold. I don't want to run into any of those storms. I'm too partial to this California sunshine."

The doctor, who was also the county coroner, came in, felt around with his stethoscope, then carefully took the folded-down sheet as Elmo watched and put it up over his grandfather's head. "He's gone," he said, "he's gone."

Elmo took the sheet down. "He's all right," he said, laying his head on his grandfather's chest.

"Your grandfather had some arrangements with one of the funeral homes in town. I'm going to send for them now, take him back with us."

"He's staying right here," Elmo shouted, sitting up. "He's going to get well. No one's going to take him anywhere." He jumped up, knocking over the chair, and put up his fists. The doctor backed out the door into the kitchen.

Greta tried to come in. "Elmo."

"No, Greta, he's staying right here."

He heard them talking and they all went out. He locked the front door and turned off the lights. "It's early," he said aloud, "but we both need our rest. We've got a big day ahead of us tomorrow." He sat down at the kitchen table to wait.

He'd been up and around when the knocking started. Had got the water just right and was shaving his grandfather. "Elmo, it's me, Joyce, I want to come in."

"Hold on," he called back. "I'm almost finished here." He wiped his grandpa's face off with a towel and went to the door.

Joyce stepped in as soon as he opened it, with James and Greta right behind her. Eddie and Claude stayed on the porch. The kitchen was crowded. James and Joyce went into his grandfather's bedroom. "I was getting him ready," Elmo said, following.

"I can give you a hand," Joyce said.

"He was awful shy," Elmo said.

"I know what you mean."

"Maybe you could brush his blue suit. It's hanging in his closet." Elmo went back and took the basin of shaving water and dumped it into the bathtub, washed out the pan, filled it with warm water.

When the suit was brushed he put it on his grandfather. Joyce knotted the tie and Elmo put it around his neck, buttoned the top button of his shirt. He couldn't think of anything else.

Eddie and Claude had gone back to the main house. James was sitting at the table with Greta. Joyce was folding the wash. When no one was looking, Elmo slipped a piece of carrot cake wrapped in wax paper into his grandfather's pocket. Then he walked out and sat on the porch, his feet up on the rail, throwing single pieces of cracked corn to the peafowl.

When the sun was over the edge of the orchard, two men came in the hearse and wheeled a gurney into the house and then came back out. Elmo didn't look. Didn't move. Kept throwing feed, making the peafowl run for it.

"Let's go up to the house, Elmo," Joyce said. "We could all use some breakfast."

"I have a few things I have to do first," he said. "I'll be up later."

"You promise," Greta said, putting her hand on his shoulder.

"I promise," he said.

He sat there a long time. His head was emptied out. There was no connection between what he could see and what was going on inside. He closed his eyes for a minute. There was no relief there. When he opened them there was the same thing. The yard, the peacocks, the road and then the orchard. If only I could imagine the hardest math problem, he thought, and then come up with the right answer, I could do anything. But there was nothing but his grandfather in his head. Inside. Like the chick inside an egg when you hold it up to the light. That was never going to come out. There was no more room to imagine an answer. He threw the last of the feed, got up and went into the house.

He found his twelve-gauge in the pantry. He took his time, loaded, filled his jacket pockets full of shells. By the time he got back to the porch there were only five or six in sight. They must know, he thought, aiming. He got those and two more behind the house.

He dragged the dead ones, as he killed them, to a big pile of tree trimmings, higher than his head, that hadn't been burned yet. He swung them up by their legs to the top. The others couldn't get away; he took his time. Where could they go? What could they become? A tree?

When James came driving up in the pickup, he was surprised. Then he realized they must have heard the shots from the main house. One took off from behind the rhododendron bush, and he blasted it. Slowly James opened the truck door. "Just stay right there, James," he yelled. "I'm attending to my own business here." James sat still, his legs hanging out the open door, his elbow on the window frame, watching.

He nailed the last ones by the clothesline. They seemed like they were waiting for him there. He emptied the rest of the five gallons of diesel around the base of the pile and capped the can. He snapped the kitchen match alight with his thumbnail, then threw it in. There was a whoosh, and the still-green plum trimmings sizzled and foamed as heat from the flames climbed up the pile. He looked for the peacock feathers, but there wasn't any sign of them that he could see.

Don Thompson

POSO FLAT

for K.D.T.

1.

We stopped
out in the field,
uphill from a tentative road

where oak stumps
surrounded us
like surplus chopping blocks

and crows,
startled or not,
vanished into the fog.

I fix this landscape in my mind

and in it
place father and son with
nothing to say.

2.

He uncrated the new chainsaw,
set it on the tailgate

and we stared
at the yellow carapace,

---

the steel tongue
and teeth

scrimshawed
with a secret alphabet:

our hands touched,
adjusting it.

3.

In that dry place
a grandfather oak,
its roots
like the talons
of an Ozymandias for crows:

Take this one,
I said,
and he nodded
teasing the starved saw
with dead wood.

The tree fell
and I heard the lock crack open
on his old sea trunk
spilling gifts
for a wife and lost daughter:

carved ivory,
silk dragons and
Sun Yat-sen dollars
so thin
your fingers blur through.

For himself
he brought home a stubborn fungus,
snapshots of the dead,

of coolies
sipping GI piss from a beer can—

and that awkwardness
when touching
his son, his hands chilled by
the frozen windows of cargo planes
crossing the Hump.

4.

When a crow perched on a stump near us,
we watched until

it flew, its caw
like a word

we had learned to speak
splitting that firewood.

Larry Levis

PICKING GRAPES IN AN ABANDONED VINEYARD

Picking grapes alone in the late autumn sun—
A short, curved knife in my hand,
Its blade silver from so many sharpenings,
Its handle black.
I still have a scar where a friend
Sliced open my right index finger, once,
In a cutting shed—
The same kind of knife.
The grapes drop into the pan,
And the gnats swarm over them, as always.
Fifteen years ago,
I worked this row of vines beside a dozen
Families up from Mexico.
No one spoke English, or wanted to.
One woman, who made an omelet with a sheet of tin
And five, light blue quail eggs,
Had a voice full of dusk, and jail cells,
And bird calls. She spoke,
In Spanish, to no one, as they all did.
Their swearing was specific,
And polite.
I remember two of them clearly:
A man named Tea, six feet, nine inches tall
At the age of sixty-two,
Who wore white spats into downtown Fresno
Each Saturday night,
An alcoholic giant whom the women loved—

One chilled morning, they found him dead outside
The Rose Café . . .
And Angel Domínguez,
Who came to work for my grandfather in 1910,
And who saved for years to buy
Twenty acres of rotting, Thompson Seedless vines.
While the sun flared all one August,
He decided he was dying of a rare disease,
And spent his money and his last years
On specialists,
Who found nothing wrong.
Tea laughed, and, tipping back
A bottle of Muscatel, said: "Nothing's wrong.
You're just dying."
At seventeen, I discovered
Parlier, California, with its sad, topless bar,
And its one main street, and its opium.
I would stand still, and chalk my cue stick
In Johnny Palores' East Front Pool Hall, and watch
The room filling with tobacco smoke, as the sun set
Through one window.
Now all I hear are the vines rustling as I go
From one to the next,
The long canes holding up dry leaves, reddening,
So late in the year.
What the vines want must be this silence spreading
Over each town, over the dance halls and the dying parks,
And the police drowsing in their cruisers
Under the stars.
What the men who worked here wanted was
A drink strong enough
To let out what laughter they had.
I can still see the two of them:
Tea smiles and lets his yellow teeth shine—
While Angel, the serious one, for whom
Death was a rare disease,
Purses his lips, and looks down, as if
He is already mourning himself—
A soft, gray hat between his hands.

Today, in honor of them,
I press my thumb against the flat part of this blade,
And steady a bunch of red, Málaga grapes
With one hand,
The way they showed me, and cut—
And close my eyes to hear them laugh at me again,
And then, hearing nothing, no one,
Carry the grapes up into the solemn house,
Where I was born.

THE POET AT SEVENTEEN

My youth? I hear it mostly in the long, volleying
Echoes of billiards in the pool halls where
I spent it all, extravagantly, believing
My delicate touch on a cue would last for years.

Outside the vineyards vanished under rain,
And the trees held still or seemed to hold their breath
When the men I worked with, pruning orchards, sang
Their lost songs: *Amapola; La Paloma;*

Jalisco, No Te Rajes—the corny tunes
Their sons would just as soon forget, at recess,
Where they lounged apart in small groups of their own.
Still, even when they laughed, they laughed in Spanish.

I hated high school then, & on weekends drove
A tractor through the widowed fields. It was so boring
I memorized poems above the engine's monotone.
Sometimes whole days slipped past without my noticing,

And birds of all kinds flew in front of me then.
I learned to tell them apart by their empty squabblings,
The slightest change in plumage, or the inflection
Of a call. And why not admit it? I was happy

Then. I believed in no one. I had the kind
Of solitude the world usually allows
Only to kings & criminals who are extinct,
Who disdain this world, & who rot, corrupt & shallow

As fields I disced: I turned up the same gray
Earth for years. Still, the land made a glum raisin
Each autumn, & made that little hell of days—
The vines must have seemed like cages to the Mexicans

Who were paid seven cents a tray for the grapes
They picked. Inside the vines it was hot, & spiders
Strummed their emptiness. Black Widow, Daddy Longlegs.
The vine canes whipped our faces. None of us cared.

And the girls I tried to talk to after class
Sailed by, then each night lay enthroned in my bed,
With nothing on but the jewels of their embarrassment.
Eyes, lips, dreams. No one. The sky & the road.

A life like that? It seemed to go on forever—
Reading poems in school, then driving a stuttering tractor
Warm afternoons, then billiards on blue October
Nights. The thick stars. But mostly now I remember

The trees, wearing their mysterious yellow sullenness
Like party dresses. And parties I didn't attend.
And then the first ice hung like spider lattices
Or the embroideries of Great Aunt No One,

And then the first dark entering the trees—
And inside, the adults with their cocktails before dinner,
The way they always seemed afraid of something,
And sat so rigidly, although the land was theirs.

Roberta Spear

ARMONA

A breeze lifts the orchard dust
and the heavy heart of fruit
lets go. Women and children
line up in the humid corridors
of the packing shed, sharing
the heat, its warped halo.
There is no talk of rest
this day, just the day itself.
And a syrup that makes
the fingers stick together,
like cups set out to gather
more coolness, more darkness,
even after the gifts of summer
are taken back.

Once, on a dare, I walked
five miles at dawn to join them
for the day's orders. If you
could slit a seam and make
the pit leap into the ringing bucket
thirty times in one minute
you had the job.
We worked quickly, rolling
casabas and plums down
the wet, splintered tables,
twisting the stems off
the crotches of fruit. In one,
a small, dark chamber where
a black widow was sleeping,
waiting for the moment
when she and her young could ride

---

that sweet river into sunlight.
She might have said *pardon me*
or *let me pass in peace,*
but she lunged straight toward me
from her sleep, pleading
like the others for more coolness,
more life. I threw my knife
into a roadside ditch and paced
the miles back to town.
For as sure as the heat had singed
each cloud out of the sky,
that kid who'd dared me here
lay chuckling in his late-morning sleep.

The workers said that once,
years ago, it had happened
much this way. But instead of a boy,
a grown man who called himself
the deacon, who kept two women
and should have known better.
One day, the women came from different farms
to these crumbling sheds
only to discover they'd been peeling
the same peach all along,
dividing its juice. Some said
his shoes were still lashed
to the rafters above us
and, when the wind slid through
the slatted ceiling, they creaked
like the two black doves,
love and death. When we were kids,
we called this town *Aroma*
for the smell of life that has come
and gone, and left its stain
under the nails long after
the black flesh of branches
has dissolved. And for the one
carried down from the rafters,
who wore gloves and a jacket
on the hottest days and loved

the aroma of money, a ripe
freestone, like a piece of ass
with the heart cut out.

How could anyone have known
that the heat would pass?
That, in the shadows, the crisp leaf
of apple, a fork of almond
were springing inch by inch
from the stubble. Or
that a harder season would follow—
One of husks and cores,
and a wind barely getting by,
lifting its soft thumb up
for a ride to another county.

Mike Cole

ENGLISH 9

In ninth grade English Donna Cramer's skirt
pulled high up her thighs under the desk,
though I almost never let my eyes rest there.
Paul Hartley sat in the back row too.
His voice changed to a deep bass
before mine, and he got a dark look in his eyes.

When my friend Eddie told Donna dirty jokes
and asked if she wore B.F. Goodrich falsies,
I stared at my desk,
and Paul pretended to be thinking.

Donna's handwriting was neat and big
with generous loops. Paul's was tight and cramped
with no margins. We got mostly A's and B's
in the back row.

In February when we were learning to make diagrams
of adverbs hanging at an angle
below the verb and to branch
compound subjects, Donna went to a vacant house
with six high school football players.
Eddie told me about it
as we got used to Donna's
empty desk. Paul listened. I pictured
the closed bedroom door.
In March we read a story
about a peasant boy who went to the city
where he ate ice cream for the first time
and swore his mouth was on fire. I was still
trying to understand how my cold

---

could be someone else's absolute hot
when Paul stopped coming to school. His mother
asked him to zip up her dress one morning,
and he buried a pair of scissors
in the white unzipped V.
I imagined a room with blanks walls,
Paul sitting at a table with his long,
clean fingers folded, a single light overhead,
a shape across from him
asking questions.

For the school play
I dressed like a father,
with Eddie as my obedient child.
Our teacher pleaded with us
to make it true-to-life.

DeWayne Rail

GOING HOME AGAIN / POEM FOR MY FATHER

Nothing here seems to welcome my return,
the dark and weeded fields, those dying trees.
Walking this ground, I see how things have turned.

Ten years ago I played here in this yard.
With sticks for mules, I mimicked every move
my father made behind that rusting plow.

And nights when he came home too tired to move,
he sat upon that stoop and slowly cursed
the land, the lack of rain, his crops, his luck.

I grew up on this farm and learned to curse
what made him hang on to his father's place,
the bank that threatened to foreclose each year,

The wind, this house and land, and his plodding ways
that kept us here, losing, year after year.
How much it marked him. Indeed, my father's face

I can still see, grown old as dirt, the lines
like furrows in the stubble of his face.

Arthur Smith

ELEGY ON INDEPENDENCE DAY

Over the balcony eave, seaside,
One after another, the rockets arc
Barely into view, each sudden thud
Rolling from behind the brickface.
We used to say the rockets "burst,"
As though speaking of someone's heart—
Star-beam, dream-light, bright spokes
Wheeling, falling in a sort of glory.

One summer, in an orchard in Manteca,
The scent of peaches was like fog,
The dust rose and settled like fog,
And both of us went waving sparklers.
You ran on, out farther, tracing
Spirals high in the air. They stayed
Long after the light went, after you
And the heavy, sweet trees were one.

Now I close my eyes and find only
Traces of those wiry figures burned
Into the night. They are fading as
They must, and as they always do.
Whatever shines, however briefly,
We tend toward and love perhaps,
Grounded as we are in the literal—
The powder, the ashes, the earth.

David Kherdian

THE WALNUT TREE TRUNK CLOTHED IN THE GREEN LEAVES OF AN EMBRACING VINE

In our garden walnut pomegranate quince loquat and apricot trees
And an unplanted nearly trunkless palm tree with fanning leaves.

I want to remember them now for I have been gone a week
And a week of California spring is more dramatic
Than the average equal interval in one man's life—
And especially now that I have been away
I can only contrast the upheaval I feel
With the easy and abundant growth that has occurred in my absence—
Teaching me that to have a sense of place
And to feel rooted in my own life
I must sit still and let my life gather about me

As the trunk of the walnut tree stands rooted in solid silence
To be clothed by the green leaves of an embracing vine.

IT BECOMES THIS FOR ME

rain-soaked leaves
that twirling fall
mark the path
the cat will walk,
as he steps from
the curb,
coming from hurt
to healing to home;
and the cars that
push waters in sprays
of reflecting colors,
everything suddenly a
metallic blue;
whirring wheels
their own special noise;
or whatever else
I see from the window
or that comes to me
from beneath the door,
because I have embraced
the silence of a slowly
turning life in the
adopted valley of
my home

Robert Mezey

THE UNDERGROUND GARDENS

for Baldasare Forestiere

Sick of the day's heat, of noise
and light and people, I come
to walk in Forestiere's
deep home, where his love never
came to live; where he prayed to Christ;
slept lightly; put on his clothes;
clawed at the earth forty years
but it answered nothing.
Silence came down with the small
pale sunlight, then the darkness.
Maybe the girl was dead. He
grew accustomed to the silence
and to the darkness. He brought
food to his mouth with both
invisible hands, and waited
for night's darkness to give way
to the darkness of day. If
he held his breath and his
eyes closed against the brown light
sifting down by masonry and roots,
he could see her spirit among
his stone tables, laughing and
saying no. When he opened
them to emptiness, he wept.
And at last he kept them closed.
Death gripped him by the hair and
he was ready. He turned and slept
more deeply.

There were many
rooms, tunnels and coves and arbors,
places where men and women
could sit, flowering plazas
where they could walk or take food,
and rooms for the tired to rest.
He could almost imagine
their voices, but not just yet.

He is buried somewhere else.

WHITE BLOSSOMS

Take me as I drive alone
through the dark countryside.
As the strong beams clear a path,
picking out fences, weeds, late
flowering trees, everything
that streams back into the past
without sound, I smell the grass
and the rich chemical sleep
of the fields. An open moon
sails above, and a stalk
of red lights blinks, miles away.

It is at such moments I
am called, in a voice so pure
I have to close my eyes and enter
the breathing darkness just beyond
my headlights. I have come back,
I think, to something I had
almost forgotten, a mouth
that waits patiently, sighs, speaks
and falls silent. No one else
is alive. The blossoms are
white, and I am almost there.

Dennis Schmitz

THE CALIFORNIA PHRASEBOOK

west of the Sierras where
the central valley
drifts on its crusts of almond
orchards the fields
die in a holiday accident
the freeways snapping
back in the dust like severed
arteries while the accomplished
doctor of silence stitches the evening
closed with stoplights which
never hold. gardens go
on their knees to the sun
all summer turning
over the brief counterfeit the rain
leaves, looking for a real
coin. in the arbors the Italian
uncles sit stirring anise-flavored
coffees, red bandanas
over their knees pulsing
with the sweating
body's rhythm like an open
chest in which the transplanted
organ of the homeland has not yet
begun to function.
in East Bay beyond the valley
towns pore to pore
the children black & brown
press out a test pattern
of veins, their faces rigid with long
division. they stand
in front of the blackboards looking

---

for their features, refusing
to draw the white mark
of a dollarsign while the mayor
waits & all the examining
board of cops waits, the correction
texts trembling in their hands.
why must we repeat our lessons,
let us go. in the alleys we rehearse
the lonely patrol of hands
over each other's bodies.
if we unfold a woman's creases
we are afraid to read
the platitudes. the black
penis is the last piece
of the puzzle we put in place
before the streetlamps have slipped
away in the wet fingers of the April
night, before the pathways
through the asphalt gardens
disgorge the feathers of the black
angel.

you who arrive late
from some forgotten Kansas
laminated of wheat & the sweet
alfalfa wasted with incurable winter
take root in the familiar
flat valley where the only
winter is overweight with rain
again & again welling up in fallen
wet fruit an early unearned
bitterness like the bum who drowses
under the indelible azaleas scrawled
against the capitol's white
walls. his life too
is a fragrant perennial.
he is less foreign than you,
but you must learn his difficult
language full of inflections for another
self palpable as the stone in spoiled

fruit. another self! the cheap
foundations of love shift—
before you always built in the quake-
proof plains where small rivers
pillow their heads in poplar
roots & turn all night
in the drought's persistent
insomnia. the cellar was dirt
still alive with roots
from which your father cut
your life in rigid board walls
incredibly steady
on the rippling floor of yellow
grain. in California the cellarless
houses sway at the slightest
tremor.

Joan Didion

NOTES FROM A NATIVE DAUGHTER

IT IS VERY EASY to sit at the bar in, say, La Scala in Beverly Hills, or Ernie's in San Francisco, and to share in the pervasive delusion that California is only five hours from New York by air. The truth is that La Scala and Ernie's are only five hours from New York by air. California is somewhere else.

Many people in the East (or "back East," as they say in California, although not in La Scala or Ernie's) do not believe this. They have been to Los Angeles or to San Francisco, have driven through a giant redwood and have seen the Pacific glazed by the afternoon sun off Big Sur, and they naturally tend to believe that they have in fact been to California. They have not been, and they probably never will be, for it is a longer and in many ways a more difficult trip than they might want to undertake, one of those trips on which the destination flickers chimerically on the horizon, ever receding, ever diminishing. I happen to know about that trip because I come from California, come from a family, or a congeries of families, that has always been in the Sacramento Valley.

You might protest that no family has been in the Sacramento Valley for anything approaching "always." But it is characteristic of Californians to speak grandly of the past as if it had simultaneously begun, *tabula rasa*, and reached a happy ending on the day the wagons started west. *Eureka*—"I Have Found It"—as the state motto has it. Such a view of history casts a certain melancholia over those who participate in it; my own childhood was suffused with the conviction that we had long outlived our finest hour. In fact that is what I want to tell you about: what it is like to come from a place like Sacramento. If I could make you understand that, I could make you understand California and perhaps something else besides, for Sacramento *is* California, and California is a place in which a boom mentality

---

and a sense of Chekhovian loss meet in uneasy suspension; in which the mind is troubled by some buried but ineradicable suspicion that things had better work here, because here, beneath that immense bleached sky, is where we run out of continent.

In 1847 Sacramento was no more than an adobe enclosure, Sutter's Fort, standing alone on the prairie; cut off from San Francisco and the sea by the Coast Range and from the rest of the continent by the Sierra Nevada, the Sacramento Valley was then a true sea of grass, grass so high a man riding into it could tie it across his saddle. A year later gold was discovered in the Sierra foothills, and abruptly Sacramento was a town, a town any moviegoer could map tonight in his dreams—a dusty collage of assay offices and wagonmakers and saloons. Call that Phase Two. Then the settlers came—the farmers, the people who for two hundred years had been moving west on the frontier, the peculiar flawed strain who had cleared Virginia, Kentucky, Missouri; they made Sacramento a farm town. Because the land was rich, Sacramento became eventually a rich farm town, which meant houses in town, Cadillac dealers, a country club. In that gentle sleep Sacramento dreamed until perhaps 1950, when something happened. What happened was that Sacramento woke to the fact that the outside world was moving in, fast and hard. At the moment of its waking Sacramento lost, for better or for worse, its character, and that is part of what I want to tell you about.

But the change is not what I remember first. First I remember running a boxer dog of my brother's over the same flat fields that our great-great-grandfather had found virgin and had planted; I remember swimming (albeit nervously, for I was a nervous child, afraid of sinkholes and afraid of snakes, and perhaps that was the beginning of my error) the same rivers we had swum for a century: the Sacramento, so rich with silt that we could barely see our hands a few inches beneath the surface; the American, running clean and fast with melted Sierra snow until July, when it would slow down, and rattlesnakes would sun themselves on its newly exposed rocks. The Sacramento, the American, sometimes the Cosumnes, occasionally the Feather. Incautious children died every day in those rivers; we read about it in the paper, how they had miscalculated a current or stepped into a hole down where the American runs into the Sacramento, how the Berry Brothers had been called in from Yolo County to drag the river but how the bodies remained unrecovered. "They were from away," my grandmother would extrapolate from the newspaper stories. "Their parents had no *business* letting them in the river. They were visitors from Omaha." It was not a bad lesson, although a less than reliable one; children we knew died in the rivers too.

When summer ended—when the State Fair closed and the heat broke, when the last green hop vines had been torn down along the H Street road and the tule fog began rising off the low ground at night—we would go back to memorizing the Products of Our Latin American Neighbors and to visiting the great-aunts on Sunday, dozens of great-aunts, year after year of Sundays. When I think now of those winters I think of yellow elm leaves wadded in the gutters outside the Trinity Episcopal Pro-Cathedral on M Street. There are actually people in Sacramento now who call M Street Capitol Avenue, and Trinity has one of those featureless new buildings, but perhaps children still learn the same things there on Sunday mornings:

> Q. *In what way does the Holy Land resemble the Sacramento Valley?*
> A. *In the type and diversity of its agricultural products.*

And I think of the rivers rising, of listening to the radio to hear at what height they would crest and wondering if and when and where the levees would go. We did not have as many dams in those years. The bypasses would be full, and men would sandbag all night. Sometimes a levee would go in the night, somewhere upriver; in the morning the rumor would spread that the Army Engineers had dynamited it to relieve the pressure on the city.

After the rains came spring, for ten days or so; the drenched fields would dissolve into a brilliant ephemeral green (it would be yellow and dry as fire in two or three weeks) and the real-estate business would pick up. It was the time of year when people's grandmothers went to Carmel; it was the time of year when girls who could not even get into Stephens or Arizona or Oregon, let alone Stanford or Berkeley, would be sent to Honolulu, on the *Lurline.* I have no recollection of anyone going to New York, with the exception of a cousin who visited there (I cannot imagine why) and reported that the shoe salesmen at Lord & Taylor were "intolerably rude." What happened in New York and Washington and abroad seemed to impinge not at all upon the Sacramento mind. I remember being taken to call upon a very old woman, a rancher's widow, who was reminiscing (the favored conversational mode in Sacramento) about the son of some contemporaries of hers. "That Johnston boy never did amount to much," she said. Desultorily, my mother protested: Alva Johnston, she said, had won the Pulitzer Prize, when he was working for *The New York Times.* Our hostess looked at us impassively. "He never amounted to anything in Sacramento," she said.

Hers was the true Sacramento voice, and, although I did not realize it then, one not long to be heard, for the war was over and boom was on and the voice of the aerospace engineer would be heard in the land. VETS NO DOWN! EXECUTIVE LIVING ON LOW FHA!

Later, when I was living in New York, I would make the trip back to Sacramento four and five times a year (the more comfortable the flight, the more obscurely miserable I would be, for it weighs heavily upon my kind that we could perhaps not make it by wagon), trying to prove that I had not meant to leave at all, because in at least one respect California—the California we are talking about—resembles Eden: it is assumed that those who absent themselves from its blessings have been banished, exiled by some perversity of heart. Did not the Donner-Reed Party, after all, eat its own dead to reach Sacramento?

I have said that the trip back is difficult, and it is—difficult in a way that magnifies the ordinary ambiguities of sentimental journeys. Going back to California is not like going back to Vermont, or Chicago; Vermont and Chicago are relative constants, against which one measures one's own change. All that is constant about the California of my childhood is the rate at which it disappears. An instance: on Saint Patrick's Day of 1948 I was taken to see the legislature "in action," a dismal experience; a handful of florid assemblymen, wearing green hats, were reading Pat-and-Mike jokes into the record. I still think of the legislators that way—wearing green hats, or sitting around on the veranda of the Senator Hotel fanning themselves and being entertained by Artie Samish's emissaries. (Samish was the lobbyist who said, "Earl Warren may be the governor of the state, but I'm the governor of the legislature.") In fact there is no longer a veranda at the Senator Hotel—it was turned into an airline ticket office, if you want to embroider the point—and in any case the legislature has largely deserted the Senator for the flashy motels north of town, where the tiki torches flame and the steam rises off the heated swimming pools in the cold Valley night.

It is hard to *find* California now, unsettling to wonder how much of it was merely imagined or improvised; melancholy to realize how much of anyone's memory is no true memory at all but only the traces of someone else's memory, stories handed down on the family network. I have an indelibly vivid "memory," for example, of how Prohibition affected the hop growers around Sacramento: the sister of a grower my family knew brought home a mink coat from San Francisco, and was told to take it back, and sat on the floor of the parlor cradling that coat and crying. Although I was not born until a year after Repeal, that scene is more "real" to me than many I have played myself.

I remember one trip home, when I sat alone on a night jet from New York and read over and over some lines from a W.S. Merwin poem I had come across in a magazine, a poem about a man who had been a long time in another country and knew that he must go home:

. . . But it should be
Soon. Already I defend hotly
Certain of our indefensible faults,
Resent being reminded; already in my mind
Our language becomes freighted with a richness
No common tongue could offer, while the mountains
Are like nowhere on earth, and the wide rivers.

You see the point. I want to tell you the truth, and already I have told you about the wide rivers.

It should be clear by now that the truth about the place is elusive, and must be tracked with caution. You might go to Sacramento tomorrow and someone (although no one I know) might take you out to Aerojet-General, which has, in the Sacramento phrase, "something to do with rockets." Fifteen thousand people work for Aerojet, almost all of them imported; a Sacramento lawyer's wife told me, as evidence of how Sacramento was opening up, that she believed she had met one of them, at an open house two Decembers ago. ("Couldn't have been nicer, actually," she added enthusiastically. "I think he and his wife bought the house next *door* to Mary and Al, something like that, which of course was how *they* met him.") So you might go to Aerojet and stand in the big vendors' lobby where a couple of thousand components salesmen try every week to sell their wares and you might look up at the electrical wallboard that lists Aerojet personnel, their projects and their location at any given time, and you might wonder if I have been in Sacramento lately. MINUTEMAN, POLARIS, TITAN, the lights flash, and all the coffee tables are littered with airline schedules, very now, very much in touch.

But I could take you a few miles from there into towns where the banks still bear names like The Bank of Alex Brown, into towns where the one hotel still has an octagonal-tile floor in the dining room and dusty potted palms and big ceiling fans; into towns where everything—the seed business, the Harvester franchise, the hotel, the department store and the main street—carries a single name, the name of the man who built the town. A few Sundays ago I was in a town like that, a town smaller than that, really, no hotel, no Harvester franchise, the bank burned out, a river town. It was the golden anniversary of some of my relatives and it was 110° and the guests of honor sat on straight-backed chairs in front of a sheaf of gladioluses in the Rebekah Hall. I mentioned visiting Aerojet-General to a cousin I saw there, who listened to me with interested disbelief. Which is the true California? That is what we all wonder.

Let us try out a few irrefutable statements, on subjects not open to interpretation. Although Sacramento is in many ways the least typical of the Valley towns, it *is* a Valley town, and must be viewed in that context. When you say "the Valley" in Los Angeles, most people assume that you mean the San Fernando Valley (some people in fact assume that you mean Warner Brothers), but make no mistake: we are talking not about the valley of the sound stages and the ranchettes but about the real Valley, the Central Valley, the fifty thousand square miles drained by the Sacramento and the San Joaquin Rivers and further irrigated by a complex network of sloughs, cutoffs, ditches, and the Delta-Mendota and Friant-Kern Canals.

A hundred miles north of Los Angeles, at the moment when you drop from the Tehachapi Mountains into the outskirts of Bakersfield, you leave Southern California and enter the Valley. "You look up the highway and it is straight for miles, coming at you, with the black line down the center coming at you and at you . . . and the heat dazzles up from the white slab so that only the black line is clear, coming at you with the whine of the tires, and if you don't quit staring at that line and don't take a few deep breaths and slap yourself hard on the back of the neck you'll hypnotize yourself."

Robert Penn Warren wrote that about another road, but he might have been writing about the Valley road, U.S. 99, three hundred miles from Bakersfield to Sacramento, a highway so straight that when one flies on the most direct pattern from Los Angeles to Sacramento one never loses sight of U.S. 99. The landscape it runs through never, to the untrained eye, varies. The Valley eye can discern the point where miles of cotton seedlings fade into miles of tomato seedlings, or where the great corporation ranches—Kern County Land, what is left of DiGiorgio—give way to private operations (somewhere on the horizon, if the place is private, one sees a house and a stand of scrub oaks), but such distinctions are in the long view irrelevant. All day long, all that moves is the sun, and the big Rainbird sprinklers.

Every so often along 99 between Bakersfield and Sacramento there is a town: Delano, Tulare, Fresno, Madera, Merced, Modesto, Stockton. Some of these towns are pretty big now, but they are all the same at heart, one- and two- and three-story buildings artlessly arranged, so that what appears to be the good dress shop stands beside a W.T. Grant store, so that the big Bank of America faces a Mexican movie house. *Dos Peliculas, Bingo Bingo Bingo.* Beyond the downtown (pronounced *down*town, with the Okie accent that now pervades Valley speech patterns) lie blocks of old frame houses—paint peeling, sidewalks cracking, their occasional leaded amber windows overlooking a Foster's Freeze or a five-minute car wash or a State Farm Insurance office; beyond those spread the shopping centers and the miles of tract houses, pastel with redwood siding, the unmistakable signs of cheap building already blossoming on those houses which have survived the first rain. To a stranger driving 99 in an air-conditioned car (he would be on business, I suppose, any stranger driving 99, for 99 would never get a tourist to Big Sur or San Simeon, never get him to the California he came to see), these towns must seem so flat, so impoverished, as to drain the imagination. They hint at evenings spent hanging around gas stations, and suicide pacts sealed in drive-ins.

But remember:

Q. In what way does the Holy Land resemble the Sacramento Valley?
A. In the type and diversity of its agricultural products.

U.S. 99 in fact passes through the richest and most intensely cultivated agricultural region in the world, a giant outdoor hothouse with a billion-dollar crop. It is when you remember the Valley's wealth that the monochromatic flatness of its towns takes on a curious meaning, suggests a habit of mind some would consider perverse. There is something in the Valley mind that reflects a real indifference to the stranger in his air-conditioned car, a failure to perceive even his presence, let alone his thoughts or wants. An implacable insularity is the seal of these towns. I once met a woman in Dallas, a most charming and attractive woman accustomed to the hospitality and social hypersensitivity of Texas, who told me that during the four war years her husband had been stationed in Modesto, she had never once been invited inside anyone's house. No one in Sacramento would find this story remarkable ("She probably had no *rel*atives there," said someone to whom I told it), for the Valley towns understand one another, share a peculiar spirit. They think alike and they look alike. *I* can tell Modesto from Merced, but I have visited there, gone to dances there; besides, there is over the main street of Modesto an arched sign which reads:

WATER — WEALTH
CONTENTMENT — HEALTH

There is no such sign in Merced.

I said that Sacramento was the least typical of the Valley towns, and it is—but only because it is bigger and more diverse, only because it has had the rivers and the legislature; its true character remains the Valley character, its virtues the Valley virtues, its sadness the Valley sadness. It is just as hot in the summertime, so hot that the air shimmers and the grass bleaches white and the blinds stay drawn all day, so hot that August comes on not like a month but like an affliction; it is just as flat, so flat that a ranch of my family's with a slight rise on it, perhaps a foot, was known for the hundredsome years which preceded this year as "the hill ranch." (It is known this year as a subdivision in the making, but that is another part of the story.) Above all, in spite of its infusions from outside, Sacramento retains the Valley insularity.

To sense that insularity a visitor need do no more than pick up a copy of either of the two newspapers, the morning *Union* or the afternoon *Bee.* The *Union* happens to be Republican and impoverished and the *Bee* Democratic and powerful ("THE VALLEY OF THE BEES!" as the McClatchys, who own the *Fresno, Modesto,* and *Sacramento Bees,* used to headline their advertisements in the trade press. "ISOLATED FROM ALL OTHER MEDIA INFLUENCE!"), but they read a good deal alike, and the tone of their chief editorial concerns is strange and wonderful and instructive. The *Union,* in a county heavily and reliably Democratic, frets mainly about the possibility of a local takeover by the John Birch Society; the *Bee,* faithful to the letter of its founder's will, carries on overwrought crusades against phantoms it still calls "the power trusts." Shades of Hiram Johnson, whom the *Bee* helped elect governor in 1910. Shades of Robert La Follette, to whom the *Bee* delivered the Valley in 1924. There is something about the Sacramento papers that does not quite connect with the way Sacramento lives now, something pronouncedly beside the point. The aerospace engineers, one learns, read the *San Francisco Chronicle.*

The Sacramento papers, however, simply mirror the Sacramento peculiarity, the Valley fate, which is to be paralyzed by a past no longer relevant. Sacramento is a town which grew up on farming and discovered to its shock that land has more profitable uses. (The chamber of commerce will give you crop figures, but pay them no mind—what matters is the feeling, the knowledge that where the green hops once grew is now Larchmont

Riviera, that what used to be the Whitney ranch is now Sunset City, thirty-three thousand houses and a country club complex.) It is a town in which defense industry and its absentee owners are suddenly the most important facts; a town which has never had more people or more money, but has lost its *raison d'être.* It is a town many of whose most solid citizens sense about themselves a kind of functional obsolescence. The old families still see only one another, but they do not see even one another as much as they once did; they are closing ranks, preparing for the long night, selling their rights-of-way and living on the proceeds. Their children still marry one another, still play bridge and go into the real-estate business together. (There is no other business in Sacramento, no reality other than land—even I, when I was living and working in New York, felt impelled to take a University of California correspondence course in Urban Land Economics.) But late at night when the ice has melted there is always somebody now, some Julian English, whose heart is not quite in it. For out there on the outskirts of town are marshaled the legions of aerospace engineers, who talk their peculiar condescending language and tend their dichondra and plan to stay in the promised land; who are raising a new generation of native Sacramentans and who do not care, really do not care, that they are not asked to join the Sutter Club. It makes one wonder, late at night when the ice is gone; introduces some air into the womb, suggests that the Sutter Club is perhaps not, after all, the Pacific Union or the Bohemian; that Sacramento is not *the city.* In just such self-doubts do small towns lose their character.

I want to tell you a Sacramento story. A few miles out of town is a place, six or seven thousand acres, which belonged in the beginning to a rancher with one daughter. That daughter went abroad and married a title, and when she brought the title home to live on the ranch, her father built them a vast house—music rooms, conservatories, a ballroom. They needed a ballroom because they entertained: people from abroad, people from San Francisco, house parties that lasted weeks and involved special trains. They are long dead, of course, but their only son, aging and unmarried, still lives on the place. He does not live in the house, for the house is no longer there. Over the years it burned, room by room, wing by wing. Only the chimneys of the great house are still standing, and its heir lives in their shadow, lives by himself on the charred site, in a house trailer.

That is a story my generation knows; I doubt that the next will know it, the children of the aerospace engineers. Who would tell it to them? Their grandmothers live in Scarsdale, and they have never met a great-aunt. "Old"

Sacramento to them will be something colorful, something they read about in *Sunset*. They will probably think that the Redevelopment has always been there, that the Embarcadero, down along the river, with its amusing places to shop and its picturesque fire houses turned into bars, has about it the true flavor of the way it was. There will be no reason for them to know that in homelier days it was called Front Street (the town was not, after all, settled by the Spanish) and was a place of derelicts and missions and itinerant pickers in town for a Saturday-night drunk: VICTORIOUS LIFE MISSION, JESUS SAVES, BEDS 25¢ A NIGHT, CROP INFORMATION HERE. They will have lost the real past and gained a manufactured one, and there will be no way for them to know, no way at all, why a house trailer should stand alone on seven thousand acres outside town.

But perhaps it is presumptuous of me to assume that they will be missing something. Perhaps in retrospect this has been a story not about Sacramento at all, but about the things we lose and the promises we break as we grow older; perhaps I have been playing out unawares the Margaret in the poem:

Margaret, are you grieving
Over Goldengrove unleaving? . . .
It is the blight man was born for,
It is Margaret you mourn for.

Richard Rodriguez

from NOTHING LASTS A HUNDRED YEARS

We arrived late on a summer afternoon in an old black car. The streets were arcades of elm trees. The houses were white. The horizon was flat.

Sacramento, California, lies on a map around five hundred miles from the ruffled skirt of Mexico. Growing up in Sacramento, I found the distance between the two countries to be farther than any map could account for. But the distance was proximate also, like the masks of comedy and tragedy painted over the screen at the Alhambra Theater.

Both of my parents came from Mexican villages where the bells rang within an hour of the clocks of California. I was born in San Francisco, the third of four children.

When my older brother developed asthma, the doctors advised a drier climate. We moved one hundred miles inland to Sacramento.

Sacramento was a ladies' town—"the Camellia Capital of the World." Old ladies in summer dresses ruled the sidewalks. Nature was rendered in Sacramento, as in a recipe, through screens—screens on the windows; screens on all the doors. My mother would close the windows and pull down the shades on the west side of the house "to keep out the heat" through the long afternoons.

My father hated Sacramento. He liked an open window. When my father moved away from the ocean, he lost the hearing in one ear.

Soon my mother's camellias grew as fat and as waxy as the others on that street. She twisted the pink blossoms from their stems to float them in shallow bowls.

Because of my mother there is movement, there is change in my life. Within ten years of our arrival in Sacramento, we would leap from one sociological chart onto another, and from house to house to house—each house larger than the one before—all of them on the east side of town. By

---

the time I went to high school, we lived on "Eye" Street, in a two-story house. We had two cars and a combination Silvertone stereo-television. My bedroom was up in the trees.

I am not unconscious. I cherish our fabulous mythology. My father makes false teeth. My father received three years of a Mexican grammar-school education. My mother has an American high-school diploma. My mother types eighty words per minute. My mother works in the governor's office, where the walls are green. Edmund G. "Pat" Brown is governor. Famous people walk by my mother's desk. Chief Justice Earl Warren says hi to my mother.

After mass on Sundays, my mother comes home, steps out of her high-heel shoes, opens the hatch of the mahogany stereo, threads three Mexican records onto the spindle. By the time the needle sinks into the artery of memory, my mother has already unwrapped the roast and is clattering her pans and clinking her bowls in the kitchen.

It was always a man's voice. Mexico pleaded with my mother. He wanted her back. Mexico swore he could not live without her. Mexico cried like a woman. Mexico raged like a bull. He would cut her throat. He would die if she didn't come back.

My mother hummed a little as she stirred her yellow cake.

My father paid no attention to the music on the phonograph. He was turning to stone. He was going deaf.

I am trying to think of something my father enjoyed.

Sweets.

Any kind of sweets. Candies. Nuts. Especially the gore oozing from the baker's wreath. Carlyle writes in *The French Revolution* about the predilection of the human race for sweets; that so much of life is unhappiness and tragedy. Is it any wonder that we crave sweets? Just so did my father, who made false teeth, love sweets. Just so does my father, to this day, disregard warnings on labels. Cancer. Cholesterol. As though death were the thing most to be feared in life.

My mother remembered death as a girl. When a girl of my mother's village died, they dressed the dead girl in a communion dress and laid her on a high bed. My mother was made to look; whether my mother was made to kiss that cold girl I do not know, but probably she did kiss her, for my mother remembered the scene as a smell of milk.

My mother would never look again. To this day, whenever we go to a funeral, my mother kneels at the back of the church.

But Mexico drew near. Strangers, getting out of dusty cars, hitching up their pants, smoothing their skirts, turned out to be relatives, kissing me on the front porch. Coming out of nowhere—full-blown lives—staying a month

(I couldn't remember our lives before they came), then disappearing when back-to-school ads began to appear in the evening paper.

Only my Aunt Luna, my mother's older sister, lives in Sacramento. Aunt Luna is married to my uncle from India, his name an incantation: Raja Raman. We call him Raj. My uncle and aunt came to the Valley before us. Raj is a dentist and he finds work where he can, driving out on weekends to those airless quonset-hut villages where farmworkers live. His patients are men like himself—dark men from far countries—men from India, from Mexico, from the Philippines.

One Sunday in summer my father and I went with my Uncle Raj and my Aunt Luna to Lodi, about thirty miles south of Sacramento. There was a brown lake in the center of Lodi where blond teenagers skied. We stopped to eat lunch in the shade of a tree and then we drove on, past dust-covered vineyards.

We stopped at an old house. I remember the look on my Aunt Luna's face. My father and my uncle got out of the car. There was some question about whether or not I should go with them. Aunt Luna fretted. "Don't be afraid of anything you'll see," she said. "It's just an old house where some men live." Aunt Luna stayed in the car.

Inside, the house is dark. The front-room windows are painted over. There are cots along the walls. On several cots men recline. They are dressed. Are they sick? They watch as we pass. We hear only the sound of our steps on the boards of the floor.

A crack of light shines from behind a door at the rear of the house. My uncle pushes open the door. A man wearing an apron is stirring a pot.

Romesh!

Romesh quickly covers the pot. He kisses my uncle on the mouth. He shakes hands with my father. Then, turning to me, he salutes: "General."

Romesh was Raj's older brother. Every Christmas, when Romesh came to my uncle's house, he called me the general. My brother was the colonel. Romesh came with his sister—"the doctor." One time he stood on his head. Every Christmas, Romesh and his sister gave me presents that either had no sex or should have gone to a girl. Once, a green cup; another year, a string of pearls. I was never sure if there was menace in Romesh.

Uncle Raj offered my father a job managing a "boardinghouse"—like the one in Lodi—where derelicts slept.

In private my mother said no, Leo, no.

My father ended up working in the back room of my uncle's office on J Street, making false teeth for several dentists. My father and Uncle Raj became closest friends.

My classmates at Sacred Heart School, two blocks from our house, belong to families with names that come from Italy and Portugal and Germany. We carry aluminum lunch boxes decorated with scenes from the lives of Hopalong Cassidy and Roy Rogers. We are an American classroom. And yet we are a dominion of Ireland, the Emerald Isle, the darling land. "Our lovely Ireland," the nuns always call her.

During the hot Sacramento summers, I passed afternoons in the long reading gallery of nineteenth-century English fiction. I took an impression of London and of the English landscape. Ireland held no comparable place in my literary imagination. But from its influence on my life I should have imagined Ireland to be much larger than its picayune place on the map. As a Catholic schoolboy I learned to put on the brogue in order to tell Catholic jokes, of grave diggers and drunkards and priests. Ireland sprang from the tongue. Ireland set the towering stalks of the litanies of the Church to clanging by its inflection. Ireland was droll. Ireland was omniscient, Ireland seeping through the screen of the confessional box.

• • •

In my own version of my life, I was not yet the hero—perhaps California was the hero, perhaps my mother. I used to lie awake in the dark and imagine myself on a train far, far away, hurtling toward the present age of my parents. Forty-six. Forty-six. I used to imagine my future as the story of the Welsh coal miner's son who leaves home to take the high road to London. But, as it was, I didn't come from Ireland or India. I was born at the destination.

In the 1950s, billboards appeared on the horizon that beckoned restless Americans toward California. Sacramento of the 1950s was the end of the Middle Ages and Sacramento growing was the beginning of London.

In those days, people were leaving their villages and their mothers' maiden names to live among strangers in tract houses and God spoke to each ambition through the GI Bill. Highways swelled into freeways. If you asked, people in Sacramento said they were from Arkansas or from Portugal. Somewhere else.

• • •

Sacramento's ceremonial entrance was the Tower Bridge, where a sign proclaimed the town's population as 139,000. From the bridge you could see the state capitol—a wedding cake topped by a golden dome. Then, for six blocks, Sacramento posted BAIL BONDS; WEEKLY RATES; JESUS SAVES. Skid Row was what remained of the nineteenth-century river town, the Sacramento one sees in the early lithographs—a view from the river in the 1850s: young trees curling upward like calligraphic plumes; wooden sidewalks; optimistic storefronts; saloons; hotels; the Eagle Theater.

In the 1950s, Sacramento had begun to turn away from the land. Men who "worked for the state" wore white short-sleeved shirts downtown. There were office buildings, hotels, senators. Sacramento seemed to me a long way from the Okie evangelists at the far end of the car radio. Except for Mexican farmworkers, I rarely saw men wearing cowboy hats downtown.

The urban progress of Sacramento in the 1950s—the pouring of cement and of asphalt—imitated, even as it attempted to check, the feared reclamation of California by nature. But in the 1950s there was plenty of nature left. On summer evenings, houses became intolerable. We lolled on blankets on the grass. We were that much closer to becoming Indians.

Summer days were long and warm and free and I could make of them what I liked. America rose, even as the grasses, even as the heat, even as planes rose. America opened like a sprinkler's fan, or like a book in summer. At Clunie Library the books which pleased me most were about boyhood and summer and America; synonyms.

I hate the summer of Sacramento. It is flat and it is dull. Though it is not yet noon, the dry heat of Sacramento promises to rise above the leaves to a hundred degrees. Just after noon, the California Zephyr cuts through town, pauses for five minutes, stopping traffic on K and L, and for those five minutes I inhabit the train's fabulous destination. But then the train sweeps aside like the curtain at the Memorial Auditorium, to reveal the familiar stage set for a rural comedy. A yellow train station.

Yet something about the Valley summer is elemental to me and I move easily through it—the cantaloupe-colored light, the puddles of shade on the street as I bicycle through. There is a scent of lawn.

When I think of Sacramento, I think of lawns—force-fed, prickling rectangles of green. Lawns are not natural to California. Even one season without water, without toil, is ruinous to a garden, everyone knows. The place that had been before—before California—would come back; a place the Indians would recognize. Lovely tall grasses of dandelion or mustard in spring would inevitably mean lapsarian weeds, tinder grass and puncture vines come summer.

On Saturdays I mow the front lawn. On my knees I trim the edges. Afterward I take off my shoes to water down the sidewalk. Around noontime, as I finish, the old ladies of Sacramento, who have powdered under their arms and tied on their summer straw hats, walk by and congratulate me for "keeping your house so pretty and clean. Whyn't you come over to my house now," the ladies say.

I smile because I know it matters to keep your lawn pretty and green. It mattered to me that my lawn was as nice as the other lawns on the block. Behind the American façade of our house, the problem was Ireland. The problem was India. The problem was Mexico.

Mexico orbited the memory of my family in bitter little globes of sorrow, rosary beads revolving through the crushing weight of my mother's fingers. Mexico. Mexico. My mother said Mexico had skyscrapers. "Do not judge Mexico by the poor people you see coming up to this country." Mexico had skyscrapers, pyramids, blonds.

Mexico is on the phone—long-distance.

Juanito murdered!

My mother shrieks, drops the phone in the dark. She cries for my father. For light.

A crow alights upon a humming wire, bobs up and down, needles the lice within his vest, surveys with clicking eyes the field, the cloud of mites, then dips into the milky air and flies away.

The earth quakes. The peso flies like chaff in the wind. The Mexican police chief purchases his mistress a mansion on the hill.

The doorbell rings.

I split the blinds to see three nuns standing on our front porch.

Mama. Mama.

Monsignor Lyons has sent three Mexican nuns over to meet my parents. The nuns have come to Sacramento to beg for Mexico at the eleven o'clock mass. We are the one family in the parish that speaks Spanish. As they file into our living room, the nuns smell pure, not sweet, pure, like candles or like laundry.

The nun with a black mustache sighs at the end of each story the other two tell: Orphan. Leper. Crutch. One-eye. Casket.

¡Qué lástima!

"Someday you will go there," my mother would say. "Someday you'll go down and with all your education you will be 'Don Ricardo.' All the pretty girls will be after you." We would turn magically rich in Mexico—such was the rate of exchange—our fortune would be multiplied by nine, like a dog's age. We would be rich, we would be happy in Mexico.

• • •

My father made false teeth for Dr. Wang. Mrs. Wang was the receptionist. Mrs. Wang sat at a bare table in an empty room. An old Chinese man, Dr. Wang's father, climbed the stairs at intervals to berate his son and his son's wife in Chinese.

Because of my mother, we lived as Americans among the middle class. Because of my father, because of my uncle from India, we went to Chinese wedding receptions in vast basement restaurants downtown, near the Greyhound station. We sat with hundreds of people; we sat in back; we used forks. When the waiter unceremoniously plopped wobbly pink desserts in front of us, my mother pushed my plate away. "We'll finish at home."

My father and my uncle worked among outsiders. They knew a handsome black doctor who sat alone in his office on Skid Row, reading the newspaper in his chair like a barber.

Sacramento was filling with thousands of new people each year—people fleeing the tanks of Hungary, people fleeing their fathers' debts or their fathers' ghosts or their fathers' eyes.

One of my aunts went back to Mexico to visit and she returned to tell my mother that the wooden step—the bottom step—of their old house near Guadalajara was still needing a nail. Thirty years later! They laughed.

My father said nothing. It was as close as he came to praising America.

We have just bought our secondhand but very beautiful blue DeSoto. "Nothing lasts a hundred years," my father says, regarding the blue DeSoto, as regarding all else. He says it all the time—his counsel. I will be sitting fat and comfortable in front of the TV, reading my *Time* magazine. My mother calls for me to take out the garbage. *Now!* My father looks over the edge of the newspaper and he says it: Nothing lasts a hundred years.

Holiday magazine published an essay about Sacramento by Joan Didion. The essay, an elegy for old Sacramento, was about ghostly ladies who perched on the veranda of the Senator Hotel and about their husbands, who owned the land and were selling the land. Joan Didion's Sacramento was nothing to do with me; families like mine meant the end of them. I so thoroughly missed the point of the essay as to be encouraged that a national magazine should notice my Sacramento.

Whenever Sacramento made it into the pages of *Time* magazine, I noticed the editors always affixed the explanatory *Calif.*, which I took as New York City's reminder. We were nothing. Still, that caliphate had already redeemed our lives. My mother, my father, they were different in California from what history had in store for them in Mexico. We breathed the air, we ate the cereal, we drank the soda, we swam in the pools.

My father was surprised by California and it interested him. It interested him that Sacramento was always repairing itself. A streetlight would burn out, a pothole would open in the asphalt, a tree limb would crack, and someone would come out from "the city" to fix it. The gringos were always ready to fix things, my father said.

In high school I worked as a delivery boy for Hobrecht Lighting Company. I delivered boxes of light fixtures to new homes on the north side of the river.

I remember standing outside a house near Auburn, waiting for the contractor to come with a key. I stood where the backyard would be. The March wind blew up from the fields and I regretted the loss of nature—the fields, the clear distance.

Yet California was elemental to me and I could no more regret California than I could regret myself. Not the dead California of Spaniards and forty-niners and Joan Didion's grandmother, but Kodachrome, CinemaScope, drive-in California—freeways and new cities, bright plastic pennants and spinning whirligigs announcing a subdivision of houses; hundreds of houses; houses where there used to be fields. A mall opened on Arden Way and we were first-nighters. I craved ALL-NEW and ALL-ELECTRIC, FREE MUGS, and KOOL INSIDE and DOUBLE GREEN STAMPS, NO MONEY DOWN, WHILE-U-WAIT, ALL YOU CAN EAT.

At a coffee shop—open 24 hours, 365 days a year—I approved the swipe of the waitress's rag which could erase history.

Through the years I was growing up there, Sacramento dreamed of its own redevelopment. Plans were proposed every few months to convert K Street downtown into a mall and to reclaim a section of Skid Row as "Old Town." The *Bee* published sketches of the carnival future—an expansive street scene of sidewalk cafés with banners and clouds and trees that were also clouds and elegant ladies with their purse arms extended, pausing on sidewalks that were made of glass.

A few years later, the future was built. Old Sacramento became a block of brick-front boutiques; some squat glass buildings were constructed on Capitol Avenue; and K Street, closed to traffic for several blocks, got a concrete fountain and some benches with winos asleep in the sun.

Never mind. Never mind that the future did not always meet America's dream of itself. I was born to America, to its Protestant faith in the future. I was going to be an architect and have a hand in building the city. There was only my father's smile that stood in my way.

It wasn't against me; his smile was loving. But the smile claimed knowledge. My father knew what most of the world knows by now—that tragedy wins; that talent is mockery. In the face of such knowledge, my father was mild and manly. If there is trouble, if there is a dead bird to pick up, or when the lady faints in church, you want my father around. When my mother wants water turned into wine, she nudges my father, for my father is holding up the world, such as it is.

My father remains Mexican in California. My father lives under the doctrine, under the very tree of Original Sin. Much in life is failure or compromise; like father, like son.

For several years in the 1950s, when one of my family makes a First Holy Communion, we all go to Sutter's Fort to have our pictures taken. John Sutter's wall against the Indians becomes a gauge for the living, a fixed mark for the progress of my mother's children. We stand in formal poses against the low white wall of the fort. One of my sisters wears a white dress and veil and a little coronet of seed pearls. Or, when it is a boy's turn, one of us wears a white shirt, white pants, and a red tie. We squint into the sun. My father is absent from all the photographs.

The Sacramento Valley was to have been John Sutter's Rhineland. He envisioned a town rising from his deed—a town he decided, after all, to name for himself—Sutterville. Sutter imagined himself inventing history. But in the Eastern cities of Boston and Philadelphia and New York, Americans were imagining symmetry. They were unrolling maps and fixing them with weights net down upon the Pacific Ocean. Newcomers—Americans—were arriving in Mexican territory.

Already there were cracks in the sidewalk where the roots of the elm tree pushed up.

My father smiled.

Ask me what it was like to have grown up a Mexican kid in Sacramento and I will think of my father's smile, its sweetness, its introspection, its weight of sobriety. Mexico was most powerfully my father's smile, and not, as you might otherwise imagine, not language, not pigment. My father's smile seemed older than anything around me. Older than Sutter's Fort.

• • •

At noontime exactly on a clear winter day, California will officially become the largest, by which we mean the most populous, state in the union. Governor Edmund G. "Pat" Brown wants fire sirens and factory whistles; he wants honking horns and church bells to detonate the hour of our numerical celebrity. The bureaucrat's triumphant tally will change California forever.

Sacred Heart Church in Sacramento has no bells, none that ring, nothing to hang in this clear blue sky. There is a brick tower, a campanile, a shaft of air. When the church was built, neighbors had complained first thing about bells. So no bells ever rang at Sacred Heart.

Five to twelve: The Irish nun stands at her blackboard. Tat. Tat. Tat. Slice. Tat. Tat. Tat. I strain to hear outside. Nothing, beyond the tide of traffic on H Street; the chain of the tether ball lifting in the wind, then dropping to lash its pole. Clank clink. Clank clink. At noontime exactly there is a scraping of chairs as we stand to pray the "Angelus" ("The angel of the Lord declared unto Mary . . .").

One hundred miles away, the governor's son was also reciting the "Angelus." Edmund G. Brown, Jr.—Jerry Brown—was a seminarian in Los Altos, studying to become a priest. We saw his picture in the newspaper every year in the Christmas-tree photo from the governor's mansion—a young man with dark eyebrows, dressed in black.

Within twenty years Jerry Brown would assume his father's office. The ex-seminarian expounded upon limits, his creed having shriveled to "small is beautiful," a catchphrase of the sixties. Like John Sutter, Jr., Jerry Brown dismantled his father's huge optimism.

"Here tragedy begins," whispers Sister Mary Regis. "The wheel of fortune creaks downward toward a word buried for centuries in the bed of a stream. . . ."

The Irish nun, an unlikely Rhinemaiden, rehearses the Protestant parable for her fifth-grade class. John Sutter sent one of his men, James Marshall, up to the Sierra foothills to build a sawmill. It was January 1848. In the clear winter stream (here Sister hikes her skirt, shades her eyes with her hand; her eyes seem to scrutinize the linoleum), Marshall spied a glint. "It looked like a lady's brooch."

In the morning, or the next morning, or the next, Sutter dispatched several men to see if they could find more of the stuff. Sutter hoped to keep the discovery secret. But several merchants in the region made it their business to take the news over the hills to San Francisco. And so, within weeks, Sutter's kingdom—the kingdom intended for John Sutter by the providence of God—was lost in a rush of anarchy. Sutter was abandoned by his men, who would be kings themselves. His herd of cattle, his horses—all were stolen. And his fort—that placid rectangle of dirt in an ocean of grass—his fort was overrun by men who heard the mystical word pronounced out loud.

GOLD.

There were stories. People vaulting from the dentist chair at the approach of my Uncle Raj. "No nigger dentist is going to stick his fingers in my mouth!" There was a price to be paid for living in Sacramento, the dark-green car, the green lawn, the big green house on 45th Street.

My Uncle Raj had three daughters. He wanted a son. I flirted with him. Every Christmas I would ask my Uncle Raj for expensive toys my parents couldn't afford. I wanted a miniature circus. I wanted a cavalry fort.

My uncle called me Coco.

Sutter's son inherited the shambles of his father's dream. John Sutter, Jr., was a tradesman and not a man of the land. John Sutter, Jr., recognized the advantages for trade in situating a new town alongside the river, two miles west of his father's fort. John Sutter, Jr., proposed giving the new river town the old Spanish name for the Valley—Sacramento. Thus Sacramento City flourished while Sutterville—which John Sutter had imagined as teeming, spired, as blessed as Geneva—Sutterville dried up and died.

In my bedroom I sketched visionary plans. A huge aquatic amusement park would parallel the river beneath an animated neon sign of a plunging diver. There would be a riverboat restaurant decked with Christmas lights and a dinner theater. (The movie *Show Boat* had recently been filmed near Sacramento.)

It was the father's failure Sacramento remembered, not the son's success. John Sutter abandoned his fort, retired to a farm. A short time later he left California for good. His bones are buried in Pennsylvania.

". . . A victim of sudden good fortune," the Irish nun pronounced over him.

Such, too, was Sacramento's assessment of Sutter's life. The town may have owed its combustion to the Gold Rush, but when, within months, the vein of anarchy was exhausted, Sacramento would turn to regard the land itself as representing God's grace. Sacramento became a farm town, the largest in the Central Valley.

Toward the end of August, the air turned foul. Farmers were burning, people said—always this speculation on the cardinal winds—just as people said "peat dust" in March or "hay fever" in May or "alfalfa" in summer.

John Sutter served as a reminder to generations of farmers who lived by the seasons (Low Church lessons of slow growth, deferred reward). For his fall was steep, a lesson suitable to the moral education of generations of flatland teenagers—farm kids who came to Sacramento every August for the State Fair. We would see their pictures in the *Sacramento Bee*, kids from Manteca or Crow's Landing in their starched 4-H uniforms, holding sunflower ribbons over the heads of doomed beasts.

After work, my father and my Uncle Raj drove home together in Raj's new green car—the Mexican and the Indian. They sat in the car with the motor running.

"Go out and bring in your father," my mother would say.

I stole only bits of the story; the rest lay on a shelf too high for me. My Uncle Raj had gotten mixed up in some kind of politics. Once, Uncle Raj had taken my father to "a meeting" in San Francisco.

That was all I knew. That, and I once overheard my mother say on the phone, "We're afraid to have Leo apply for the job. There may be some record."

My father, as everyone knew, would have liked a job at the post office. But he never applied.

Alongside copies of *Life* and *Saturday Evening Post* on the mahogany table in Raj's waiting room were "international" pamphlets.

Was Raj a Communist?

Communists were atheists. Raj was a Catholic. He converted when he married my aunt. He went to mass every Sunday.

One afternoon a dental-supply salesman—his sample case and his hat were in his hand—came out into the waiting room with Raj. It was nearing six o'clock. I sat turning the pages of a magazine. My father was taking down the garbage. I could hear the trash barrel bumping down the stairwell. The man crossed to the table next to me, picked up the stack of pamphlets, and stooped to place them into the waste basket my father had just emptied. "Let's just get rid of these, Dr. Raman." His tone was not friendly but he smiled. "Why look for trouble, I always say," holding out his hand to Uncle Raj. Uncle Raj stared down into the waste basket.

Had my Uncle Raj conspired against paradise?

My uncle's eyes began to cloud at their perimeters. I heard them say he had a heart bother. I heard them say he was worried he might get deported. My uncle had been in some kind of trouble with the Immigration Service. There had been "a trial."

My uncle put an Adlai Stevenson sticker on the back fender of his new green car.

But there were afternoons when my uncle's heart tightened like a fist at the prospect of losing California. In the back room of my uncle's office was a maroon sofa and, over the sofa, a painting of a blond lady being led by a leopard on a leash. Beneath the painting my uncle lay prostrate. At such times my father applied warm towels to the back of my uncle's skull.

Raj had come on ships all the way from Bombay when he was a boy. Now Raj hated the ocean. The swirling ocean was a function of the Immigration and Naturalization Service. The ocean would surely carry him back to India.

What was India, I pondered, but another Mexico. Indians in both places. "Cattle freely roam the streets of Bombay," according to the *World Book Encyclopedia*.

• • •

My father is a man nearly as old as the century. As a boy he saw Halley's Comet and he gauged his life by the sighting. He said he would live to see the comet's return and now he has surpassed Halley's Comet. My father understands that life is as surprising as it is disappointing. He left Mexico in his late twenties for Australia. He ended up in Sacramento in a white coat, in a white room, surrounded by shelves of grinning false teeth. Irony has no power over my father.

Our last house on "Eye" Street was across from an old cemetery. No memory attached to it. The grass was watered and cut once a month by the city. There were no scrolls or wrought-iron fences; no places to put flowers. There were granite plaques level with the ground. Early dates. Solitary names. Men. Men who had come early to California and died young.

No grandsons or granddaughters came forward in the 1950s when Sacramento needed the land to build a school, a new Sutter Junior High School. A plywood fence was hammered up around the cemetery and, within that discretionary veil, bulldozers chugged and grunted, pulling up moist hairy mounds of what had once been the light of day; trucks came to carry it all away.

In early November, white tule fog rises from the valley floor. My father is easy with this ancient weather reminiscent of the sea. My father is whistling this morning as he scrambles two eggs. My mother turns away from the window, pulling her blue bathrobe closer around her throat. I am sitting at the kitchen table. I am sixteen years old. I am pouring milk onto Sugar Frosted Flakes, watching milk rise in the bowl. My parents will die. I will die. Everyone I know will someday be dead. The blue parakeet my mother has taught to say "pretty boy" swings upon his little trapeze, while my mother pours coffee.

I can no longer remember the cold. In my memory, it is always summer in Sacramento; the apricot tree in the backyard is heavy; the sky is warm and white as a tent.

One summer, my uncle was beautiful. His skin was darker than Mexico. His skin wore shade. It was blue. It was black.

When I was seven years old, my girl cousins threw me into the lake at Lodi and, with several islands to choose from, I swam toward the island of my uncle. His eyes were black and so wide with surprise they reflected the humor of the water. His nipples were blue and wet black fur dripped down his front and floated in the water at his waist.

In the family album, Raj yet lifts me upside down by my legs. I am confident in my abandon as the trees whirl by. My aunt backs away from the camera, regret blurring from her eyes.

"Don't . . . Please. Put him down now, Raj."

Comedy and tragedy merged when I was sixteen. My uncle died.

He was extracting a tooth. He had just begun to tighten the clamp; the water swirled in the expectoral basin (within the patient's skull the awful grating, as of sepulchral stone), when my uncle began to sweat.

"I don't think I'm going to be able to finish. . . ."

The clamp banged down on the metal tray.

The receptionist's scream brought my father from the back lab, still drying his hands with a towel. My father placed the towel under my uncle's head. My father took my uncle's hand, where my uncle lay on the dark-green linoleum; my father easing my uncle down into the ocean.

Jose Montoya

RESONANT VALLEY

When I was
Young among
The pregnant
Vineyards of
All the un-domed
Capitols of
The raisin
Industry—
Musically chiming
Charming towns like
Fowler
Reedley
Del Rey
Selma
Clovis
Parlier
Kingsburg and Sanger—

I was lazy

Me!

From a family
Of clean pickers
The pride of
Any Fijikawa or
Saroyanesque Krikor—

Quinienteros of

Jose Montoya

Five hundred trays
And the day barely
Two-thirds along

But everyone said I was

Lazy.

I knew. But how I knew!

Why I was easy
On the clusters—
Careful with
The leaves,

Slow.

I was too quick to
Sadden at the sight
Of the green, iridescent worm
Scorching itself in the
Hot, planed-for-trays, sand.
And knowing I had something
To do with its death

I wept.

And rather than
Repeat the

Senseless carnage
I remained lazy

Sitting under the vines
Imagining what my reactions
Would be to some similar
Onslaught.

Panic!

Intolerable panic!
With both
Hands
Upon my head
My eyes
Shut tight
Flashing
Stabs of color
On the roof
Of my skull-and
A child-young urge
To roll
Naked
Upon the burning sand,
I remained lazy.

And the family
Of quinienteros
Didn't make as much

Money
That summer
In the valley of the San Joaquin.

But the worms, the wasps and the
Black widow spiders—for a short
Time, at least—frolicked
cooly in that green-leaf world

Beneath the sun.

GABBY TOOK THE 99

Gabby took the 99
The 99 highway
El highway 99

EL JAIGUEY
The ninety nine
Donde se requió el troque
Con aquel jentío
The HIGHWAY
Down the 99
Up the 99
Ahi, por todo el 99

El noventa y nueve
As you leave the 99—

WHO LEAVES???

. . . who ever leaves the
Cold
    ugly
    dry
    hot slippery
    bloody

Dirty
    foggy
    sleek
    powerful
Ninety nine?
NOT RAZA, Okies, Arkies,
Chapos, Armenians—not even
THE MCHIGGENBOTHOMS!

That Impersonal 99. And
Not even crosses for the
Dead alongside this road
Of the timeless vanishing
Point.

And in that infinity
You dared to dream, Gabby.

The riches that passed
As you turned trays, loco,
Y el 99 te iba a llevar
Pa ya, Gabby . . .
El Gabby took the 99—
La salvación!!
. . . and he died in Viet Nam
of an overdose, pa'
Cabala de chingar!

And Visa laughs
And Dina laughs
And Goshen and Cutler
Y el 99 laughs
And all do the
Tecato tattoo taps.

Philip Levine

RICKY

I go into the back yard
and arrange some twigs
and a few flowers. I go alone
and speak to you as I never could
when you lived, when you
smiled back at me shyly.
Now I can talk to you as I talked
to a star when I was a boy,
expecting no answer, as I talked
to my father who had become
the wind, particles of rain
and fire, these few twigs
and flowers that have no name.

Last night they said a rosary
and my boys went, awkward
in slacks and sport shirts,
and later sitting under the hidden
stars they were attacked and beaten.
You are dead. It is 105,
the young and the old burn
in the fields, and though they cry
enough the sun hangs on
bloodying the dust above the aisles
of cotton and grape.

This morning they will say a mass
and then the mile-long line of cars.

Teddy and John, their faces swollen,
and four others will let you
slowly down into the fresh earth
where you go on. Scared now,
they will understand some of it.
Not the mass or the rosary
or the funeral, but the rage.
Not you falling through the dark,
moving underwater like a flower
no one could find until
it was too late and you had gone out,
your breath passing through dark water
never to return to the young man,
pigeon-breasted, who rode
his brother's Harley up the driveway.

Wet grass sticks to my feet, bright
marigold and daisy burst in the new day.
The bees move at the clumps
of clover, the carrots—
almost as tall as I—
have flowered, pale lacework.
Hard dark buds
of next year's oranges, new green
of slick leaves, yellow grass
tall and blowing by the fence. The grapes
are slow, climbing the arbor,
but some day there will be shade
here where the morning sun whitens
everything and punishes my eyes.

Your people worked
for some small piece of earth,
for a home, adding a room
a boy might want. Butchie said
you could have the Harley
if only you would come back,
anything that was his.

A dog barks down the block
and it is another day. I hear
the soft call of the dove,
screech of mockingbird and jay.
A small dog picks up the tune,
and then *tow-weet tow-weet*
of hidden birds, and two finches
darting over the low trees—
there is no end.

What can I say to this mound
of twigs and dry flowers, what
can I say now that I would speak
to you? Ask the wind, ask
the absence or the rose burned
at the edges and still blood red.
And the answer is you
falling through black water
into the stillness that fathers
the moon, the bees ramming into
the soft cups, the eucalyptus
swaying like grass under water.
My John told me your cousin
punched holes in the wall
the night you died and was afraid
to be alone. Your brother
walks staring at the earth.
I am afraid of water.

And the earth goes on
in blinding sunlight.
I hold your image
a moment, the long
Indian face
the brown almond eyes
your dark skin full
and glowing as you grew
into the hard body
of a young man.

And now it is bird screech
and a tree rat suddenly
parting the tall grass
by the fence, lumbering
off, and in the distance
the crashing of waves
against some shore
maybe only in memory.

We lived by the sea.
My boys wrote
postcards and missed you
and your brother. I slept
and wakened to the sea,
I remember in my dreams
water pounded the windows
and walls, it seeped
through everything,
and like your spirit,
like your breath,
nothing could contain it.

ASHES

Far off, from the burned fields
of cotton, smoke rises and scatters
on the last winds of afternoon.
The workers have come in hours ago,
and nothing stirs. The old bus creaked
by full of faces wide-eyed with hunger.
I sat wondering how long the earth
would let the same children die day
after day, let the same women curse
their precious hours, the same men bow
to earn our scraps. I only asked.
And now the answer batters the sky:
with fire there is smoke, and after, ashes.

You can howl your name into the wind
and it will blow it into dust, you
can pledge your single life, the earth
will eat it all, the way you eat
an apple, meat, skin, core, seeds.
Soon the darkness will fall on all
the tired bodies of those who have
torn our living from the silent earth,
and they can sleep and dream of sleep
without end, but before first light
bloodies the sky opening in the east
they will have risen one by one
and dressed in clothes still hot
and damp. Before I waken they are
already bruised by the first hours
of the new sun. The same men
who were never boys, the same women
their faces gone gray with anger,
and the children who will say nothing.
Do you want the earth to be heaven?
Then pray, go down on your knees
as though a king stood before you,
and pray to become all you'll
never be, a drop of sea water,
a small hurtling flame across the sky,
a fine flake of dust that moves
at evening like smoke at great height
above the earth and sees it all.

Luis Valdez

QUINTA TEMPORADA

First Performance: At Delano, California, during a grape strikers' meeting in Filipino Hall, 1966.

Characters:

CAMPESINO	FALL
DON COYOTE	SPRING
PATRÓN	THE UNIONS
WINTER	THE CHURCHES
SUMMER	LA RAZA

The farm labor contractor, satirized as the archetypical DON COYOTE *in this acto, is one of the most hated figures in the entire structure of agri-business. He is paid by the growers for having the special skill of rounding up cheap stoop labor in the barrio and delivering it to the fields. The law stipulates that he must provide safe transportation and honest transactions. The sorrowful reality is something else again, ranging from broken down buses that are carbon monoxide death traps to liquor and meager lunches sold at exorbitant prices to the workers. In the field,* DON COYOTE *sits in his air conditioned pickup while the workers suffer the blistering heat or freezing cold of inclement weather. He originally appeared in this acto with the name of real-life contratistas.* DON COYOTE *has earned the unrelenting hatred of the campesino, but it is ultimately agri-business that condones and protects him. The only solution to the injustices of the farm labor contracting system is the union hiring hall.*

In addition to DON COYOTE, *the seasons appear in "Quinta Temporada" as characters. It is necessary to emphasize the effect of summer, fall, winter and spring on the survival of the farmworker. If it rains he is out of a job, and there is no unemployment compensation. Winter or "El Invierno" is thus almost a living, breathing creature to the campesino—a monster, in fact bringing with him humiliation, starvation and disease. If the strikers laughed at winter in this acto, it was because of the real hope offered by the United Farmworkers Organizing Committee, which created a new "fifth" season.*

Enter FARMWORKER *to center stage from* S.L. *He addresses audience.*

WORKER: Oh, hello—quihúbole! My name is José. What else? And I'm looking for a job. Do you have a job? I can do anything, any kind of field work. You see, I just got in from Texas this morning and I need to send money back to my familia. I can do whatever you want—pick cotton, grapes, melons. *(*DON COYOTE *enters while* FARMWORKER *is talking. He smiles and comes toward the* FARMWORKER.*)*

COYOTE: My friend! My name is Don Coyote and I am a farm labor contractor.

WORKER: En la madre, ¡me rayé! Un contratista. *(The* FARMWORKER *kisses the contractor's outstretched hand.)*

COYOTE: So you want work, eh? ¿Busca jale? Bueno, véngase pa'ca un momento. *(*COYOTE *pulls* FARMWORKER *over to* S.R.*)* Mire, this summer is coming fat, fat! Covered with money! Dollar bills, five dollar bills, ten, twenty, fifty, a hundred dollar bills and all you have to do is . . . *(*COYOTE *gestures above* FARMWORKER's *head as if holding a wad of money which he now releases.)* catch! *(*FARMWORKER *pretends to catch money in his hat.* COYOTE *moves downstage center.)* Well, what do you say? Will you work for me?

WORKER: ¡Oh, sí, patroncito! ¡Sí, señor! *(Approaches* COYOTE's *hand out.)*

COYOTE: *(Grasping hand, shaking it.)* A deal is a deal. *(The* PATRONCITO *enters on* S.R., *stomps downstage smoking a cigar.)*

PATRON: Boy! *(*DON COYOTE *shoves* FARMWORKER *aside and leaps toward the* PATRON, *landing at his feet and kissing his boots. He rises dusting off the* PATRON.*)* Like your patron, eh, boy?

COYOTE: ¡Oh, sí, patrón!

PATRON: Good. You got my summer crew ready, boy?

COYOTE: Sí, señor. *(He motions to* FARMWORKER, *who hesitates, then comes over to* PATRON. COYOTE *points to his hat.)* El sombrero, tonto. *(The* FARMWORKER *removes his hat and stands beside the contractor, both smiling assininely toward the* PATRON.*)*

PATRON: Well, I don't much care what he looks like, so long as he can pick.

COYOTE: Oh, he can pick, patrón! *(The* PATRON *stomps over to* S.L. COYOTE *elbows the* FARMWORKER *and makes a gesture, holding his hands widely apart as if describing how fat summer will be. The* PATRON *at* S.L. *calls in* SUMMER.*)*

PATRON: Summer! Get in here. *(*SUMMER *is a man dressed in ordinary workshirt and khaki hat. His shirt and hat, however, are completely covered with paper money: Tens, twenties, fifties. He walks in with his arms outstretched, and continues across the stage at a normal pace.)*

SUMMER: I am the Summer.

WORKER: ¡Ajúa! ¡El jale!

COYOTE: ¡Entrale, mano! *(The* FARMWORKER *attacks the* SUMMER, *and begins to pick as many dollar bills as his hands can grab. These he stuffs into his back pockets.* DON COYOTE *immediately takes his place behind the* FARMWORKER *and extracts the money from his back pockets and hands it over to the* PATRON, *who has taken his place behind the contractor. This exchange continues until* SUMMER *exits. The* PATRON *then moves to* S.R., *counting his money.* DON COYOTE *takes the* FARMWORKER *to* S.L. *Enthusiastically.)* ¡Te aventastes! Didn't I tell you we're going to get rich? Didn't I tell you? *(*DON COYOTE *breaks off abruptly and goes over to his* PATRON'S *side.)* How'd we do, boss?

PATRON: Terrible! We're going to have to ask for a federal subsidy. *(The* FARMWORKER *searches his pockets for money and panics when he can't find a single dollar bill. He spots the* PATRON *with handfuls of money and his panic turns to anger.)*

WORKER: *(To* DON COYOTE.*)* Hey! Where's my money?

COYOTE: What money?

WORKER: Pos, what? The money I work for all summer.

COYOTE: You know what's wrong with you? You're stupid. You don't know how to save your money. Look at my patrón, how come he always has money?

WORKER: *(Lunging toward* PATRON.*)* That's my money!

COYOTE: *(Stops him.)* No! I know who has your money. Come here. *(He takes* FARMWORKER *to* S.L. *again.)* It's . . . *(He points out toward the audience, making a semi-circle from* S.L. *to* S.R., *finally stopping at the* PATRON *and pointing at him inadvertantly.)*

PATRON: Hey!

COYOTE: No! Not my patrón! It's Autumn! Autumn has your money.

WORKER: Autumn?

COYOTE: El otoño.

WORKER: Puras papas. I don't believe you.

COYOTE: You don't believe me? *(Faking his sincerity.)* But I swear by my madrecita! *(Pause.)* Still don't believe me, eh? Okay. Do you want to see the truth in action? Well, here's the truth in action! *(*DON COYOTE *makes a flourish with his arms, and spits on the floor, then stomps vigorously on the spit with his foot. All in a grandiose manner.)*

WORKER: That's it?

COYOTE: La verdad en acción.

WORKER: Well, here goes mine! *(*FARMWORKER *spits at* DON COYOTE'S *foot, but* COYOTE *pulls it back just in time. He retaliates by spitting toward*

FARMWORKER'S *foot.* FARMWORKER *pulls his foot back just in time, as* DON COYOTE *stomps toward it. The* FARMWORKER *now catches* DON COYOTE *off guard by spitting on his face.)*

COYOTE: *(Retreats momentarily, decides to suppress his anger.)* No matter. Look, mano, this autumn is coming FAT! Fatter than last summer. You go to work for me and you'll be rich. You'll have enough money to buy yourself a new car, a Cadillac! Two Cadillacs! You'll be able to go to Acapulco! Guadalajara! You'll be able to send your kids to college! You'll be able to afford a budget! You'll be middle-class! You'll be Anglo! You'll be rich! *(The* FARMWORKER *responds to all of this with paroxysms of joy, squeals of delight.)* So, what do you say? Will you work for me?

WORKER: *(Suddenly deadpan.)* No.

COYOTE: *(Turns away, goes to* D.S.C.*)* Okay, no me importa! I don't give a damn. Anyway, winter's coming.

WORKER: *(Suddenly fearful.)* Winter?

COYOTE: El invierno!

WORKER: No! *(He rushes toward the contractor, hand outstretched.* DON COYOTE *grabs it quickly, before the* FARMWORKER *can think twice.)*

COYOTE: Lío es lío, yo soy tu tío, grillo.

PATRON: *(At* S.R.*)* Boy!

COYOTE: *(Whirling around.)* Yes, patrón?

PATRON: *(Stuffing money into his pockets.)* Is my fall crew ready?

COYOTE: Sí, patrón. *(*DON COYOTE *motions the* FARMWORKER *over to* S.R. *The* FARMWORKER *steps forward, hat in hand, with a smile on his face. The* PATRON *moves forward with a grunt and the* FARMWORKER *steps in front of him. The* PATRON *tries to move around him and the* FARMWORKER *moves in front of him again. The* PATRON *finally shoves the* FARMWORKER *aside and goes* S.L. *The* COYOTE *yells to* FARMWORKER.*)* A un lado, suato!

PATRON: Fall, come in here, boy. *(*FALL *comes in. He is a thinner man than* SUMMER. *His work shirt is covered with money, though more sparsely than* SUMMER'S.*)*

COYOTE: ¡Entrale, mano! *(With a shout, the* FARMWORKER *leaps to his work, picking money off the shirt that* FALL *wears. The same* FARMWORKER-DON COYOTE-PATRON *arrangement is used until* FALL *is almost off stage at* S.R. *At this point, the* FARMWORKER *reaches back and accidentally catches* DON COYOTE'S *hand in his back pocket. Spotting this, the* PATRON *rapidly crosses to* D.S.L.*)*

WORKER: Hey! That's my money! You're stealing my money! Pos, mira, qué hijo de . . . *(*FARMWORKER *strikes at contractor.* DON COYOTE *knocks him down and kicks him three times. The* PATRON *stands watching all of this, then finally calls out.)*

PATRON: You, boy!

COYOTE: *(In a sweat, fearful of reprimand.)* ¿Sí, patroncito? I didn't mean it, boss. *(Pointing to his foot.)* Mira, rubber soles, patrón. *(*DON COYOTE *obsequiously slides over to the boss. The* PATRON *is expansive, beaming, pleased.)*

PATRON: I like the way you do that, boy.

COYOTE: You do? Oh, I can do it again, patrón. *(He runs over to the* FARMWORKER *and gives him one final kick in the ribs. The* FARMWORKER *groans.)*

PATRON: *(With corporate pride.)* Beautiful! If there's anything we need in our company, boy, it's discipline and control of our workers!

COYOTE: Sí, señor, disciplina, control de los mexicanos!

PATRON: And just to show you our appreciation for what you do for the business, the corporation, I am going to give you a little bonus. *(Above the flat behind* PATRON *and contractor, a hand appears holding a huge bone with big black letters spelling out the word "bonus." The* PATRON *picks this up and hands it to the contractor.)*

COYOTE: *(Overcome with emotion.)* ¡Oh, patrón! ¡Un hueso! *(There is a loud rumbling noise backstage. Snowflakes come tumbling over the flats.* COYOTE *runs to* S.R.*)* Winter is coming! *(The* FARMWORKER *picks himself up off the floor and cowers at* U.S.C. PATRON *stands* S.L.*, undisturbed by the advent of* WINTER. *With a final rumble* WINTER *leaps into the scene around the corner of the flat at* S.L.*)*

WINTER: I am Winter and I want money. Money for gas, lights, telephone, rent. *(He spots the contractor and rushes over to him.)* Money! *(*DON COYOTE *gives him his bonus.* WINTER *bites the bone, finds it distasteful, throws it backstage over the flats. He whirls around toward the* PATRON.*)*

COYOTE: Money!

PATRON: *(Remaining calm.)* Will you take a check?

WINTER: *(Rushing over to him.)* No, cash!

PATRON: Okay, here! *(Hands him a small wad of bills.)* Well, that's it for me. I'm off to Acapulco 'til next spring. *(Exits* S.L.*)*

COYOTE: And I'm off to Las Vegas. *(Exits* S.R.*)*

WORKER: And I'm off to eat frijoles! *(*WINTER *nabs the* FARMWORKER *as he tries to escape.)*

WINTER: Ha, ha, Winter's got you! I want money. Give me money!

WORKER: I don't have any. I'm just a poor farmworker.

WINTER: Then suffer! *(*WINTER *drags the* FARMWORKER *D.S.C., kicking and beating him, then dumps snow on him from a small pouch. The* FARMWORKER *shivers helplessly.* SPRING *enters at* S.L.*, singing a happy tune.)*

SPRING: *(Skipping in.)* La, la, la, la, la. *(Stops, sees* WINTER *maltreating* FARMWORKER.*)* What are you doing here?

WINTER: Mamasota, who are you?

SPRING: I am Spring, la primavera, but your time is past. You have to go!

WINTER: Some other time, baby.

SPRING: Aw, come on now, you've had your turn. You've got to leave. *(*WINTER *ignores her with a grunt.)* Get the hell out of here!

WINTER: All right, I'm going for now, but I'll be back again next year, campesino. *(Exits* S.R.*)*

SPRING: *(Crosses to* FARMWORKER *and helps him to get* up.*)* There, there, you poor, poor farmworker, here, now, get up. You mustn't let this happen to you again. You've got to fight for your rights!

WORKER: You mean I've got rights?

SPRING: Sure!

WORKER: Ahora, sí. I'm going to fight for my rights like Pancho Villa, like Francisco I. Madero, like Emiliano Zapata . . . *(*SPRING *startles him by touching his shoulder.)* Ta-ta-ta! *(From backstage is heard the cry: Campesino!)*

SPRING: Oh, my time has come . . . *(Crosses in front of* FARMWORKER.*)* I must go now. But, remember, fight for your rights! La, la, la, la. *(Exits* S.R., *singing and skipping.)*

WORKER: She's right! From now on I'm going to fight for my rights, my lefts, and my liberals. *(*DON COYOTE *enters* S.L.*)*

COYOTE: Amigo . . .

WORKER: *(Turns, frightened. Runs to* S.R. *after* SPRING.*)* Pri . . .

COYOTE: Pri, what?

WORKER: Pri . . . prepare yourself! You robbed me!

COYOTE: No! No, I'm your friend.

WORKER: Ni madre! You're a thief!

COYOTE: No, soy tu amigo. ¡Somos de la misma raza!

WORKER: ¡Simón, eres rata! *(He swings at* DON COYOTE.*)*

COYOTE: ¡Calma, hombre! ¡Ahí viene mi patrón!

WORKER: Que venga ese cab . . .

PATRON: *(Enters* S.L.*)* Boy!

COYOTE: *(Running over to him.)* Yes, boss?

PATRON: You got this year's summer crew ready?

COYOTE: *(Hesitating, hat in hand.)* Well, you see, patrón, it's this way . . .

PATRON: Well?

COYOTE: *(With a forced smiled.)* Sure, boss, it's all ready.

PATRON: Good! *(He turns and crosses to corner of flat at* S.L., *anticipating the entrance of* SUMMER. DON COYOTE *rushes to* FARMWORKER *at* S.R.*)*

COYOTE: ¡Andale, mano! You got to work. Haven't I always give you work? Don't I always treat you good?

WORKER: ¡No!

COYOTE: Andale, hombre, be a sport! Do it for old times sake!

WORKER: ¡No, te digo! *(He spots* SUMMER *coming in at* S.L.*)* ¡Estoy en huelga! *(He squats.)*

PATRON: What's going on? Why isn't he working?

COYOTE: He says he's on strike.

PATRON: Strike? But he can't be! Summer's going by! What does he want?

COYOTE: *(To* FARMWORKER.*)* ¿Qué quieres?

WORKER: Un contrato bien firmadito.

COYOTE: He wants a signed contract!

PATRON: He's crazy! We need some more workers! Find me some more workers! Find me some more workers! Summer's passing! *(To audience.)* Five hundred workers! I need five hundred workers! *(Meanwhile,* SUMMER *continues to cross the stage and finally exists* S.R. *The* PATRON *is frantic, hysterical. He ends up following* SUMMER *off stage. There is a silence. The* PATRON *re-enters in shock and disbelief.)* He's gone. Summer's gone. My crop! Ahhhhhhhh! *(He leaps and snorts like an animal.)*

COYOTE: *(Fearfully.)* ¡Patrón! ¡Patrón! *(The* PATRON *is on the floor, kicking and snorting like a wild horse.* DON COYOTE *leaps on his back and rides him like a bronco until the* PATRON *calms down and settles on all fours, snorting and slobbering incoherently.* COYOTE *pats the side of his head like a horse.)* Chihuahua, cada año se pone más animal mi patrón. It's okay, boss. He can't last, because he's getting hungry. (FARMWORKER *doubles over with pangs of hunger.)* And anyway, here comes Autumn! (AUTUMN *crosses the stage and the* FARMWORKER *approaches him with one hand on his stomach and his other arm outstretched.)*

WORKER: Con esto me compro un taco.

COYOTE: *(Slapping his hand down.)* None of that! Put it here first! *(Stretches out his hand.)*

WORKER: No, I can't. I'm on strike!

COYOTE: No work, no eat! Put it here!

WORKER: No, I . . . *(He hesitates. He is almost ready to take the contractor's hand.* SPRING *enters* S.L. *dressed as a nun representing the churches.)*

CHURCHES: Wait! *(Crosses* to FARMWORKER.*)* I am the Churches. I bring you food and money. *(She hands him some cash and fruit.)*

PATRON: *(Back to his senses.)* You . . . you lousy contractor! You lost me my summer crop and my fall crop. You're fired! And you, you communist farmworker. *(Points to nun.)* You, too, you Catholic communist! *(A rumbling noise backstage. The* FARMWORKER *is frightened. He tries to run but the nun holds him. The* PATRON *cowers* U.S.C. SUMMER *enters dressed as "unions" and carries a contract and an oversized pencil.)*

UNIONS: *(D.S.C.)* I am the Unions. We're with you, brother! Keep fighting! *(Crosses to* FARMWORKER *and shakes his hand and stands by his side. There is another rumbling noise backstage.* FALL *re-enters dressed as a Mexican revolutionary representing "la raza."*

LA RAZA: La raza está contigo, mano. Sigue luchando. *(He also joins the ranks around the* FARMWORKER. *One final gigantic rumble from backstage. With snow spilling over the flats,* WINTER *enters with a vengeance.)*

WINTER: ¡Llegó el lechero! And my name ain't Granny Goose, baby! Money, give me money! *(He charges toward the* FARMWORKER *and is repulsed by the* CHURCHES, LA RAZA, *and the* UNIONS *who shout "No!")* That's what I like, spunk! *(He tries again and is repulsed a second time.)* God damn!!! *(He tries one final time, making himself as big and as frightening as possible, but he fails again. He asks them.)* Who has money? *(*CHURCHES, UNIONS *and* LA RAZA *point at* PATRON *and shout: "he has." With a gleeful shout,* WINTER *assails the* PATRON, *demanding money. The* PATRON *pulls out money from all of his pockets, wads and wads of it, until he runs out.)* More!

PATRON: That's all I have!

WINTER: More!

PATRON: I don't have any more. Except what I have in the bank. *(With a savage look in his eyes,* WINTER *takes a step backward and gets ready to leap at the* PATRON'*s throat. The* PATRON *is transfixed with fear. He is unable to move until* WINTER *grabs him by the throat and drags him* D.S.C.*)*

PATRON: But I don't have any money.

WINTER: Then freeze to death! *(*WINTER *kicks and beats the* PATRON *and pours snow all over him. The* PATRON *shivers and looks up toward the* CHURCHES, UNIONS *and* RAZA.*)*

PATRON: Help me!

UNIONS: *(With* RAZA *and* CHURCHES.*)* Sign a contract!

WORKER: ¡Firma un contrato!

PATRON: *(After a pause.)* All right! *(*UNIONS *hand the* FARMWORKER *the contract and the pencil. The* FARMWORKER *comes forward and hands them to the* PATRON. *In panic,* DON COYOTE *comes around and kneels beside his boss.)*

COYOTE: No, patrón, don't sign! I'll be out of a job. I brought you wetbacks. They're communists. Nooooo! *(The* PATRON *signs the contract and hands it to the* FARMWORKER *who looks at it in disbelief.)*

WORKER: $2.00 an hour . . . rest rooms in the fields . . . vacations with pay . . . GANAMOS!!!! *(The* FARMWORKER'S *supporters give out a cheer and pick him up on their shoulders and carry him out triumphantly. The* PATRON *crawls out on his hands and knees in the opposite direction.* DON COYOTE *tries to sneak out with the crowd, but* WINTER *catches him.)*

WINTER: Ah-hah! Winter's got you!

COYOTE: *(Bluffing.)* Winter? Hah! Winter's already past! *(*WINTER *slaps his forehead stupidly.* DON COYOTE *laughs and starts to walk out. Then suddenly* WINTER *snaps his fingers as if realizing something.)*

WINTER: The fifth season! I'm the fifth season!

COYOTE: What fifth season? There are only four!

WINTER: *(Tearing off the top layer of the sign hanging from his neck, revealing a new sign underneath.)* ¡La justicia social!

COYOTE: Social justice? Oh, no! *(*WINTER *kicks* DON COYOTE *offstage, then turns toward the audience.)*

WINTER: Si alguien pregunta que pasó con ese contratista chueco, díganle que se lo llevo la quinta chin . . . ¡¡¡La quinta temporada!!! *(Exits* S.L.*)*

Greg Pape

¡VIVA LA HUELGA!

There was resentment and pride
in the way they waved the red flag
so that the blocky black wings
of the symbol that stood for the eagle
flapped and pushed at the angry red air,
trying to fly.

I thought I was just going to work,
finally, after all those mornings
standing in line to earn again
my weekly check, twenty-nine dollars,
after all those forms and phone calls
and interviews in which the questions
and the answers feel like lies
even when they are true, and the one
who asks and the one who answers
are both liars by omission.
That year it seemed the whole San Joaquin,
my valley, my state, my country,
my own eyes were all liars by omission.
The woman who loved me was exempt,
and each morning when I rose to
and from her body, my own body
was a gift she had given me.

That year I read some great books,
burned my draft card, and lamented
all the deaths that weren't mine
and the one that was. That year
I got a job working for an entrepreneur
farmer, inventor of a mechanical peach harvester.

For two weeks we trained for the day
of its demonstration. Each morning when the sun
rose over the Sierras, we tucked our pants
into our boots, tied bandanas over our faces,
pulled the gloves over our cuffs, grabbed hold
of the clamp handles on the hydraulic arms,
and swung our shakers down the rows
like machine gunners attacking the trees.
Clouds of fuzz rose as peaches fell in the heat,
and sweat meant to cool the body
turned to mold. By noon each day,
we had shaken the peaches from a dozen rows
of trees and all the thoughts out of our minds.

When the big day arrived
and we drove slowly up the dirt road
trailering the harvester behind the truck
between the rows of solemn peach trees
and angry people waving the red flag
with the black eagle, shouting ¡Viva la Huelga!
I stared at them like the daydreamer I was,
a young man dreaming his lover next to him
in a summer orchard who leans over
and looks into his eyes, but instead
of parting her lips for the deep kiss
that leads to heaven, she slaps him hard
in the face to awaken him to shame.

George Keithley

'TIME ACCOMPLISHES FOR THE POOR WHAT MONEY DOES FOR THE RICH'

—Cesar Chavez

They sit on the windy sidewalk waiting
to enjoy a joke. No matter how many times
they've seen this they never grow
tired of watching the ritual
which is performed every morning.

Twin limousines glide together, stopping
traffic momentarily in both lanes.
Governor Reagan looks grim as he leans
from the rear door of one limo, then
he climbs into the other, changing cars...

Partly superstition, partly precaution.
From the front seat his guard
observes the Chicanos who sit, laughing,
on the walk outside the bars
and furniture stores on Folsom Boulevard.

Under swollen sycamores and tall palm trees
both vehicles continue toward the capitol,
pursuing one another through the traffic.
Since they're identical it's impossible
to say in which one he will arrive today

until he emerges and his fine hair is disturbed
by the wind, the wind that blows tomorrow
and the next night and
the next day when we will all eat
the bread in our hands and we will all drink wine.

THE RED BLUFF RODEO

1

They travel the ranch roads that connect
with paved roads feeding into the freeway
to enter town by car or truck
or on horseback,
the main street blocked by traffic—
Not one empty seat in a bar and grill
for a mile around the fairgrounds.

"Let me have coffee and three eggs."
"How do you *like* your eggs?"
"I like 'em fine!"

2

Clowns in costume (cow-
boys, ranch hands)
fall off
a plow horse
too old to mind the jeers
of spectators. One clown
climbs a cow, flings
his feet in the air,
slaps her rear and lifts
his eyes to the sky,
asking why she won't buck—
Little laughter now
people slide close,
settling onto the wooden slats.

3

They roar for the first rider—
What's caught the eye of the crowd
is the swift stride of his horse
and the queer stiff legs of the calf.
His lasso loops her neck
to stop her short.

He drops beside her in the dust,
ties her legs tight with one length
of rope, and one twisted sweep of his arm.
Throws his hands in the air to show he's done,
like a man surrounded, forced to surrender.

Applause rattles the grandstand from its full height;
man and horse trot toward the gate.

The clowns come for the calf; hurry off.

The p.a. announces another number. The chute opens.
Out flies a fresh horse, a hat, hands
too quick for the calf
rolling helpless
in the grip of the rope.

4

Girls squirm in the sun.
Small boys turn their eyes
to women unwrapping
sandwiches. Wives
worried if a man might earn his entry fee
lean heads together
ignoring the noise, the nervous heat.

> "Did he come home sober last night?"
> "In the dark he tripped over a chair."
> "What did you do?"
> "I laughed so hard I fell out of bed."

5

The last man on a saddle bronc
provokes a rough ride
to impress the judges.
Jabbing flesh,
his spurs urge
the bronc to kick
three ways at once—

He flies from his mount
in mid-air, tossed
free. Falls
like a sack of meal in the dust.
The throng disapproves and boos.
On hands and knees he crawls away from the hooves.

"Some sort of fun."
"I'm glad it's done. All day
I been dizzy as a squirrel."

The crowd staggers out of the stands,
pressing into the parking lot.

Evening dresses the country in cool stars.

Far in the night a fitful line of light
searches the length of the valley
where families follow one another home
beyond the lost barns and deserted fields.

Leonard Gardner

from FAT CITY

Hundreds of men were on the lamplit street, lined for blocks with labor buses, when Billy Tully arrived, still drunk. He had been up most of the night, as he had nearly every other night since the loss of his cook's job; and he had been fired because of absences following nights out drinking. It had been agony getting up after three hours' sleep. After the night clerk's pounding, Tully had remained motionless, shaken, hearing the knocking at other doors, the same hoarse embittered summons down the hall. It had been so demoralizing that he had taken his bottle out with him under the morning stars. In the other pocket of his gray zipper jacket were two sandwiches in butcher paper. He had eaten no breakfast.

The wine calmed his shivering as he passed the dilapidated buses, the hats and sombreros and caps of the men inside silhouetted in the windows. The drivers stood by the doors addressing the crowds.

"Lettuce thinners! Two more men and we're leaving."

"Onion toppers, over here, let's go."

"Cherries! First picking."

"They ripe?"

"Sure they're ripe."

"How much you paying?"

"A man can make fifteen, twenty dollars a day if he wants to work."

"Shit, who you kidding?"

"Pea pickers!"

The sky was still black. Only a few lights were on in the windows of the hotels, dim bulbs illuminating tattered shades and curtains, red fire-escape globes. Under the streetlights the figures in ragged overalls, army fatigues, khakis and suit coats all had a somber uniformity. They pushed to board certain buses that quickly filled and rolled away, grinding and backfiring, and in these crowds Billy Tully jostled and elbowed, asking where the buses were going and sometimes getting no answer. He crossed the street, which

---

was crossed continuously by the men and the few women and by trotting preoccupied dogs, and stopped at a half-filled sky-blue bus with dented fenders and a fat young man in jeans at the door.

"Onions. Ever topped before?"

"Sure."

"When was that?"

"Last year."

"Get on."

Tully climbed into the dark shell, his shoes contacting bottles and papers, and waited amid the slumped forms while the driver recruited outside. "If these onions were any good," Tully said, "looks like he could get him a busload."

"They better than that damn short-handle hoe."

"Maybe I ought to go pick cherries."

"You make more topping onions, if we can get this man moving."

The stars paled, the sky turned a deep clear blue. Trucks and buses lurched away. The crowd outside thinned and separated into groups.

"Let's get going, fat boy," Tully yelled.

"Driver, come *on*. I got in this bus to top onions and I want to top onions. I'm an onion-topping fool."

The bus rattled past dark houses, gas stations, neon-lit motels, and the high vague smokestack of the American Can Company, past the drive-in movie, its great screen white and iridescent in the approaching dawn, across an unseen creek beneath ponderous oaks, past the cars and trailers and pickup-truck caravans of the gypsy camp on its bank and out between the wide fields. Near a red-and-white-checkered *Purina Chows* billboard, it turned off the highway. Down a dirt road it bumped to a barn, and the crew had left the bus and taken bottomless buckets from a pickup truck when the grower appeared and told them they were in the wrong man's onion field. The buckets clattered back into the truckbed, the crew returned to the bus, and the driver, one sideburn hacked unevenly and a bloodstained scrap of toilet paper pasted to his cheek, drove back to the highway swearing defensively while the crew cursed him among themselves. The sky bleached to an almost colorless lavender, except for an orange glow above the distant mountains. As the blazing curve of the sun appeared, lighting the faces of the men jolting in the bus—Negro paired with Negro, white with white, Mexican with Mexican and Filipino beside Filipino—Billy Tully took the last sweet swallow of Thunderbird, and his bottle in its slim bag rolled banging under the seats.

They arrived at a field where the day's harvesting had already begun, and embracing an armload of sacks, Tully ran with the others for the nearest rows, stumbling over the plowed ground, knocking his bucket with a knee in the bright onion-scented morning. At the row next to the one he claimed knelt a tall Negro, his face covered with thin scars, his knife flashing among the profusion of plowed-up onions. With fierce gasps, Tully removed his jacket and jerked a sack around his bottomless bucket. He squatted, picked up an onion, severed the top and tossed the onion as he was picking up another. When the bucket was full he lifted it, the onions rolling through into the sack, leaving the bucket once again empty.

In the distance stood the driver, hands inside the mammoth waist of his jeans, yelling: "Trim those bottoms!"

There was a continuous thumping in the buckets. The stooped forms inched in an uneven line, like a wave, across the field, their progress measured by the squat, upright sacks they left behind. In the air was a faint drone of tractors, hardly audible above the hum that had been in Tully's ears since his first army bouts a decade past.

He scrabbled on under the arc of the sun, cutting and tossing, onion tops flying, the knife fastened to his hand by draining blisters. Knees sore, he squatted, stood, crouched, sat, and knelt again and, belching a stinging taste of bile, dragged himself through the morning. By noon he had sweated himself sober. Covered with grime, he waddled into the bus with his sandwiches and an onion.

"You got you a nice onion for lunch," a Negro woman remarked through a mouthful of bread, and roused to competition, an old, grizzled, white man, with the red inner lining exposed on his sagging lower lids, brought from under his jacket on the seat his own large onion.

"Ain't that a beauty?" All the masticating faces were included in his stained and rotting smile. "Know what I'm going to do with it? I'm going to take that baby home and put it in vinegar." He covered it again with his jacket.

Out in the sun the scarred Negro at the row beside Tully's worked on in a field now almost entirely deserted.

Through the afternoon heat the toppers crawled on, the rows of filled sacks extending farther and farther behind. The old grizzled man, half lying near Tully, his face an incredible red, was still filling buckets though he appeared near death. But Tully was standing. Revived by his lunch and several cupfuls of warm water from the milk can, he was scooping up onions from the straddled row, wrenching off tops, ignoring the bottom fibrils where sometimes clods hung as big as the onions themselves, until a sack was full. Then he thoroughly trimmed several onions and placed them on top. Occasionally there was a gust of wind and he was engulfed by sudden rustlings and flickering shadows as a high spiral of onion skins fluttered about him like a swarm of butterflies. Skins left behind among the discarded tops swirled up with delicate clatters and the high, wheeling column moved away across the field, eventually slowing, widening, dissipating, the skins hovering weightlessly before settling back to the plowed earth. Overhead great flocks of rising and falling blackbirds streamed past in a melodious din.

In the middle of the afternoon the checkers shouted that the day's work was over.

Back in the bus, glib and animated among the workers he had surpassed, the Negro who had topped next to Tully shouted: "It easy to get sixty sacks."

"So's going to heaven."

"If they onions out there I get me my sixty sacks. I'm an onion-topping fool. Now I mean onions. I don't mean none of them little pea-dingers. Driver, let's go get paid. I don't want to look at, hear about, or smell no more onions till tomorrow morning, and if I ain't there then hold the bus because I'm a sixty-sack man and I just won't quit."

"Wherever you go there's always a nigger hollering his head off," muttered the old man beside Tully.

"Just give me a row of good-size onions and call me happy."

"You can have them," said Tully.

"You want to know how to get you sixty sacks?"

"How's that?"

"Don't fool around."

"You telling me I wasn't working as hard as any man in that field?"

"I don't know what you was doing out there, but them onions wasn't putting up no fight against me. Driver, what you waiting on? I didn't come out here to look at no scenery."

They were driven to a labor camp enclosed by a high Cyclone fence topped with barbed wire, and as the crew rose to join the pay line outside, the driver blocked the way. "Now I want each and every one of those onion knives. I want you to file out one by one and I want every one of those knives."

"You going look like a pincushion," said the sixty-sack Negro.

The crew handed over the short, wooden-handled knives, and the driver frowned under the exertions of authority. "One by one, one by one," he repeated, though the aisle was too narrow for departing otherwise.

Tully stepped down into the dust and felt the sun again on his burned neck. Standing in the pay line behind the old man, he looked down the rows of whitewashed barracks. A pair of stooped men in loose trousers, and shirts darkened down the backs with sweat, passed between buildings. In the brief swing of a screen door Tully saw rows of iron bunks. A Mexican with both eyes blackened crossed the yard carrying a towel. Tully moved ahead in the line. The paid were leaving the window of the shack and returning to the bus, some lining up again at a water faucet

"Is that all you picked?" the paymaster demanded of the old man. "What's the matter with you, Pop? If you can't do better than that tomorrow I'm going to climb all over you."

"Well, it takes a while to get the hang of it," came the grieving reply.

Two dimes were laid on the counter under the open window. "Here's your money."

The old man waited. "Huh?"

"That's it."

The creased neck sagged further forward. Slowly the blackened fingers, the crustaceous nails, picked up the dimes. The slack body showed just the slightest inclination toward departing, though the split shoes, the sockless feet, did not move, and at that barely discernible impulse toward surrender, three one-dollar bills were dealt out. With a look of baffled resignation the man slouched away, giving place to Billy Tully, who stepped up to the grinning paymaster with his tally card.

As the bus passed out through the gate, Tully saw, nailed on a whitewashed wall, a yellow poster.

BOXING
ESCOBAR
VASQUEZ

The posters were up along Center Street when the bus unloaded in Stockton. There was one in the window of La Milpa, where Tully laid his five-dollar bill on the bar and drank two beers, eyeing the corpulent waitress under the turning fans, before taking the long walk to the lavatory. He washed his face, blew his dirt-filled nose in a paper towel, and combed his wet hair.

On El Dorado Street the posters were in windows of bars and barber shops and lobbies full of open-mouth dozers. Tully went to his room in the Roosevelt Hotel. Tired and stiff but clean after a bath in a tub of cool gray water, he returned to the street dressed in a red sport shirt and vivid blue slacks the color of burning gas. Against the shaded wall of Square Deal Liquors, he joined a rank of leaners drinking from cans and pint bottles discreetly covered by paper bags. Across the street in Washington Square rested scores of men, prone, supine, sitting, some wearing coats in the June heat, their wasted bodies motionless on the grass. The sun slanted lower and lower through the trees, illuminating a pair of inert legs, a scabbed face, an outflung arm, while the shade of evening moved behind it, reclaiming the bodies until the farthest side of the park had fallen into shadow. Billy Tully crossed the sidewalk to the wire trash bin full of empty containers and dropped in his bottle. Over the town a dark haze of peat dust was blowing from the delta fields.

He ate fried hot dogs with rice in the Golden Gate Café, his shoes buried in discarded paper napkins, each stool down the long counter occupied, dishes clattering, waitresses shouting, the cadaverous Chinese cook in hanging shirt and spotted khaki pants piled over unlaced tennis shoes, slicing pork knuckles, fat pork roast and tongue, making change with a greasy hand to the slap slap of the other cook's flyswatter.

Belching under the streetlights in the cooling air, Tully lingered with the crowds leaning against cars and parking meters before he went on to the Harbor Inn. Behind the bar, propped among the mirrored faces in that endless twilight was another poster. If Escobar can still do it so can I, Tully thought, but he felt he could not even get to the gym without his wife. He felt the same yearning resentment as in his last months with her, the same mystified conviction of neglect.

At midnight he negotiated the stairs to his room, its walls covered with floral paper faded to the hues of old wedding bouquets. Undressing under the dim bulb, he stared at the four complimentary publications on the dresser: *An Hour With Your Bible. El Centinela y Heraldo de la Salud. Signs of the Times—The World's Prophetic Monthly. Smoke Signals—A Renowned Anthropologist Marshals the Facts on What Smoking Does to Life Before Birth.* He wondered if anyone ever read them. Maybe old men did, and wetbacks staying in off the streets at night. And was this where he was going to grow old? Would it all end in a room like this? He sat down on the bed and before him on the wall was the picture of the wolf standing with vaporized breath on a snow-covered hill above a lighted farm. Then the abeyant melancholy of the evening came over him. He sat with his shoulders slumped under the oppression of the room, under the impasse that was himself, the utter, hopeless thwarting that was his blood and bones and flesh. Afraid of a crisis beyond his capacity, he held himself in, his body absolutely still in the passing and fading whine and rumble of a truck. The blue and gold frame, the long cord hanging from the molding, the discolored gold tassel at its apex, all added to the feeling that he had seen the picture in some room in childhood. Though it filled him with despondency he did not think of taking it down, or of throwing out the magazines and pamphlets and removing from the door the sign

IF YOU SMOKE IN BED
PLEASE LET US KNOW
WHERE TO SEND YOUR ASHES.

It did not occur to him that he could, because he did not even feel he lived here.

In the dark he arranged himself with tactical facility in the lumpy terrain of the mattress. When the pounding came again on the door, he lunged up in the blackness crying: "Help!"

Out in the hall the hoarse voice warned: "Four o'clock."

Susan Kelly-DeWitt

YUBA CITY ORCHARDS, 1961

Heat was everywhere.

It crept down the ladders
of leaves, to where brown
men bent; it sent out runners
of sweat along their necks.

The message was clear:
This heat would be here
all summer—nailing them
with its spike to that
monotonous place.

There were orders and the men
worked a little faster. They climbed
and stretched and stooped

until the shade
finally took them.

Susan Kelly-DeWitt

FEBRUARY HEAT

Fifth drought spring:
one could easily imagine the whole parched fabric
of the world was ending, unraveling from dun

colored tapestries of dust: Grit, chemicals
fertilizers thicken the air
in a stifling layer. In the garden, havoc: dead things

left skeletoned by January's freeze—The frazzled
stump of a fern: Will it come back?
The rare and tender lemon

bougainvillea reduced
to a monotony of twigs.
Too soon! too soon! the bird of summer

sings. Along the Interstate,
tumbleweed contradicts
neatly furrowed fields, pasteled willows—but hawks

plummet without thinking in human terms;
cry out, exultant, in hawkish speech. And drunk
with light, hundreds of field sparrows

tilt in a cyclone-shaped mist, a tornado of birds
darkening the bone disc
of the horizon—

Wings beating the still
February air, peppering dusty valley
gusts which sift and settle

over cattle
standing like ceramic cream
pitchers, on mounds of dung

in the incendiary sun.

Don Thompson

MEDUSA

In this heat the hills
Can neither hold still nor move
beyond their shimmering

The granite I stand on breathes
and the air is like stone
in my lungs

I can never learn.

Jean Janzen

SOMETIMES

Sometimes
when the sun
is overhead
and we lie
in its great outpouring,
our skins evaporate
and our bones become
light and airy.

We are alone in this.
The garden continues
to drink the earth,
stalks leaning heavily
with blooms,
sycamores extending
and thickening into
silent shade.

Only we can see
that the sky
is coming nearer,
that it is wearing
a new iridescence
at its rim,
and we know
with a certainty
that we are not made
for earth,
a feeling
that already
with hair burning
we rise.

TOWARD THE END OF THE CENTURY

Worst is the ivy
    which works its way
through a window crack when
    I turn my head, then pushes
a pale arm through the drapery.
    Good, proper English ivy
turned to forced entry.
    But also, the creeping fig,
at first so demure,
    has grown hungry and large,
etches the brick and window glass
    with a million hands.
I open the door for air
    and the moths swarm in.
One night a baby possum
    at the doorstep, one paw up, courteous.
Even after the garden shears
    and the locked doors, honeybees
continue to claim the chimney,
    generation after generation
filling it with their tiny rooms,
    snarling angrily when we try
to gas them out.
    And when I play Chopin's Prelude in A Flat
some fly in to the sudden sun
    of the piano light, scorching
their wings as I repeat
    the one essential tone,
the search for resolution.

Peter Everwine

BACK FROM THE FIELDS

Until nightfall my son ran in the fields,
looking for God knows what.
Flowers, perhaps. Odd birds on the wing.
Something to fill an empty spot.
Maybe a luminous angel
or country girl with a secret dark.
He came back empty-handed,
or so I thought.

Now I find them:
thistles, goatheads
the barbed weeds
all those with hooks or horns
the snaggle-toothed, the grinning ones
those wearing lantern jaws
old ones in beards
leapers in silk leggins
the multiple pocked moons
and spiny satellites, all those
with juices and saps
like the fingers of thieves
nation after nation of grasses
that dig in, that burrow, that hug winds
and grab handholds
in whatever lean place.

It's been a good day.

THE HATCH

At dusk they come swarming from the grass,
columns and clouds, almost invisible
like dark feathery seeds, the air
whining with their constant song.
Soon they will lie wrecked in corners,
burnt-out, drifted in the dust they resemble.
But now they are all around me, fanning
like curtains in the air, in spirals
entering the secret places
of my body. They batter urgently
my eyes and whirl in my nostrils.
Their grit breaks against my tongue.
They are butting and strangling
in my fine hairs.
My beard sings like a hive.
The ends of my fingers live, I dance
in deep columns of living cells.
Sweet Jesus! I am the King of Bugs!
And I flap my arms home, home
under the fluttering stars, trailing
the endless robes of my flesh.

Archie Minasian

EARLY SUMMER THROUGH LOS BANOS

From off the fields
abstract designs in crazy patterns
appear and assemble
like an instant gallery
in greens and mustard-yellow.

Two eyes,
two legs, two wings,
a twisted head,
and suddenly a burst,
like mushy grapes or tinted snow,
life instantly immobilized
and lost forever.

No lengthy summer sun,
no wide expanse to flit across,
no promise of a winged destiny—
fulfillment of a frenzied afternoon,
relinquished all to progress of the road
and mounted in some grotesque bas-relief.

HOW FOOLISH THE SWEATING MEN

Ha for a long review of everything
now that old Summer plans to leave.

How comforting the fields we pass
with the slanting hay;

How pure the air we breathe
and our thoughts;

How inviting the lushy meadows,
the wandering girls by roadsides;

How jubilant each hedge
with the bird;

How dear the farmhouse half hidden
with gardens adjoining and vines sloping;

How cool the broad canal
and the lanky weeds and the weir;

How distant the brown bare hills,
the solitary trees;

How still the air,
how dense the views we pass;

How foolish the men sweating in orchards,
shaking the peach and the last ripe plum.

DeWayne Rail

PICKERS

Scattered out like a handful of seeds across
The field, backs humped to the wind,
Faces like clumps of dirt in the white rows,
Their hands keep on eating cotton.
The long sacks fill and puff up tight,
Like dreams of money they're going to make
Or of what they'll have for supper.
They stagger with the weight to the wagon and scales,
Dragging their tracks out in the dust,
Faces bent close to the ground,
Their hands ragged from the cotton bolls.

THE FIELD

This field is the last refuge of squirrels,
Jackrabbits, and mice. Deserted, left
To its own devices, it has taken years
To grow a thick cover of weeds that tangle

And arch over long tunnels. A dozen kinds
Of birds feed here on sprays of seeds
That hang from dried stalks, and tracks lead
Everywhere. At dark, the animals push up

From their holes and look cautiously out.
The hours between now and dawn belong to them.
They sniff the air, or nibble
At young shoots of grass, and each one moves

Slowly away from his den. They are free
To wander the length of the field,
Never hunted, as each night rolls
Easily into the next. Some stay on here

For years. Others die, drawn
To the highway that loops the edge of the field,
With its hiss of cars whose headlights
Shine briefly into their lives, then flash on.

Joyce Carol Thomas

TRACY POEM

We stood in the
Fields loving
Even though we didn't
Know it then
Green stained jeans
Where picking crops on
The knees is a constant
Prayer and a truck ride
On long handled hoes
Is a benediction
To young girls who believe

We crooned up and down
the rows each harvest
A different noted rhythm
Melody where the saxophone
Is a nibble in the ear

And she all ebony
And sometimes taunted
And sought after by
Stubby chinned boys, later
Beaten blue by
Bossy brothers
Grew to bend a bit
But not too much

He totes her tomato box
And one so bold is
A possible lover
But only in blameless sleep

Slow simmered pinto beans
Catch flavors from the air
And a home is a home
Wherever a mother is
It's my turn to spend the night
Mama, may I
Mama, may I
Well, if it's all right . . .

All right to stay awake
Spilling out grown up dreams
And wondering
Where can our heroes be

Somebody Stockton hopped
And country girls only knew
How to dance slow and never
Could understand why
Their partners sweated so

And we stood in the fields
Loving
One another
And we didn't even know

Gary Soto

HOEING

During March while hoeing long rows
Of cotton
Dirt lifted in the air
Entering my nostrils
And eyes
The yellow under my fingernails

The hoe swung
Across my shadow   chopping weeds
And thick caterpillars
That shriveled
Into rings
And went where the wind went

When the sun was on the left
And against my face
Sweat   the sea
That is still within me
Rose and fell from my chin
Touching land
For the first time

HARVEST

East of the sun's slant, in the vineyard that never failed,
A wind crossed my face, moving the dust
And a portion of my voice a step closer to a new year.

The sky went black in the 9th hour of rolling trays,
And in the distance ropes of rain dropped to pull me
From the thick harvest that was not mine.

Art Coelho

EVENING COMES SLOW TO A FIELDHAND

Evening comes slow
to a fieldhand.
Dust of diesel cats
owned like the smell
of his own sweat
down endless barley checks.
Fingers cracked from
water-syphoning pipes,
and exhausted with
the sunup to sundown.

"Don't you tell me
there's a better day
for us a-comin'.
How come there
ain't enough food
to go around?"

You think the moon
feels for an aching back,
or that a cool shower
in a clapboard house understands?
You think a migrant child
doesn't know every solitary shingle
of a falling down shack
is making her daddy an old man?

Evening comes slow to a fieldhand.

David St. John

THE OLIVE GROVE

Never lost in its false, dense weather
The man in the olive grove sits

Dreaming as the fog drifts
And smokes through the lattice of the branches.

There is nothing about his life
He does not know. He knows that even the split

Parings of his bones will at last gleam
In the wet earth. He knows the buckle

Of muscle will fasten his heart, as the coronas
Blaze out of their saints. But where, he thinks,

Is that blessing pale as a summer?
Where is the reaper beneath his shoulders of wheat?

The bark of the trees flakes in long curls.
The man in the olive grove crouches a moment

Beside the turned roots, their shoots
Like the hair pulled from the moon's harvest shag.

He shakes out his pockets—
Photos torn from a newspaper, a letter from

A friend blaming him, lint, a ring of keys
To a chain of rooms, phone numbers smudged into

Names, a tan cigar, matches turned to clay, clay.
What is visible at last in the dark

Leaves guttered in the road's throat? Or
Does the wind remember the last room of the valley?

He thinks *he* must not remember the business
Of a man in the olive grove, though he has never

Before lied. He told what he saw.
He was milk to everyone; the certain, pale medium

Of desire. But the heart of the window breaks—
Now, the peacocks sweep through the grove.

Someone steps out onto the path
Bowing, as if beginning a dance he does not know.

Everything is over. O*nce more.* The worst is past.

James D. Houston

IN SEARCH OF OILDORADO

CALIFORNIA'S BEST-KNOWN EXPORTS nowadays are not things. They are images, composed of such unnatural resources as life-style frontiers and the shapes of leisure—not the bottled wine itself but the chilled chardonnay in the hot tub; the kid on the trail bike, on the rim of the bluff, at sunset. The images depict all varieties of the West Coast adventure in Life with a capital L, the Living and Exploring and Spending and Expanding and Exploding. They are exported on film, on record and tape, via the covers and inside pages of the *National Enquirer, Road and Track, TV Guide, People, Self,* and *Us.* Meanwhile, more traditional resources, such as timber, cotton, cattle, hogs and poultry, crude oil and natural gas, come to the public attention when some feature of the environment has been violated or is about to be swamped. But for the most part, though they fuel and finance the rest of it, these tend to exist in the shadow of The Great Post-Industrial Experiment, which dazzles us with such a blinding light.

To see this other part of California, the resource-full part, it helps to get away from the coast from time to time, and head inland, which is why I found myself on the road to Bakersfield again, over there in Kern County. About 7 percent of the cotton grown in the United States comes out of Kern County. According to Bill Rintoul, it is also the nation's fourth largest oil-producing region.

"If Kern seceded from the rest of California," says Bill, "which of course it is not planning to do, at least not right away, but if it were separated off, this county all by itself would be running fourth in oil, after Texas, Alaska and Louisiana. For that matter, it is the eighteenth largest oil producer in the world."

Bill is a Kern County patriot. I do not mean he would defend his county's honor with guns and knives. But he likes the place, he has spent most of his

---

life there, and takes its flaws along with its virtues. He actually prefers the unrelieved flatness of the landscape. We were talking once about the heavy groves and wooded canyons characteristic of the northern coast, and he said, "You know, it's funny, but there is something about that kind of country that just doesn't feel right. Those trees all around you. And the rain it takes for that kind of growth, the way it drips down through the trees. The way the mountains rise up. Half the time you can't see the sun. I suppose it's just what a person gets used to. You get used to a certain idea of what the world is supposed to look like. I'd just rather see the sun and know where I'm going."

Bill makes his living as a petroleum journalist. He writes columns for the Bakersfield *Californian* and the Tulsa *Daily World*. When I called and told him I might be heading his way, there was a pause. He is not a fast talker. He thinks things over. After a moment he said, "Well, if you time it for next weekend, you could be here for some of Oildorado. I am going to be the Grand Marshal in the parade this year. Maybe you could ride along with me in the limousine."

I had heard of Oildorado, but didn't know much about it. The fact is, on the day I called him, everything I knew about oil in California could fit easily into the spare can I carry around in the back of my Mercedes, which has a diesel engine, by the way, an old 1960 190D. It will give me thirty-five miles to the gallon on a trip like this, where the roads are straight and flat.

Naively I asked, "Does the parade run through the oil fields?"

"No, it runs right through downtown Taft. On second thought, maybe you'd be better off watching from the sidewalk. That way you'll be sure to see all the floats and the Oildorado Queen and the Maids of Petroleum. If you want to see the fields themselves, my suggestion is to drive down Highway Thirty-Three. There is no other road quite like it in the eleven western states."

In literature, Kern's finest moment comes in that scene midway through *The Grapes of Wrath*, when the Joad family stops at Tehachapi Pass to take a first long and thirsty look at this land they have struggled so hard to reach. Ruthie and Winfield, the youngest, are awestruck by the sight, "embarrassed before the great valley," is the way Steinbeck described it:

> The distance was thinned with haze, and the land grew softer and softer in the distance. A windmill flashed in the sun, and its turning blades were like a little heliograph, far away. Ruthie and Winfield looked at it, and Ruthie whispered, "It's California."

The vista Steinbeck chose, in 1939, to flesh out the dream these immigrants carried with them from Dust-Bowl Oklahoma is a long way from the world most Californians inhabit now. The largest cities, the densest networks of subdivisions, mobile home parks and retirement towns are found along a coastal strip, some forty miles wide, between Sonoma County and San Diego. By and large, this is where The Big Experiment is going on. It is also a zone of intense tourism. People living in or near the coastal communities often find themselves caught in that mind-boggle between how a place once looked and felt and how its packaging looks once it has become merchandise on the international travel circuit.

In Kern County you do not have to put up with much of this. In the 1979 *Atlas of California*, on the page where "Major Tourist Attractions" are marked with circles and dots of various colors, Kern is blank. Gray Line buses do not linger in Bakersfield or Oildale or Taft. Movie stars and sports celebrities seldom buy homes there, though they might well invest in the land. People don't visit, as a rule, unless they have business there, or relatives, or have come searching for Oildorado.

Heading east out of Paso Robles, I cross the county line halfway through a lonesome dip in the Coast Range called Antelope Valley. The hills along here are so dry and brown they shine in the sunlight as if ready to burst into flame. Just as the county sign flashes past, I hear Conway Twitty on the radio, station KUZZ out of Bakersfield coming in clear now. His voice rich with stoic remorse, Conway tells some lost sweetheart she is standing on a bridge that just won't burn. It seems perfect. This land could torment you for years without ever quite killing you. And the road signs for what lies up ahead don't seem to promise much relief: Bitterwater Valley, Devils Den, Lost Hills.

The first blur of color is startling, almost uncanny, when this narrow passage opens out into cotton fields, hundreds, perhaps thousands of acres, with bolls white and ready for picking. A couple of miles go by, and one side of the road turns from cotton into a long orchard of dusty almond trees. As the last slopes level out, where Antelope Valley joins the broad San Joaquin, the first grapevines appear, their leaves half green, half rusty brown after harvest. To the south the rows look about a mile long, stretching across to the base of the Temblor Range. To the north there is no telling how far the rows extend, no limits visible up that way. Vines merge toward the horizon.

A few more miles go by, and the vines give way to another stand of almond trees, older and thicker, bearing well, and then a peach orchard, and now, across this highway, facing the orchards you can see what all this land looked like once, and would look like now without the aqueduct. The contrast is spectacular. In this landscape almost nothing grows naturally. No shrubs, no trees, no houses. No people. Out this way there aren't even any beer cans. Without the aqueduct that intersects this highway a few miles up ahead, there would be nothing here but tumbleweed and sagebrush and the diesel rigs powering past on their way to Interstate 5.

Water is one of three resources that have shaped Kern County. Oil and country music are the other two. The water is imported from rivers farther north. The music is what you might call a hybrid product, transplanted southwestern and Okie energy finding new roots here, giving Bakersfield its nickname, Nashville West. The oil, however, is indigenous. While the guitars and the fiddles and the gospel quartets float through the airways a few feet above the ground, and while the piped-in water taps the riches in the first foot of earth, thousands and thousands of wells suck up the riches farther down, planted there fifty or sixty million years ago when uncountable generations of plankton sifted downward through the fathoms of this one-time inland sea and left tiny skeletons behind to be transformed into crude.

When we talked on the phone Bill told me Kern County is now producing more oil than some of the OPEC nations. He said this with such genuine pride in his home region, I felt obliged to ask him which OPEC nations he was referring to.

Again there was a pause, as if he had forgotten. He had not forgotten. Bill is a living encyclopedia of petroleum lore, but he will hesitate like this, as if the facts are elusive and hard to pin down. "Oh, I think Qatar is one," he said. "Gabon is another. Ecuador is in there somewhere."

On the radio Willie Nelson is singing, "Whiskey River, don't run dry." Out here on Highway 46 the crops have disappeared for a while. Sand and sagebrush stretch away on both sides of the road. It's odd to be comparing this alkaline wasteland with Ecuador.

Standing all alone in the sand and the wind, where Highway 46 meets 33, there is a cafe with a couple of gas pumps called Blackwells Corner. I swing right, as Bill instructed me to do, heading south. Within a few miles I am surrounded by walking beams, steam generators, derricks and fields of grimy pipe. No crops at all grow along this side of the San Joaquin. The soil is parched. Animals are scarce. The only movement in this moon-like realm comes from the pumps, their metal beams nodding with the motion and profile that has stirred several dozen writers to compare them to praying mantises. I now see why. This type of field pump resembles nothing in the world as much as a mantis on a string. What you see from the road are hundreds of praying heads, painted orange or yellow or black and connected by cable to something underground that seems to pull each one by the nose, so that they are all, on their various cycles, silently, ceaselessly bowing.

I will soon learn from Bill that if there was a Guinness book of financial records, this oil field would be in it. Late in 1979 Shell bought it for what was said at the time to be the largest sum ever to change hands in a corporate transaction. The field, called Kernridge, which now produces fifty thousand barrels each day, was sold for three billion, six hundred and fifty-three million, two hundred and seventy-two thousand dollars. Shell's geologists estimate three hundred and sixty-four billion cubic feet of natural gas are waiting underground here, along with five to six hundred million barrels of oil, most of which had long been considered too expensive to get at or too viscous to pump. Soaring prices changed that view.

South of this field, and near the village of McKittrick, I enter a much larger and more valuable oil field, a legendary field that has created five towns and numerous fortunes. It is called the Midway-Sunset. It is over twenty miles long. As Bill is soon to tell me, it is among the twelve largest fields in the United States. It was the fourth in the history of the country to deliver a billion barrels—as of 1967—and they are still a long way from the bottom.

I have some trouble with figures like a billion barrels, or three hundred sixty-four billion cubic feet. I have some trouble with the scale of this whole business. It is almost too much to grasp. These fields and Bill's almanac memory are filled with numbers that simply bring the imagination to a standstill. Standard Oil of California grossed $42 billion in fiscal 1980. Kern County produces nearly four million barrels of oil per week. The United States still imports about five million barrels per day. A typical reason, or symptom: there are 5.1 million automobiles in Los Angeles. These cars alone burn fifteen million gallons of gasoline every day. There are forty-two gallons in a barrel. By 1967 this field I'm driving through produced a billion barrels. In other words, forty-two billion gallons, a hundred and sixty-eight billion quarts. But where are they now? And how big a cavity does that leave below? Is it bigger or smaller than Carlsbad Caverns? And how much bigger can it get before the roof caves in?

I am beginning to wonder if it helps to drive out through these oil fields. Even here, right in the middle of it, along Highway 33, which displays the most elaborate collection of field equipment west of Oklahoma, there is not much to see, not much to hold to, nothing nearly as immediate as that vast field of cotton with its bolls like a field of white eyes along the roadside watching you pass.

I think it was easier in the old days, when they had gushers that would blow the tops off the derricks, puddle the earth with lakes of oil and fill the air for miles around here with an oily haze that would stain the laundry and cloud the sun. The most dramatic of these, Lakeview Number One, blew in on March 14, 1910, with such force and volume it could not be brought under control, and it was never brought under control. People still talk about the soaring column of oil and sand and rock. Unable to cap or channel it, they used timbers and sandbags to construct great dikes and holding sumps. Of the nine million barrels said to have come pouring forth during the year and a half it gushed and geysered, some four million were trapped and processed. The rest just ran free and seeped back into the earth from whence it came. This went on for five hundred and forty-four days without restraint. Then one day it stopped as suddenly as it had begun. The bottom fell in, due to some shift of underground pressure, and that was the end of the Lakeview Gusher, but the beginning of flush times in the Midway-Sunset, as well as the true launching of Taft, the largest town among these west-side fields. (The sign at the edge of town listing weekly luncheon times for Kiwanis, Lions, Optimists, and Rotary, claims a population of 18,500, but a plumbing contractor Bill introduced me to confessed that the true population is closer to twelve thousand. "People will throw these numbers around," he said. "They will try to rope in Fellows and Tupman and some of the outlying areas, but as long as I've been here, which is forty-four years, the population of the town itself hasn't changed that much one way or the other.")

Taft happened to incorporate in the same year Lakeview Number One burst forth, and it looks back upon that event as a grand and almost supernatural announcement of the town's arrival. Emblazoned across the 1980 souvenir T-shirts being sold in Taft, where Oildorado originated, is the phrase SEVENTY YEARS OF BLACK GOLD.

Seventy years. Even in the foreshortened history of California this place is very, very young. My house in Santa Cruz is older than any two-by-four in Taft. In this part of the world the coastal mission towns, like Santa Cruz, go back as far as towns go. The central valley had to wait until after statehood. And Taft's side of the valley had to wait even longer. As late as 1900 there was still nothing here but sagebrush. In 1902 Southern Pacific ran a spur line out this far, to service the new wells. The end of that spur line gathered a dusty collection of shacks and tents and converted boxcars. The town is still built close to the ground, in a hollow between hills studded with producing wells. The view in all directions is of dry, oil-bearing hillsides, their bare slopes defined by a few wooden derricks that survive from the early days—the same kind of definition trees provide.

Hills like these are uncommon in the San Joaquin, which by nature is as flat as a football field and about the same shape. Geologists call them anticlines. Where they rise from the plain, oil from sand and shale layers farther down has been gathered upward, within easy reach of the surface. Just out of sight, a few miles east, there lies another low range called Elk Hills, an unassuming cluster of tawny ridges that offer almost nothing to the passing eye, yet Elk Hills happens to be California's number-one producing field. Since 1912 it has belonged to the federal government, as a naval petroleum reserve. For a while in the 1920s, it was famous for its role in the most notorious oil scandal of all time though never quite as famous as Wyoming's Teapot Dome field—another naval reserve—which gave the scandal its name.

In 1921 Albert Fall, then Secretary of the Interior, brought the control of these fields into his department and promptly leased them out to high-rolling cronies. Elk Hills went to Edward Doheny, the original California oil baron and L.A. entrepreneur, in exchange for "a personal loan" of $100,000. After the dealings were exposed, Fall became the first cabinet officer in American history to be convicted of a felony and sent to prison. Doheny was acquitted. Elk Hills spent the next fifty years rather quietly, as a low-production reserve administered by the navy. It was not until the Arab oil embargo of 1973, and the new pressure to develop domestic fuel supplies, that the idea of reopening Elk Hills for commercial use was seriously considered. Bill Rintoul says a hundred and sixty thousand barrels a day come out of there now, and two hundred million cubic feet of gas.

At a small shop called Oildorado Headquarters, on the main street in downtown Taft, I pick up a couple of the black and gold T-shirts, an official program, and the special issue of the Taft *Daily Midway Driller.* In a back room, beyond the cash register, bright lights are shining. Life-size costume boards have been set up, with notches at the top for neck and chin. The costumes are old-time and turn-of-the-century Western. People are waiting in line to have their pictures taken standing behind these boards, as souvenirs of this festival which is already three days along. It started Wednesday with the queen contest and the official opening of the Westside Oil Museum. Things will begin to peak tomorrow with the big parade and hopefully climax with the World Championship Welding and Backhoe Races at Franklin Field.

Outside this shop, where I am meeting Rintoul, numerous sheriffs are passing by, numerous vests and cowboy hats, bonnets and gingham dresses. Bill is easy to spot. He is not in costume. A man of simple tastes, he never overplays his hand. As Grand Marshal of tomorrow's parade he could get away with almost anything, but he shows up in checkered slacks and a short-sleeved sport shirt. No string tie. No turquoise. He has a brand-new cowboy hat, which he has left in the car. He pretends to be worried that the parade committee might get the wrong idea.

"If they see that hat they might start talking about horses. Then I might have to tell them to forget the whole thing." He grins a weathery grin, as if he is turning into a heavy wind to look at me. "I think the Grand Marshal deserves a limousine, don't you?"

Whatever they give him to ride, Bill is going to do well as Oildorado Marshal. His heart is in exactly the right place. He is an oil fields aficionado, a man fascinated with every feature of the way this business works. He grew up here in Taft, joined the army during World War II, came back for a few years in the fields before going off to Stanford for a degree in journalism. Secretly I suspect he writes his daily columns and his feature pieces for *Pacific Oil World, Well Servicing,* and *The Drilling Contractor* so he can continue to roam among these anticlines at will. He loves all of it, the mathematics, the geology, the look of a drilling rig, the lore of the roughneck, the gusher legends, the way a late sun tints steam plumes rising from the generators. He jumped at the chance to meander once again through his own home territory, to take me on a little tour, which now begins, in the Veterans Hall, where the Oildorado Civic Luncheon is being served, and where the Grand Marshal's attendance is expected.

We walk in moments before the invocation. Bill moves directly to the head table, while I find a spot, a vacant folding chair, at one of the long tables lined up in rows across the hall. Maybe two hundred people are here, merchants and their wives, the civic leaders of Taft, dressed as cowhands, schoolmarms, desperadoes. The brightest outfits are black and gold, worn by a dozen young girls dressed as saloon dancers, 1890s style, with high fringed skirts and high-heeled shoes. These are the girls who ran for queen, sponsored by such groups as the Taft Rotary, the Desk and Derrick Club, the Moose Lodge. The winner, sponsored by Veterans of World War I, Barracks 305, is a slim and pretty senior from Taft High. She wears a glossy beauty-contest banner that says OILDORADO QUEEN. The others each wear a banner saying MAID OF PETROLEUM. Later, two hours from now, after speeches by Oildorado presidents, mayors, and council members past and present, after all the testimonials to the community of Taft and the Oildorado tradition, these fourteen girls will dance a cancan routine to taped music, a high-stepping, side-kicking, skirt-lifting dance that will end with their backsides first to the speakers' table and then to the crowd. They are giddy with anticipation, jumping up and sitting down and hurrying out to the lobby. Now they all come to a temporary halt as the M.C. calms the room and asks us to stand while a portly minister intones the blessing.

Still standing, we put our hands over our hearts, face the flag, and say the pledge of allegiance. Then we sing "God Bless America." It has been a long time since I sang "God Bless America" before lunch. It feels good. One thing I will say about the people of Taft: They are not cynical. They do not intend the phrase "Maid of Petroleum" to have more than one meaning. They genuinely want America to be blessed by someone. And though relatively little of the profits from these mammoth oil reserves trickles into town, they are thankful when the fields come back to life, as they have in these past few years, because then Taft comes back to life. The mood on this particular afternoon in Veterans Hall is one of carefully nurtured prosperity.

During lunch—paper plates of fried chicken, potato salad, carrot-and-raisin salad, coffee or iced tea from pitchers—I talk to the plumbing contractor who moved here forty-four years ago. He wears a rodeo shirt and cowboy hat and has let his beard grow out for the whiskerino contest. Business has been good, but it is a mixed blessing. He complains about the hard time he has finding qualified help. "They all want to go work in the oil fields now. Out there they start at seven-fifty, and move up to nine, nine-fifty right away. Your best workers are going to head for the fields. So I've always got a new man to break in."

After lunch we drive out to the site of the Lakeview Gusher, south of town, on Petroleum Club Road. The distance is ten kilometers, a figure I remember from the program. Early Saturday morning there is to be a footrace, an event called The Lakeview Gusher 10K starting where the original derrick stood before the gush of oil and gas reduced it to splinters, and ending downtown.

The race is another tribute to that great explosion. The place where those racers will assemble, "The Site," is a built-up pit perhaps forty yards across. Its walls were originally made of timbers, sandbags, and dirt. A wetter climate would long ago have flattened these walls, but seventy years from now it will probably look pretty much the same. Sandbags can still be seen, tattered strips of burlap show through the dirt and the grimy boards. In the special edition of the *Midway Driller* there is a photo of two grinning, oil-stained men standing on an oil-encrusted wooden raft floating in a shiny lake of oil. They are poling from one side to the other. This caked and sandy pit Bill and I are standing in used to be that lake.

While we wander to the far side he is telling me the story, one he heard firsthand from an uncle who worked this field in the earliest years. Eventually both of us fall silent, kicking at the shards of oily sand. Something eerie hovers here, something reverent in the breeze across these scarred dunes, in the near-absence of motion or life where once there had been such a swarm, such frenzy.

Between the pit and roads stands a monument, with a plaque affixed, which says, "AMERICA'S MOST SPECTACULAR GUSHER." It's a California Registered Historical Landmark. If you squint, it could be a gravestone. Around here they talk about this gusher as if it had a life and an identity, the great creature who sprang forth, lived its wild existence for a year and a half, and suddenly died, of subterranean causes. Seventy years later there are photos everywhere—in the Oildorado program, in the special issue of the *Driller,* in shop windows downtown, and on permanent display in the new Westside Oil Museum—the lakes and rivers of oil, the oily workers rafting through it, the shattered derrick, the black spurting geyser of oil. They commemorate the early time, which was also the time of wildness, before this piece of earth had been quite tamed, and the sky-high gusher who could not be contained burst forth, made a huge and glorious mess, then disappeared or perhaps just retreated back into its cave.

I don't know why, but when you're driving around in the central valley, the songs from these country and western stations always seem to be providing some ironic comment on the landscape or the general situation. As we pull away from the gusher site, Bill switches on KUZZ, and it's an old Buck Owens arrangement called "Today I Started Loving You Again."

Buck is singing it. Merle Haggard wrote it. They both happen to be Kern County heroes. As we near the oil field village of Maricopa, Bill says in passing that he knows someone who went to high school with Merle in the days before he married Bonnie Owens. Everyone in Kern County seems to know someone who knows Buck Owens or Bonnie or Merle or all three. "She was married to both of 'em, you know. Of course, not at the same time."

By pure coincidence the one thriving business in Maricopa is called Buck's Steakhouse. No connection. "Best place to eat, for as long as I can remember," says Bill. From the look of things, it is the only place to eat. Taft and Maricopa started even, back in 1910, when they both incorporated. By 1911 Maricopa had its own opera house. In the mid-1970s, when they were filming *Bound for Glory,* Woody Guthrie's life story, Maricopa was chosen as the location for some early scenes set in the wind-blown west Texas town of Pampa during the 1930s. Very little had to be changed. It still has that sanded, worn-down Western patina. Only the cars are new, and some of the pickups outside Buck's.

Maricopa sits near Midway-Sunset's southern edge. We head due east now, along the base of the Temblor Range, running almost parallel to the California Aqueduct, which also swings eastward here, with its long flow from the Delta passing through these lowlands before making the salmon's leap over the Tehachapis and down again, to water Southern California. Along here the aqueduct waters more cotton fields. A few miles out of Maricopa we are driving through one of those tracts so vast your eyes burn trying to see the end of it. And right out in the middle of this small continent of cotton, about half a mile off the road, stands the drilling rig we have come looking for. With no pumps or derricks in view, the rig looks like some intergalactic vehicle that has landed in the wrong place.

It's part of Bill's beat to see how they are doing, how deep the hole is, and whether they've had any show. This is a wildcat well, he says, in a part of the valley that hasn't been drilled before. "There's probably some oil down there, but ten years ago you wouldn't have found anybody drilling that deep."

He takes a side road, looking for access, finds it, and we are easing along a dirt track between cotton rows toward a clearing where maybe half an acre has been opened up to make room for the platform, the caravan of trailers.

Bill has two hard hats in the trunk. We don these and climb the metal ladder. Everything is made of metal and painted battleship gray. The steel platform is thirty feet above the cotton, and rising a hundred thirty feet above the platform is the bolted network of struts and pulleys they call the mast. Climbing aboard you have the feeling of boarding a great vessel, anchored in the invisible waters of the inland sea.

Five men are working, the standard crew, all wearing hard hats and T-shirts, and smeared with grease and oil. Bill introduces me to Terry the tool pusher, the crew leader, who grins when I offer to shake his hand and shows me the palm thick with grease. Something about the way he stands would tell you he is in command, even if you hadn't been told in advance. His face is lean, his hair black and straight. He could be part Cherokee. Early thirties. He wears cowboy boots and jeans, no shirt. Without being muscular, his body looks powerful, whip-like. He stands with one foot forward, like a sailor on a rolling deck.

Bill mentions a man who died a couple of days ago, on another rig in some other part of the county. "I read in the paper that the cable crushed his chest."

Terry laughs and shakes his head, nods toward the draw works, the broad metal drum that winds and unwinds the cable that feeds up to the top of the mast, then down toward the center of the platform where the lengths of pipe are lowered or raised.

"I read that story," Terry says. "There is no way this cable is gonna crush your chest. If it breaks and comes whipping out of that drum it might knock you around some. It got me once. But it's not gonna crush your chest. What I figure is, they were pulling on the pipe, and it was the pipe broke loose and come free and swung out and got him. That pipe is what can crush you."

Any of this stuff could crush you. This is what Terry is ready to roll with, not an ocean, but the great chunks of moving metal that surround him. The drilling pipe comes in thirty-foot sections. They are slung from the cable on a lobster-shaped hook the size of a VW bus, which hangs directly overhead. Next to us, another large piece of thick steel is hanging loose, about the size of a Harley-Davidson. When I ask Terry about it he says, "All this is here is a great big pipe wrench. We just clamp it onto the pipe there to tighten the fit."

He shrugs it off, makes it simple. And it is simple. You put a bit on the end of a pipe, and you start cutting a hole in the ground. After awhile you screw on another length of pipe, and you cut a little deeper. It isn't the act that's impressive. Like everything else in the oil business, it's the magnitude. This rig down here at the absolute bottom edge of the San Joaquin Valley is running a pipe that is now twelve thousand feet into the ground, chipping and grinding through the next inch or foot of sand or shale or ancient fossil layer. What we have here is a brace and bit over two miles long, which is deep, Bill says. An average well in Kern County runs four to five thousand feet. Lakeview Number One came in at 2,225 in 1910.

Retracing our route through the rows of cotton we head back toward Taft, where I left my car. From there I follow Bill to Bakersfield and our final stop, a Basque place downtown called The Pyrenees, established in the days when Basque shepherds roamed the foothills east of here. The food is served family style, and the folks who run the place pile it on the table as if everyone who walks in is a shepherd just back from a month in the mountains or a roughneck coming off a seventy-two-hour shift.

This is Bill's favorite restaurant. The house beverage, Picon Punch, warms his county patriotism, stirs to life an epilogue for today, a prologue for tomorrow's parade.

"Kern is a kind of headquarters, you see. This is the biggest drilling year in the history of the state of California, and two-thirds of the new wells are here in this county—over two thousand wells. Meanwhile, oil people fan out from here in all directions. The contractor Terry works for has twenty-one rigs like that one we climbed. His main yard is outside town, but his rigs are trucked to Nevada, Wyoming, and into the Rockies where all the new exploration is going on. You take Terry himself. He grew up here in Bakersfield, started out as a roughneck, worked his way up to tool pusher. He just got back from Evanston, Wyoming. Couple of years ago he was working in Alaska. It's typical. Some of those fellows over in Taft, you would never know it to see them walking along the street, but they have been to Peru, to Arabia, Iran, and Venezuela, and they wind up right back home again."

The next morning we start out in separate cars, planning to meet at the reviewing stand. I am ten miles south of Bakersfield, whizzing along to Loretta Lynn's version of "I've Never Been This Far Before," and near the village of Pumpkin Center, when I smell the smell of warm rust, glance down and see my temperature needle heading off the gauge.

I pull over, pop the hood open, and gingerly ease the cap loose. Steam pours out but no water. There isn't much left. The top seam on my radiator has split. I make it to a phone and call the first place listed in the Yellow Pages that claims to be open on weekends. I cadge some water from a Freddy Fast-Gas and limp back to this radiator shop on the outskirts of Bakersfield, in the middle of a district that seems coated with rubber dust and filmed with oil, not from wells but from the generations of cars that have moved through its grimy jungle of transmission shops, upholstery parlors, abandoned service stations, warehouses, and wrecking yards.

I pull in with steam billowing around my hood and a trail of what could be diluted blood. The radiator man is sympathetic, a fellow in dark coveralls whose eyes cannot quite open, as if the lashes have been coated with honey. He says that since it is Saturday and he plans to close at 1:00 P.M. he will do the repairs only if I pull the radiator myself and install it again. I agree, making it clear that I am truly in a rush since my friend is going to be the Grand Marshal in the Oildorado parade. I emphasize this, figuring the local reference might enhance our relationship. He stops me right there.

"The what parade?"

"Oildorado."

"What the hell is that?"

Well, I think, this is curious. Here is a fellow whose entire livelihood depends on the internal-combustion engine, the heat it generates while burning gasoline, heat that must be cooled with water, which must circulate and sooner or later spring a leak, which he then is qualified to repair. And this fellow has never heard of the event I have driven halfway across the state to attend, the celebration of an industry that keeps not only this radiator man but a good part of California, if you will pardon the expression, solvent.

"You mean to tell me," I say, "that you work on cars all day long, every day, and you have never heard of Oildorado over there in Taft?"

Something in my tone unnerves him. We are both on edge. For a moment his eyelashes pull apart. I see indignation in there. "Hey," he says, "you want this goddamn radiator fixed?" Before I can reply he says, "Taft is thirty-five miles away. How the hell am I for Christ sake supposed to know what's going on over in Taft!"

I don't argue. It is important for the two of us to get along, at least for the next hour, which we do, working side by side, sometimes eyelash to eyelash, above and below the fittings.

By the time I reach Taft it is mid-afternoon. I have missed the parade. I have missed the Lakeview Gusher 10K Run. I have missed the Fly-in at Taft Airport and the barbecue at The Petroleum Club. Searching for Bill, I stop at Franklin Field, where the World's Championship Backhoe Contest is scheduled. It hasn't started yet. I happen to catch the most intense tug-of-war I have ever witnessed. These are not college kids pulling each other across a mudhole at the Spring Fling. These are some of the largest men in the state—truckers, hay-buckers, roustabouts, and derrick-men—thick men bursting through their T-shirts and playing for keeps.

One team is fully outfitted with matching blue and white jump suits, paratrooper boots for sure footing, and ball caps that say "Duval Sporting Goods." They have a coach, also in uniform, who paces back and forth in a half-crouch, muttering instructions before the match begins. They have some moves worked out, some hand and voice signals. They have prepared for this moment, and it is sad to watch those thick boot soles sliding through the sand as they are dragged across the line in less than thirty seconds by some ragtag group who have evidently organized at the last moment, who wear Adidas running shoes, whose T-shirts do not even match, but whose pecs and biceps would bring tears to the eyes of many regulars at Santa Monica's Muscle Beach. After a few rebel yells the winners swagger off toward the Budweiser truck, while the team from Duval Sporting Goods stand there frowning at each other, gazing at their shoes in bewilderment.

In the middle of the bare acreage called Franklin Field, sixty backhoes are lined up, their earth-scraping bulldozer blades drawn in low at the back, their crane-arms crooked high in front, scoops at the ready, waiting for the championship to begin. The cabs are empty. On one windshield a sticker says "Iran Sucks." In a half-circle a thousand fans wearing cowboy hats stand around sipping Budweiser and waiting too. This is where I finally find Bill. I recognize him by his bare head. He is carrying his hat. I still have not seen him wearing it.

He doesn't take my absence personally. When I explain what happened he laughs and says it reminds him of the old Merle Haggard tune "Radiator Man from Wasco," which is set in a Kern County town north of Bakersfield, up Highway 43.

"Sounds like you've been reliving Merle's song," he says.

I ask him if the parade committee had made him ride a horse.

"Nope, a fellow here in town provided a Cadillac, which would have been real comfortable if the top was down. It wasn't a convertible, though. The roof was so low I couldn't get my cowboy hat on. In that respect I guess I lucked out. My wife, Frankie Jo, was with me, so we all three sat hunched in the front. The fact is, I am weary. Waving to so many people, trying to keep a smile on for an hour and a half, and then not sure anybody can even see you, under a low roof like that."

Bill grins all through this account, amused by the parade, and narrows his eyes as if faced with heavy weather, which at the moment happens to be true. It is clouding up. Warm heavy clouds have filled the sky.

"I'm glad I don't have to do this again," he says. "Five years from now, it'll be somebody else's turn. They only have Oildorado every five years, you know. For a while it was annual. But people were running out of time for anything else."

I have a couple of beers and wait for the backhoe drivers to mount their rigs. I can only guzzle so much Bud, since I plan to start north this afternoon, sooner or later, and there are mountains to climb. I decide on sooner. It is looking more and more like rain, and the backhoes are still sitting like tanks on D day waiting for the signal from Eisenhower. Something has delayed the contest. No one is certain what, and no one much cares as long as the Bud holds out.

I say good-bye to Bill and start out 33 through the Midway-Sunset, making one last quick stop at Fellows, another oil-field village five miles away. Fellows reminds me of the pueblos in New Mexico, the ones you pass driving out of Albuquerque—no roadside billboards in any direction, no pitch to the motorist, no hotels or motels or fast-food neon, no Rotary lunch. It's just a village, a cluster of low buildings a mile or so off the road, out there all by itself in the high desert. Pumping wells dot the slopes beyond town, where the Temblor Range begins to rise. More wells decorate the plains spreading south. There is no main street in Fellows, no grocery store. Where the houses stop, I find what I have left the highway for, another gusher site Bill has recommended, another monument stone, another plaque:

> THE FIRST GUSHER
>
> Midway Field 2-6, which made the Midway Oil Field famous. Blew in over the derrick top, November 27, 1909, and started the Great California Oil Boom. At its peak it produced 2000 barrels a day.

Lakeview was the biggest and wildest. Midway 2-6, coming four months sooner, gets credit for being first. The granite block stands by itself inside a low fence. The air and land nearby is strangely quiet under the lowering sky, punctuated by a faint creaking from the nearest well, where the cable rubs once in every cycle. No wind this day, no dogs, no cars. Perhaps everyone who lives in Fellows went off to Franklin Field. In the hills and plains around the monument where the first gusher blew in, nothing moves but the bowing pumps, the silent praying of a thousand mantises near and far across the western San Joaquin.

William Rintoul

THE GANGPUSHER

JAKE LUTTRELL the gangpusher reminded Chris Olmsted of a comic-strip version of the foreman. He had a thick neck that was beginning to get stringy, a barrel chest that had slipped and an earnest face that showed no appreciable enlightenment. Bi-focals added the only incongruous touch. They made him look benevolent.

Olmsted kept his thoughts to himself in the security of the maintenance superintendent's office, where he had just been introduced to Luttrell. In Luttrell's roustabout gang, he would begin the first phase of the six months' training program that was supposed to prepare engineering graduates for management.

There would be a predictability, Olmsted supposed, about the gangpusher's method of dealing with any who crossed him, his blind loyalty to Conestoga American Oil, Inc., probably even his swearing. When it came time to write the bi-monthly report that was required of trainees, he would have little trouble finding something to write about. If he kept his eyes open, he'd probably be able to suggest some ways the gang could do its work better, which shouldn't hurt his standing with superiors.

As he followed Luttrell out of the office, Olmsted couldn't help contrasting the gangpusher with the company representative who had come to college to interview seniors. That man had smoked a pipe and worn an expensive suit. He was very informal and very interested in everything Olmsted had done. He spoke of the work men he called "our people" were doing: of subterranean fires they set to boil oil out of the ground, of closing loops with automation, of completing oil wells on the ocean floor. He had made it sound as if it would be the finest thing in the world to be on the same team with "our people."

That man would be off now smoking his pipe on another campus, Olmsted supposed. He was the recruiting sergeant; Luttrell the drill sergeant. As Olmsted approached the black-top where an awkward-looking truck with A-frame waited, he wished his work clothes did not look so new, his gauntlet gloves so obviously unused.

The gang truck carried him in company with Luttrell, the driver and two roustabouts over a rough road to an oil field, where the first stop was beside a Rube Goldberg maze of valves and gauges that Olmsted recognized as the surface controls for a newly-drilled well.

Olmsted watched Luttrell expectantly.

The gangpusher walked slowly to the well, taking care not to step in soggy patches where rotary mud and water remained from the drilling operation. He turned a valve, looking at the same time toward the end of a pipe that ran from the well-head to a shallow earthen pit. A stream of watery brown fluid sputtered from the pipe, raining into the pit. Apparently satisfied, Luttrell closed the valve and looked at distant pump units that gave the appearance of so many grazing horses, rhythmically dipping their heads. He stared at the steel mast of a drilling rig, more than a hundred yards away, watching puffs of exhaust issue from the diesels. He began walking toward the rig.

Perplexed, Olmsted looked back at the truck. He was startled to see the driver and the two roustabouts running toward the bin full of pipe fittings. The roustabout whom the others called Sallisaw took the lead, separating fittings into two piles, one larger than the other. He pushed the larger pile toward one of two vises on the back of the truck. He looked at Olmsted. "Give Vernon a hand," he said, pointing to the smaller pile. "He knows what to do." Sallisaw looked sharply at the roustabout. "Scrubber, gate, check-valve. Remember?"

Olmsted looked at the drilling rig in time to see someone wave at Luttrell.

Sallisaw followed Olmsted's gaze. "You're lucky," he said. "They could of put you in Albert's gang. He breathes down your neck." Sallisaw looked at Luttrell's retreating figure. "Jake's the best pusher they got." He turned abruptly to the fittings. "Let's make up the header."

The job, Olmsted found, involved screwing together in series the valves and other fittings that would handle separate flows of oil and gas. He wondered at the way the others worked, hurrying as if at any given moment they expected someone to blow a whistle and shout "time." It seemed only a short while before he and Vernon finished their series, Sallisaw and the driver, whose name seemed to be Broyles, theirs. Sallisaw took charge of putting the two series together in parallel. The four of them, virtually running, carried the header to a position on the ground several feet from the well. Sallisaw completed the hook-up.

Olmsted was surprised to see Vernon put down his pipe wrench, climb onto the truck bed and lie back against the rag bag as if he meant to take a nap. Broyles opened his lunch pail and took out a paperback book whose cover pictured a bosomy girl who had lost her brassiere. He sat in the cab. Sallisaw looked at a distant Chinaberry tree around which the ground was littered with what appeared to be the remnants of a house foundation and the discarded debris of departed occupants.

"Last time," Sallisaw said, "I found a collector's item. A pill bottle shaped like a heart, right in the middle of a bunch of rusty cans. The sun had turned it the prettiest violet you ever saw." He started toward the tree.

Alone, Olmsted nervously sat on the ground, resting his back against a truck wheel. It seemed almost an hour before he saw the gangpusher returning. He quickly stood up, wondering if he should sound an alarm.

"Childress fell from a derrick at Round Mountain," Luttrell said somberly, walking toward the water can. "Oscar's wife had a baby. A girl." He filled a paper cup. "He's pretty proud. The boys told him, hell, that's nothing. He had the model right in front of him."

Sallisaw approached, carrying a small banjo-shaped bottle that was tinted a delicate violet.

"Whiskey bottle," Luttrell said. "Thirty, forty years since they used that kind." He peered into the cab. "About through with the chapter?" As if by afterthought, he walked slowly to the header, casually inspecting it. He turned suddenly. "Let's pick up the tools," he said to no one in particular. "Let's wind 'er up."

In the days that followed, Luttrell's gang made up headers, set pump units, shoveled oil sand, repaired leaks, dug deadman holes and ran a go-devil through a choked pipe line.

Olmsted kept a list of jobs, making notes at night on what the gang had done and why they did it. He balanced the exhaustion he felt after shoveling oil sand by telling himself he was lucky to work at a variety of jobs, the problem of cleaning crude oil from himself and his clothes by reminding himself it was valuable experience that would give him important insights when he became a member of the management team.

Increasingly as he read his notes he felt he'd left out the most interesting parts. The job at Bitterwater Creek, for instance. Before Olmsted finished digging the deadman hole to anchor the guy line from a drilling mast, the noon whistle blew. The gang ate their sandwiches in the shade of a wooden derrick. After they were through, Olmsted idly asked Luttrell how long it would have taken to build the derrick.

Luttrell looked meditative. He started talking. He told of how men built derricks, of how the company bought a suit for each man in the crew if the well was good, of how far the tongs would knock a man (he said he had the scars to prove it), of wild wells that blew oil over the countryside for days on end. When he stopped, it was forty-five minutes past time to go back to work. Sallisaw and Vernon, who had lain on the rough planking of the derrick floor with their eyes shut, looked gratefully at Olmsted.

Olmsted wondered if he was supposed to include things like that in the report.

A job took the gang to a cluster of ancient looking tanks. A spider's web of pipelines converged from surrounding wells on a manifold through which oil funneled to the tanks. Nearby a dilapidated boiler stood under a tin-roofed lean-to.

Luttrell told Broyles to replace a leaky manifold valve which had saturated the ground below with oil. He sent Sallisaw and Vernon behind the tanks to shovel oil sand into rabbit holes that riddled the earthen dike. He told Olmsted to install a new safety on the boiler.

As Olmsted fit the jaws of a pipe wrench on the union below the safety he was to replace, he was not surprised to have Luttrell take up a position nearby. The gangpusher frequently seemed to give him jobs that separated him from the rest of the gang. On such occasions Luttrell would delay his solitary wandering to explain the job and other facets of oil work. Olmsted was grateful, not only for the chance to absorb the gangpusher's schooling but also because Luttrell picked such times when there was not an audience.

"Gunk," Luttrell said disparagingly.

"That's what this oil is. It'd come up in chunks if they didn't heat it." He pointed at a glass tube on the outside of the safety. "Samples the heat. When it gets too hot, the thermostat opens a pop-off valve."

"What's wrong with the safety?"

"It won't work."

"How would the boiler go? Straight up?"

"The stack might. The chamber's made to sail out backwards. That's why it's pointed away from the tanks."

Olmsted resolved to remember what the gangpusher told him. It would make his reports sound knowledgeable.

"Wasn't all that long ago," Luttrell said, "one blew on Section Thirty-One. They found the boiler a quarter of a mile from the tanks." He pointed at the safety Olmsted was preparing to install. "All because of a defective safety." He looked sharply at the safety, scowling.

Luttrell pointed to a beaded weld mark on the back. "Reclamation," he said. "Cost them a boiler and they sent it back." He angrily put the safety on the ground, turned and walked toward the truck.

Olmsted watched the gangpusher rummage through the metal tool chest. Luttrell took out the sledge hammer.

"You send 'em in," Luttrell said, returning, "they patch 'em up and send 'em back." He raised the sledge hammer and brought it down. There was a shattering sound as glass scattered over the ground. Luttrell raised the sledge hammer and brought it down again.

Olmsted looked sideways at the battered safety. How had he meant to describe its workings?

The day before the report was due, the gang truck pulled up on the edge of a dusty scar on the earth where a crane and pickup waited. The shadow of the crane's boom fell across a partially-crated pump unit. In the background, there was a well over which the pump unit obviously was meant to be installed. A wiry man in greasy coveralls climbed out of the pickup.

Luttrell approached the crane, scowling. He joined the man beside the pickup. The two talked. Olmsted could not hear what they said, but it looked as if they were arguing. The man climbed into the pickup and drove away, turning to follow the road that paralleled the company's private telephone line.

Luttrell looked scornfully at the crane. "They must think my elevator don't run all the way to the top," he said.

The dust behind the pickup had not settled long before another vapor trail of dust appeared on the road. The dust spurted up so fast as to leave no doubt the driver was in a hurry. The car that pulled up was a business coupe with a company emblem prominent on the side. Olmsted recognized the maintenance superintendent.

"What the hell's the difficulty?" the superintendent shouted.

"Do you expect me to lift a heavy piece of equipment with that?" Luttrell pointed at the boom.

"It's adequate," the superintendent said.

"Ever seen a boom buckle?" Luttrell paused as if for effect. "Do you know what happens to the boys guiding the load?"

"Jake, do you think I'd tell you to do a job that's not safe?"

"Send a bigger crane."

"How the hell's it going to look at home office when they get the bill?

How's it going to look if I have to attach a memorandum saying one of our gangpushers refused to do his job?"

"Be sure and send a copy to Mr. T.C. Archibold."

Olmsted remembered Mr. Archibold was the company president.

The superintendent looked exasperatedly at the roustabouts as if he were seeking support. His eyes met Olmsted's. Olmsted looked away, flustered.

"Pete's on another job," the superintendent said, looking curiously calm. Pete operated the company's large crane. "He'll be over when he's through." There was a foreboding sound to his voice.

That night Olmsted sat down to write his report. The first part would be easy, he thought. He would simply recount the jobs he had worked on and the relation each bore to producing oil. He stared at the patternless linoleum on the floor of the bunkhouse room. What should he write after that? Should he report the gangpusher wasted time, smashed company equipment and refused to do what the superintendent ordered?

Olmsted studied the limp shoe string some previous occupant of the room had tied on the grill high on the wall as a gauge of whether air was entering the room.

What if he wrote an innocuous report? What would the superintendent think if he saw a report that ignored the way the gangpusher acted about the crane? Would the superintendent pass along the word that Olmsted, the new trainee, was not a company man? What if it was well known how loosely Luttrell pushed his gang? Maybe there was more than met the eye to putting a trainee in Luttrell's gang. Maybe management wanted to find out early how observant he was. Or whose side he was on.

Olmsted looked at the blank paper. He didn't have a choice. To hell with the superintendent. To hell with Luttrell. To hell with his own chances. There was only one way to handle the situation. Be honest.

He picked up the pen and began to write. The place to begin was the first day when Luttrell walked off, leaving the gang to its devices. The day Sallisaw found the whiskey bottle. The collector's item. Good-hearted Sallisaw. He'd called Luttrell the best pusher the company had. What would Sallisaw do if he had to work for a real company man?

Olmsted stopped.

What would Sallisaw do? Would he work as hard and fast as he did? Or would he dog it? Take more time to get the job done than he took now, including the time for loafing?

Olmsted crumpled the paper, feeling vastly relieved.

Wilma Elizabeth McDaniel

RETURN TO TAGUS RANCH

The old man wanted
    to see again
whatever was left
of Tagus Ranch
before my eyes are gone
    he wrote his sons

And they came from Lodi
and Santa Barbara
    for Father's Day
and took him
in Derrel's Mercedes

To a place of bewilderment
bypassed by a freeway

where is the big white house
like a southern mansion
where we got our pay
                in scrip
the old man begged to know

Someone said it burned down
Dad, several years ago
but the old man wouldn't hear

He looked out the window
of the Mercedes
and mused to himself

I picked so many peaches here
in 1936
my blood tested peach brandy
at Tulare County Hospital

Afterward
Derrel told his brother
Bobby Gene
we made a hell of a mistake
bringing Dad out here
I feel depressed myself.

VIEWING KERN COUNTY DESOLATION

Uncle Phylo
made a mournful
face
and sounded even
sadder
when he said Them
builders has ruined
this land
not even room
for a jackrabbit to hide

Gerald Haslam

OILDALE

TO GAZE DEEP WITHIN myself, I walk the streets of Oildale. On December 23, 1985, a frigid tule fog obscuring all but the nearest yards and houses, I wander up McCord Street—no sidewalks here, but many trailer parks in an area where once Dust Bowl migrants built hovels—and I notice two banty roosters scratching and jerking on a patch of brown bermuda grass, eyeing one another but ignoring, as nearly as I can tell, two wild doves feeding with them. I stop and for a few moments stand motionless, watching while my breath bursts white, then one dove flies with a muffled whir and the other poses warily. Both roosters pause, fighters awaiting the bell.

With a hollow warble, the second dove climbs swiftly into the gray surrounding us, leaving me on the sandy border between lawn and street. I notice then two pickups parked on the far border of the bermuda, one ancient and huge, the other new, small and crowned with four yellow fog lights. On the older vehicle's blotched surface someone has painted "Trust Jesus." A bumper sticker on the new one says, "If you love something, set it free. If it returns, it is yours. If it doesn't, hunt it down and kill it." The smaller pickup is metallic red, as are the roosters.

The neighborhood through which I wander is called Riverview because before a dam was constructed in the mountains east of town, you could view the Kern River flowing through your kitchen during wet winters. It was once considered the least prosperous section in this unaffluent and unincorporated community. Today, although new housing developments sprout on Oildale's outskirts, many of the same unpainted shanties I used to see here as a kid in the 1940s remain erect and apparently unchanged—like tribal elders, reminders of our collective past. Two or three rickety lawn chairs—poor people's air conditioners—sit in front of many such resi-

---

dences, as do cars or trucks whose grandeur often contrasts sharply with the setting.

Such older houses are now bracketed by ubiquitous trailer parks and what realtors like to call upgraded houses. The former feature everything from modern mobile homes complete with metallic awnings and metallic porches, to geriatric travel trailers, faded and frayed. Most of the upgraded houses are carefully painted and their yards may be tended by dark-skinned men from other communities; moreover, there seem to be more and more signs announcing security systems, indicating perhaps a seige mentality, the long shadow of hard times past.

Now contiguous with Bakersfield, Oildale grew north of the larger town during the early years of this century. Lawrence Clark Powell worked in the area in the early 1920s and even then, he tells me, "We learned to leave the Oildale guys alone, thank you." It was an enclave of oil-company camps, attracting a disproportionate number of males who did hard, physical labor, and pursued rough, masculine diversions. Except that agriculture and not petroleum was the principal lure, this was also the pattern for much of the San Joaquin Valley, where waves of migrants have been attracted since the 1870s not by gold nuggets or movie careers but by the availability of what can only be called toil: Chinese, Japanese, East Indians, Mexicans, Filipinos, Blacks, plus many varieties of Whites.

Despite their resulting heterogeneity, most agricultural towns have by no means been racially integrated but have at least hosted residents of varied colors, whereas the oil industry, unofficially but actually, did not welcome nonwhites. So towns like Taft, Coalinga and Oildale developed racist reputations. In my youth there was even said to have been a sign—which no one ever saw but everyone talked about—on the outskirts of town: "Nigger, don't let the sun set on you here."

In any case, my hometown's renown as a rough section intensified following the so-called Dust-Bowl migration of the 1930s when large numbers of Southwesterners settled here. This is Merle Haggard's home town; Buck Owens' Enterprises is a major local business. Today, fifty years after that migration began in earnest, and now boasting its own hospital, its own high school, its own civic organizations, Oildale is nonetheless spoken of by local liberals as a redneck enclave.

A close friend of mine—a Bakersfield boy who understands well what my hometown *means* locally—mentioned the other night that his therapist had suggested that Oildale is a crucible for fascism, which might simply mean that people here voted for Ronald Reagan, hyperbole being what it is. But my friend guessed that the three R's—racism and rowdiness and the right to bear arms—were troubling her, so he rattled the therapist's cage, telling her about the night a group of us, all high school pals, had driven to my house from Bakersfield. In our exaggerated sense of adventure, we had suggested that the one black among us duck as we entered Oildale—he'd laughingly complied—then we had dashed from the car into my place where we'd spent the night.

While our dramatics were unnecessary, they symbolize an aspect of Oildale's lingering reputation among those who do not live here: it is said to be an environment unconducive to notions as diverse as affirmative action, gun control, cigarette warnings, and seat belts. More to the point, Oildale has been to Bakersfield as Bakersfield has been to California, a scapegoat; "You're from *Oildale?*" I've heard at genteel parties, tone saying it all.

It is also what thin-wristed experts like to call a working-class area, and it remains predominantly white. Because so many of Oildale's citizens over the years have been fair-skinned Southwesterners, lovers of country music and the self-serving version of patriotism it posits, the community has been assigned a gothic Southern stereotype. This has been aided by the more important fact that many white migrants were poorly educated, products of generations of yeomanry, so they had to compete with nonwhites for jobs on nearby farms or work in the now-integrated oilfields. More than a little local pontification on matters racial has been in fact an expression of economic fear.

Racism, as well as other narrowness, hangs on most desperately among the desperate, but in this state it cannot be easily separated from issues of social class, for the latter, usually unspoken but as real and as certain as the surging of sex, often triggers racist regression. Here in California some nonwhites—the number need not be large, only visible—have been able to take advantage of the state's educational system to escape chronic destitution and assumed inferiority. While the society as a whole benefits from such a development, to whites stranded on that same desperate level, the underclass, even the slightest gain for nonwhites is clear evidence that something is wrong with America: *This is white country but a damn Mescan's bossin' me. Shee-it!* This is one reason why racist organizations tend to contain so many marginal members rather than men and women of accomplishment; the former are the threatened ones.

In the court of $225-a-month houses across the street from my folks' place, I see fair-skinned young men with long unkempt hair, bearded, disheveled, angry after three beers at a world that does not offer them well-paid jobs or much prestige, but does provide drug dealers to rip them off and does provide candy-bar and soda-pop lunches. They carry homemade tattoos on their knuckles, and their shoulders are splendid with murals of nude women on horses, but few high school diplomas grace their mantels: school sucks, man. If asked, they will often reveal that they are about to tell someone off or to kick someone's ass.

Christmas Eve, 1984, I walked out of my mother and father's small house and heard howls and screams from the courts: one young man was beating another in the street while two women shrieked and a third man yelled encouragement. The puncher was shouting over and over at the punchee something like "Take my fuckin' money!" It was not a new scene to me, but that night it struck me: those cries should have echoed through the halls of Congress, or through that therapist's office, because they were battling each other in lieu of opponents they could neither see nor understand. And I, raised on this street, having seen my father fight here and having bled here myself, fresh from a comfortable Christmas celebration with my family, I knew exactly what was going on, and was swept by imprecise guilt along with enormous gratitude for my own good fortune. But I did not allow myself to say, "There but for the grace of God go I," for I did indeed go there, or at least some part of me did. A moment later, a sheriff's cruiser pulled up and I returned to the warmth of family within my parents' house, reminded as my hometown frequently reminds me of the proximity and the possibility of poverty, and of its consequences.

Not only young men dwell in those $225-a-month units. My own grandmother lived in one—$50 a month then—and those small houses have long been refuges for the impecunious old. Many young women reside there too, more all the time it seems, often single mothers, also poorly educated and often tattooed—a small butterfly on a shoulder, a rose on one breast. In their too-young, too-fleeting primes they may combine bad teeth with bodies that make men gnaw chrome, but their boyfriends gnaw something else and soon babies ride on each hip; with them come food stamps, hours watching daytime television and, usually, revolving males who cannot support themselves, let alone families; and with them may come the unfocused outrage that accompanies an erosion of hope.

As is true of people floundering at the bottom, these young and old, women and men, tend not to see over the rim to reality, so they remain frustrated by and angry at a world that offers them only blue-light specials. And when things go wrong, as they so persistently do, someone must be blamed: mother dies and the damned doctors are responsible; the car doesn't run and the dirty Japs are guilty; I don't get the job because that other bastard has suck. Niggers cause this, and Jews, and slopes, but mostly niggers because blaming blacks has long been an acceptable way for lower-class whites to vent general grievances. Anyway, Rambo or Jimmy Swaggert or the Klan will save us—white men banded together. And no niggers better move into Oildale because this is white people's territory; at least we're better than niggers. It is an irrational, probably unavoidable stance, one held with the desperate, uncritical grip of divine revelation: it *must* be true.

My own parents, for reasons I've never fully understood—his better education, probably, and her Latin attitudes—but for which I remain constantly grateful, did not indulge in such delusion. Instead, they taught me to accept people as individuals; my dad's dictum, for example, was "Is he a good guy?" not "Is he colored?" Thus Quincy Williams and Freddie Dominguez and John Takeuchi slept and ate at our house, just as did Raymie Meyer and Ernie Antongiovanni and Tommy Alexander. If my folks had only given me that I would have been well served.

Poverty and race and class churn a bitter stew: history dictates that a much larger proportion of blacks than whites inhabits the dungeon of unabating want, so race has been and remains an effective camouflage for our system's inability to reach many endemic poor. Because it is convenient to keep the hungry fighting one another, racism is frequently *part* of poverty, just as idealized egalitarianism is *part* of liberalism: default assumptions.

I must challenge my pal's therapist: Oildale is not a breeding ground for fascism, but poverty certainly is, poverty and ignorance and hopelessness so deep that education and government programs cannot deflect it. My hometown is a place where low-rent housing and the rumor of jobs for the unskilled has traditionally attracted whites on the bottom hoping to struggle toward the middle.

Those who do make that transition may carry lower-class fears and prejudices longer than is conscionable, but the real problem is that others never make it. In Oildale you cannot be unaware of this nation's class system because this is a cusp where hopelessness and hope, or at least the *hope* of hope, abut. When even that slimmest of threads frays, despair engulfs and violence erupts, in Oildale as elsewhere.

Therein lies the rub. Without this community and its unpretentious styles, many people would be utterly abandoned by a society that fears fascism or attacks socialism but dreads and distrusts losers most of all, would be ignored by an elite often willing to love the poor only as it imagines them, not as they are: hopeless as well as hopeful. Since the poor exist, they must exist somewhere, and Oildale predictably harbors a proportionate number of losers: *Who's got the dope, man?*

But the losers constitute a small if visible minority here. I stride past the Assembly of God on Wilson Street, and from within I hear voices raised to a God who can accept, and I hear people—some of them from the courts, grown broad-hipped and repentant—praying for me, their brother burdened by sins unknown to them, but burdened, surely burdened, because that is the human condition and with God's help it can be endured. In that congregation now as in the neighborhood, one might find a sprinkling of dark faces reflecting the slow erosion of stereotypes as well as the enduring truth that class profoundly influences the acceptance of nonwhites—too rich or too poor excluded. Unfortunately, the continued lack of blacks also illustrates the persistence of America's most heinous racist illusion.

What fearers of fascism forget is that most of Oildale is populated by folks who have established themselves in the middle class by dint of hard work, survivors whose daughters now aim for honor roll and university, whose sons play football and fight wars. Oildale's citizens pay their taxes, frequently resent welfare and shake their heads at punk rock, at "Fit 'n' Forty" medallions, at sprout sandwiches, but accept the churning present anyway. When I stroll through town, nearly everyone I meet says hello, often with a Southwestern drawl. Folks I've never seen before discuss the weather—"This *dern* fog . . ."—while waiting for traffic to pass. Their children have no drawls at all.

I notice other things on my visits home, little glimpses I no longer take for granted. On North Chester Avenue, aging men and women who struggled here from Oklahoma now drink coffee in a McDonald's that is identical with several in Tulsa. There seems to be an inordinate number of "Beware of Dog" signs, especially on those blocks populated by folks who have edged into the middle class. And on winter mornings if fog doesn't obscure the world, steam plumes rise from the hills north where, on the leases that lured so many here, leases where my father and I worked, heavy oil is liquified by the hot vapor then pumped to the surface.

It seems to me now that older women always boast hair care comparable to matrons in Palm Springs, their tresses freshly dyed, curled, piled and sprayed. Coiffures are taken seriously even if bodies are not. The other morning, at about 9 A.M., I passed a gal unlocking the front door of a beer bar, her hair deep red and elaborately swirled high above a leathery face imprinted "Texas, 1914." A cigarette dangled from her crimson lips when she smiled at me. I smiled back.

And there is an aggressive angle at which many men wear hats—billed caps, straw Stetsons, but no plaid snapbrims or berets—that advises you not to let your mouth overload your ass. Short sleeves may be rolled up, biceps bulging, and beer drinking after work seems sacramental. Cars and guns are icons taken seriously, offering reasons for taking oneself seriously.

It is finally the mix of people hereabouts that most compels: friendly, plain-spoken, conservative, protestant in work ethic if not religion, scarred but not embittered by hard times, they constitute what I like to call a "front porch" society. In the days before air conditioners tamed the long, scorching summers, it was common for neighbors to gather on porches, drink iced tea, perhaps play cards or checkers, occasionally sing to the accompaniment of a guitar or banjo, but usually just talk. Many older folks still do these things, such gatherings livening balmy summer evenings, while kids clatter up and down sidewalks on skateboards now rather than skates.

Oildale is changing as the rest of California is, but its reputation is not altering apace. There is a sidewalk in front of my parents' place, built by the WPA in 1941, or so says the inscription on the corner. There is also a small parking strip there with a tree that has been so brutally trimmed that scar tissue knots it like tumors—a peculiar local style of arboreal coiffure that seems more ritual maiming than practical necessity. Below it, on summer mornings, runoff water from lawn irrigation settles in the gutter to make a small pond, and every morning if I arise early I can sip coffee with my dad and watch doves drinking out there—an unchanging reality—dipping their fawn heads, bobbing their white-splashed tails. They rise with a whir when a truck bounces past on its way to the oilfields. Then the birds return, drink again and occasionally call—a haunting, hollow sound that says "Home."

Lawson Fusao Inada

MEMORY

Memory is an old Mexican woman
sweeping her yard with a broom.
She has grown even smaller now,
residing at that vanishing point
decades after one dies,
but at some times, given
the right conditions—
an ordinary dream, or practically
anything in particular—
she absolutely looms,
assuming the stature
she had in the neighborhood.

This was the Great Valley,
and we had swept in
to do the grooming.
We were on the move, tending
what was essentially
someone else's garden.
Memory's yard was all that
in miniature, in microcosm:
rivers for irrigation,
certain plants, certain trees
ascertained by season.
Without formal acknowledgment,
she was most certainly
the head of a community, American.

Memory had been there forever.
We settled in around her;
we brought the electricity

---

of blues and baptized gospel,
ancient adaptations of icons,
spices, teas, fireworks, trestles,
newly acquired techniques
of conflict and healing, common
concepts of collective survival . . .

Memory was there all the while.
Her house, her shed, her skin,
were all the same—weathered—
and she didn't do anything, especially,
except hum as she moved;
Memory, in essence, was unmemorable.

Yet, ask any of us who have long since left,
who have all but forgotten that adulterated place
paved over and parceled out by the powers that be,
and what we remember, without even choosing to,
is an old woman humming, sweeping, smoothing her yard: Memory.

Gary Soto

HISTORY

Grandma lit the stove.
Morning sunlight
Lengthened in spears
Across the linoleum floor.
Wrapped in a shawl,
Her eyes small
With sleep,
She sliced papas,
Pounded chiles
With a stone
Brought from Guadalajara.

After
Grandpa left for work,
She hosed down
The walk her sons paved
And in the shade
Of a chinaberry,
Unearthed her
Secret cigar box
Of bright coins
And bills, counted them
In English,
Then in Spanish,
And buried them elsewhere.
Later, back
From the market,
Where no one saw her,

She pulled out
Pepper and beet, spines
Of asparagus
From her blouse,
Tiny chocolates
From under a paisley bandana,
And smiled.

That was the '50s,
And Grandma in her '50s,
A face streaked
From cutting grapes
And boxing plums.
I remember her insides
Were washed of tapeworm,
Her arms swelled into knobs
Of small growths—
Her second son
Dropped from a ladder
And was dust.
And yet I do not know
The sorrows
That sent her praying
In the dark of a closet,
The tear that fell
At night
When she touched
Loose skin
Of belly and breasts.
I do not know why
Her face shines
Or what goes beyond this shine,
Only the stories
That pulled her
From Taxco to San Joaquin,
Delano to Westside,
The places
In which we all begin.

LIKE MEXICANS

My grandmother gave me bad advice and good advice when I was in my early teens. For the bad advice, she said that I should become a barber because they made good money and listened to the radio all day. "Honey, they don't work como burros," she would say every time I visited her. She made the sound of donkeys braying. "Like that, honey!" For the good advice, she said that I should marry a Mexican girl. "No Okies, hijo"—she would say—"Look my son. He marry one and they fight every day about I don't know what and I don't know what." For her, everyone who wasn't Mexican, black, or Asian were Okies. The French were Okies, the Italians in suits were Okies. When I asked about Jews, whom I had read about, she asked for a picture. I rode home on my bicycle and returned with a calendar depicting the important races of the world. "Pues si, son Okies tambien!" she said, nodding her head. She waved the calendar away and we went to the living room where she lectured me on the virtues of the Mexican girl: first, she could cook and, second, she acted like a woman, not a man, in her husband's home. She said she would tell me about a third when I got a little older.

I asked my mother about it—becoming a barber and marrying Mexican. She was in the kitchen. Steam curled from a pot of boiling beans, the radio was on, looking as squat as a loaf of bread. "Well, if you want to be a barber—they say they make good money." She slapped a round steak with a knife, her glasses slipping down with each strike. She stopped and looked up. "If you find a good Mexican girl, marry her of course." She returned to slapping the meat and I went to the backyard where my brother and David King were sitting on the lawn feeling the inside of their cheeks.

"This is what girls feel like," my brother said, rubbing the inside of his cheek. David put three fingers inside his mouth and scratched. I ignored them and climbed the back fence to see my best friend, Scott, a second-generation Okie. I called him and his mother pointed to the side of the house where his bedroom was a small aluminum trailer, the kind you gawk at when they're flipped over on the freeway, wheels spinning in the air. I went around to find Scott pitching horseshoes.

I picked up a set of rusty ones and joined him. While we played, we talked about school and friends and record albums. The horseshoes scuffed up dirt, sometimes ringing the iron that threw out a meager shadow like a sundial. After three argued-over games, we pulled two oranges apiece from his tree and started down the alley still talking school and friends and record albums. We pulled more oranges from the alley and talked about who we would marry. "No offense, Scott," I said with an orange slice in my mouth, "but I would never marry an Okie." We walked in step, almost touching, with a sled of shadows dragging behind us. "No offense, Gary," Scott said, "but I would *never* marry a Mexican." I looked at him: a fang of orange slice showed from his munching mouth. I didn't think anything of it. He had his girl and I had mine. But our seventh-grade vision was the same: to marry, get jobs, buy cars and maybe a house if we had money left over.

We talked about our future lives until, to our surprise, we were on the downtown mall, two miles from home. We bought a bag of popcorn at Penneys and sat on a bench near the fountain watching Mexican and Okie girls pass. "That one's mine," I pointed with my chin when a girl with eyebrows arched into black rainbows ambled by. "She's cute," Scott said about a girl with yellow hair and a mouthful of gum. We dreamed aloud, our chins busy pointing out girls. We agreed that we couldn't wait to become men and lift them onto our laps.

But the woman I married was not Mexican but Japanese. It was a surprise to me. For years, I went about wide-eyed in my search for the brown girl in a white dress at a dance. I searched the playground at the baseball diamond. When the girls raced for grounders, their hair bounced like something that couldn't be caught. When they sat together in the lunchroom, heads pressed together, I knew they were talking about us Mexican guys. I saw them and dreamed them. I threw my face into my pillow, making up sentences that were good as in the movies.

But when I was twenty, I fell in love with this other girl who worried my mother, who had my grandmother asking once again to see the calendar of the Important Races of the World. I told her I had thrown it away years before. I took a much-glanced-at snapshot from my wallet. We looked at it together, in silence. Then grandma reclined in her chair, lit a cigarette, and said, "Es pretty." She blew and asked with all her worry pushed up to her forehead: "Chinese?"

I was in love and there was no looking back. She was the one. I told my mother who was slapping hamburger into patties. "Well, sure if you want to marry her," she said. But the more I talked, the more concerned she became. Later I began to worry. Was it all a mistake? "Marry a Mexican girl," I heard my mother say in my mind. I heard it at breakfast. I heard it over math problems, between Western Civilization and cultural geography. But then one afternoon while I was hitchhiking home from school, it struck me like a baseball in the back: my mother wanted me to marry someone of my own social class—a poor girl. I considered my fiancee, Carolyn, and she didn't look poor, though I knew she came from a family of farm workers and pull-yourself-up-by-your-bootstraps ranchers. I asked my brother, who was marrying Mexican poor that fall, if I should marry a poor girl. He screamed "Yeah" above his terrible guitar playing in his bedroom. I considered my sister who had married Mexican. Cousins were dating Mexican. Uncles were remarrying poor women. I asked Scott, who was still my best friend, and he said, "She's too good for you, so you better not."

I worried about it until Carolyn took me home to meet her parents. We drove in her Plymouth until the houses gave way to farms and ranches and finally her house fifty feet from the highway. When we pulled into the drive, I panicked and begged Carolyn to make a U-turn and go back so we could talk about it over a soda. She pinched my cheek, calling me a "silly boy." I felt better, though, when I got out of the car and saw the house: the chipped paint, a cracked window, boards for a walk to the back door. There were rusting cars near the barn. A tractor with a net of spiderwebs under a mulberry. A field. A bale of barbed wire like children's scribbling leaning against an empty chicken coop. Carolyn took my hand and pulled me to my future mother-in-law who was coming out to greet us.

We had lunch: sandwiches, potato chips, and iced tea. Carolyn and her mother talked mostly about neighbors and the congregation at the Japanese Methodist Church in West Fresno. Her father, who was in khaki work clothes, excused himself with a wave that was almost a salute and went outside. I heard a truck start, a dog bark, and then the truck rattle away.

Carolyn's mother offered another sandwich, but I declined with a shake of my head and a smile. I looked around when I could, when I was not saying over and over that I was a college student, hinting that I could take care of her daughter. I shifted my chair. I saw newspapers piled in corners, dusty cereal boxes and vinegar bottles in corners. The wallpaper was bubbled from rain that had come in from a bad roof. Dust. Dust lay on lamp shades and window sills. These people are just like Mexicans, I thought. Poor people.

Carolyn's mother asked me through Carolyn if I would like a *sushi*. A plate of black and white things were held in front of me. I took one, wide-eyed, and turned it over like a foreign coin. I was biting into one when I saw a kitten crawl up the window screen over the sink. I chewed and the kitten opened its mouth of terror as she crawled higher, wanting in to paw the leftovers from our plates. I looked at Carolyn who said that the cat was just showing off. I looked up in time to see it fall. It crawled up, then fell again.

We talked for an hour and had apple pie and coffee, slowly. Finally, we got up with Carolyn taking my hand. Slightly embarrassed, I tried to pull away but her grip held me. I let her have her way as she led me down the hallway with her mother right behind me. When I opened the door, I was startled by a kitten clinging to the screen door, its mouth screaming "cat food, dog biscuits, *sushi*. . . ." I opened the door and the kitten, still holding on, whined in the language of hungry animals. When I got into Carolyn's car, I looked back: the cat was still clinging. I asked Carolyn if it were possibly hungry, but she said the cat was being silly. She started the car, waved to her mother, and bounced us over the rain-poked drive, patting my thigh for being her lover baby. Carolyn waved again. I looked back, waving, then gawking at a window screen where there were now three kittens clawing and screaming to get in. Like Mexicans, I thought. I remembered the Molinas and how the cats clung to their screens—cats they shot down with squirt guns. On the highway, I felt happy, pleased by it all. I patted Carolyn's thigh. Her people were like Mexicans, only different.

John and Annie Bidwell seated on the south porch of the Bidwell Mansion, ca. 1893. Courtesy of Special Collections, Meriam Library, CSU Chico, and Bidwell Mansion State Park.

Japanese American farmer in the Fresno area, ca. 1920. Courtesy of Fresno County Free Library, Japanese American Project.

Filipino contractor taking a crew of field workers to ranch from Stockton. Photograph by J.J. Billones. Courtesy of Bank of Stockton.

Migrant youth in potato field, Kern County, 1940. Photo by Rondal Partridge, courtesy of the photographer.

Drilling action on a rig near Kettleman Hills, 1971. Photo by Bill Rintoul, courtesy of the photographer.

Sikh farm worker, San Joaquin County, 1960. Courtesy the Bank of Stockton Photo Collection.

United Farm Workers Union members picketing, 1973, Arvin-Lamont area. Courtesy of Kern County Museum.

Grape arbor. Photo by Hammond, Porterville, 1938. Courtesy of CSU Fresno, Madden Library, Special Collections.

Robert Vasquez

AT THE RAINBOW

for Linda, Theresa, and Phyllis

At fifteen, shaving by then, I passed
for eighteen and got in, in where alcoves
breathed with ill-matched lovers—
my sisters among them—who massed
and spun out their jagged, other selves.
I saw the rhythmic dark, year over

year, discharge their flare: they scored
my memory, adrift now in the drifting place.
Often I watched a slow song empty
the tabled sidelines; even the old poured
out, some dragged by wives, and traced
odd box shapes their feet repeated. *Plenty*

and *poor:* thoughts that rose as the crowd
rose—my sisters too—in the smoked air.
They rise on. . . . They say saxophones
still start up Friday nights, the loud,
troubled notes wafting out from where
I learned to lean close and groaned

into girls I chose—no, took—and meant it.
In the Rainbow Ballroom in Fresno
I sulked, held hands, and wheeled among
the deep-bodied ones who reinvented
steps and turns turned fast or slow,
and this body sang, man to woman, song to song.

Robert Vasquez

BELIEF

for Jon and Ernesto

There are those—old aunts, far-off
godparents—whose houses seem ready
for ghosts: east of Madera, bricked
and scarred in the last century,
is the house my mother's aunt
lit candles in (though wired
for lights in '38; her parlor
nightly sent forth a dozen
"flamedrops"). A Stevens .22 pump
was found in the cellar, and a woman
in a black nightgown cried and walked
from shadows to the outhouse.
She was dead; she rose
some nights the way clouds
of insects lift from long grass—dark
breath of the earth, brief, wind-broken. . . .
Take care, they warn, these
death-long witnesses of grief
any hour can dole out, unresolved,
and flood the soul. Aunt Esther
gestured back with wick yarn alone.
"But don't worry," she said. "Most
never glimpse one; even more
think you're nuts if you do." But
belief can start anywhere, even here
in a plum orchard, wind-stirred
and radiating leaf-spurs, or out beyond
the old washroom where "someone
you can't touch, some stranger
doing eternal chores, touches you."

Ernesto Trejo

AUTUMN'S END

It begins when the TV mentions the name of my street,
saying in passing that the woman next door has died.
Yes, the one whose name I never knew, the name
that even now escapes me. I will watch
the leaves from her ash trees pile up all winter.

Now deer start to come down from the high country
to a place between snow and this valley lost in fog.
And my shaggy dog scuttles between rooms.

Then there's the ants. When winter stumbles on them
they go under into their caves, tunnels,
and immense corridors.
And what happened to mosquitoes? Where have they gone
with all the blood collected?

Now there's a long peace in corners and basements
where we won't dare to step in,
black widows nest there with their young.
Outside my window a few leaves hang on.
Doubting so many things I wait for winter.
Watergrass is sprouting everywhere, even on the ground
where the nameless woman hides from winter.

Jean Janzen

WHITE FIELD

for Ernesto Trejo (1950-1991)

After your death I think *black rose,*
but write *white field,* the way you would,
a light in your eyes. The field near my house,
feathery with wild oats and blowing in May.
I write *field under full moon,* dry,
like a cough or hair bleached and falling,
what you left us, plow's blade cutting in
and the air filled with dust. It's rootedness
that lies exposed and flies. It's old
mother earth, her suck and murmurings
lifted with a groan and given up.
Not the elegance I try to shape
from her rough roots. It's these shifting,
blowing lines, you in the air I breathe.

George Keithley

AUTUMN

Autumn, old mother, you bless with lightning bolts.
Thunder shudders caprock, rolls down the canyon.
Storm clouds break above the foothills, swelling
Horseshoe Lake, flooding Butte Creek—
Out of Mad Woman Meadow a sleek dog trots home,
burrs knotted in its long ears, its switching tail.

In town the streets run red with rain,
pistachios bleed under our feet.
Persimmon trees, wind-plundered, clutch their bare fruit.
Irascible oaks shake dark fists. What is this
silver light? It glosses chiseled grey
tombstones tucked behind an iron fence—

The rusted red truck swerves to avoid
the dog struck dead in the road
beside the cemetery. Ducking into the rain,
a man and woman leave their pickup, lift the lean dog
into the open bed, drape a painter's tarp over it.
They drive down the spattered road without a word.

Mother of rain, grass, dusk. You teach us
the end of life. Now at the last
daylight hour the storm surges, dies out. Gone
is the green scent of hay drifting across the highway.
Beanfields, sodden, blue-black, glimmer like the sky.
Under ghostly sycamores the creek rolls, tumbles

down its rockbed: the voice of my father before his stroke,
his merriment irrepressible as water. Do trees listen
to our words as we attend to their thrumming music?
I want them to hear the spring concert solo

Jean played, her shy eyes proud of the blue
tunic and gold braid of the school band.

The clarinet coal-black in her pale hands. Head bowed
to applause. Her bell of brown hair by summer's end
lost to leukemia. Two months after the funeral
her sister caught by the midnight train
far from town, her boyfriend's car
stalled on the crossing above the flooded fields.

Mother of all that I remember—
Your rains wash my eyes,
I see the blue earth breathing low and still.
In this soil you gather my family,
my friends. They wait for me
as the wind waits and the waters settle.

After the flashing storm, under the turning maples,
in your streams the blood of light spills, gleams.
In you I learn the rhythm that rises
and fails in my heart. Here
in the deep fields, in crimson water—
In the slow sacred blood of our mother.

Susan Kelly-DeWitt

RICE FIELDS AT DUSK

for Marie DeSouza Corbett, in memory

The sun drops
like a single bright bead

in a rosary for
The One.

Rice stalks lean,
listening for the hawk's cry

circling West
Catlett Road.

I follow the solid
yellow line, then

the broken one
to your hospital bed.

(All my old ones
are vanishing.)

Lillian Vallee

HAWK AT THE DELTA MENDOTA CANAL

To the memory of my father,
Tadeusz Wereda

So you thought it was life-giving
This glittering river in a concrete trough
Pipe, dam, sluice
Bridge and ladder
Iron and steel
Nothing riverine about it, my fierce brother
Nothing but hills of rusting windmills
Reaping acres of sterile wind

Given a human body, you had to work
Emptying trash, mopping floors
No wingèd fantasies for you
Now an *Untermensch* again
Doing other people's laundry
Airing their fetid sheets
While they danced a frantic aerobic
Not a cottonwood in sight, my weary brother
To muffle the emptiness of the western shore
The river could change its course
But humankind was a steady race
Achieving perfect control
Oblivious to shadows
Still grazing the rancherias

You pass over
The restless Coast Ranges hail your last flight
And your toothless cry, *Eloī Eloī*
Is lost in the dissolving mists

Your wings flash clean against leaden skies
Though they tow a heavy baggage of cliff
Had you seen it all along?
Barren river, wailing skull
Always the same unending sacrifice
Fueling the steadiness of a reckless race
That wraps in concrete and ribbons of aqueduct
Its bottomless misery
Its Gorgon face?

Yes, you saw
Your vision was unblinking
But this was the only earth you were given
And so in the dying that came too soon
You touched the heads of your unworthy children
And blessed them

LOS BANOS RESERVOIR, NEW MELONES, ETC., ETC.

1.

At dusk
The oaks release their bats
Sorrow rises from the rock
Cries out as they strike the seed
Whose work, whose work
Whose digging sticks
Uprooted corms, tidytips
Tule mats melt into grass

At night
The heart fumbles
With the excess of death
Not even one is imaginable
Not a single infant
Returning to dust

Snug in her buckskin stays
Not a single mother
With singed black hair
Woven now into oriole nests
Gobbled by hungry roots
Whose life, whose life
Whose breath
Was taken from the heartsick earth

Bass in underwater villages
Behind earthfilled dams
That burst in your dreams
Nibble on bones
No one is spared
Whose eyes, whose eyes
Stare back
Whose hands flutter like moths
Rustle like lizards
In the fallen leaves

2.

At dawn
Your body becomes the earth
Licking its wounds
Each sapling remembers
The aches of its elders
Each river continues
Addressing the rocks
Snow boils down the mountains

Then suddenly you hear it:
Music pouring from elderberry stalks
Rattles, clappers, gourds and flutes
Hoofs, dewlaps, elkskin drums
Filament after filament of sound
Fills the throbbing empty spaces
Rocks the hammock of the world
And you know:
They are back to forgive us

Wendy Rose

NOVEMBER: SAN JOAQUIN VALLEY

Tumbleweed sticks to the stunned
and stripped valley
recalling the drought or the day
we turned the wind around
and slept dreamless one year
only to dream too much the year following.
Field after field there are invisible mountains
beyond the cotton baled into tightly bound trucks;
the white mounds mark the boundaries of farms
like nations backed up against each other
forever on the brink of war.
Red-winged blackbirds and magpies
lean anchored to the barbed wire,
stiffly move in the stillness.
Peach trees have slimmed and dropped their leaves,
hardening brown for winter
and stretching their hands
between the smudge pots.

I run the length of this valley
flatfooted into its visions

with eyes closed tight
on the western horizon,

white with fatigue,
holding my breath, remaining deaf.

Like infants dressed in baptismal robes
I gather the dreams back to me in this place

and promise again
to protect what's left.

THANKSGIVING ON THE SAN JOAQUIN DAILY BETWEEN FRESNO AND MARTINEZ

Done with summer, sunset finds
deathly red the vines
grown crooked with fruit
that will ferment in vats of money,
sugary balls for Italian wines
that are profit and kickback, pollution
having nothing to do
with the bare black branches of buckeye
gnarled in the fog, nearly
the willing accomplice
of the pious and smug
old thieves of the valley.

Everyone sees the pile of brush.
Tangled mess where beavers build
I heard one man tell an invisible boy.
But I see it too on the midnight slough
not the messy tangle of tules and twigs
but carefully engineered moon,
perfect half-sphere on the water,
elegant geometry of animal rising
from black and brackish water,
luminous eyes reflected
in the light of our passing.

Imperialists,
we lean from the train
in the tangled mess we made,
tracks made toxic with money,
missiles in the ground
ready to slash open
the innocent marsh,
ghosts of battleships
lined up in the river,
old dress with silver bric-a-brac
and threads flowing loose
in the current.

Egrets stand,
obnoxious and proud.
Coots assemble, drop to the water,
pass by. Irrigation canals churn poison
between claws and fins
entering the green north end of the bay.
Garnet eyes in electric light
watch us run, looking for a way
back to the boulders, crazed escape
from the holiday feast,
the bitterly cold faces, the massacre.
Blackbirds chatter and laugh.
How well they know.

COALINGA

rattles
its new ruins
picking them up
like old shells

full of pioneer steam,
of the S.P. and morning crows

dusty from the cottonfields,
touching their feet
to that thirsty place,
that dry earth;

according to the sign
on the spine of the highway
these hills are lost

yet merely recede
in a native stretch
west and west;

irrigation canals
have broken the bones
of foothill and valley,
have tied together
the old woman's wrists.

Do the songs return
in the stretch of your flesh
Grandmother, the step
of your dancing foot?

Lost Hills, California
1983 (just after the Coalinga earthquake)

Lee Nicholson

LEAVING THE MALL

Turn. Let us exit another way, leave
This funeral home and with Ishi light
A fire. Burn it down: Indians would grieve
For their dead like this too. First feel the bright

Flames and then wait a bit for Brother Time.
Grass will be the first to come back here, green
Wet pieces of God, a natural rhyme
To everlasting Life, the Holy Queen.

Falling from her Spirit World come her gifts:
Feral gardens, sedge beds, tules, rain-gray
Willows, elderberries, herbals in drifts,
Fragrant mists, sweet silences that lay

Around important beauty. Let us please
The wild roses. Let us turn. Let us ease.

WATER, WEALTH . . .

High over city traffic with much grace
Curves a metal arch which supports a sign.
Its message hanging there in open space
Reads WATER, WEALTH, CONTENTMENT, HEALTH—a fine

Combination of abstraction and pride.
They all depend in this hot, arid land
On one element—WATER—the blue bride
Of sky. For each daily wedding this band

Of iron surrounds with ceremony all
Those other essences—earth, sunlight, air.
A witness to this cosmic dance and brawl
It stands. All lovers, fighters too, should stare.

Look up. See through. Like some great Buddha's ring
An arch can show a space or everything.

Eye Street looking east, Modesto, 1948. Photo courtesy of McHenry Museum.

Richard Dokey

BIRTHRIGHT

THE PROCESSION was ready to leave. In front was the hearse that Arnold Suttman had bought back in '49 when he had taken the mortuary over from his father. He had had the hearse painted a dark grey to match the Cadillac limousine he had been able to afford in '74. Arnold didn't like black, and he didn't like driving the hearse either, and so, since '74, he had driven the limo and had had Billy Rylands, his assistant, drive the dead. It was all right, he always said, to handle people after they had just died, but on the day of a funeral he wanted to be as close to the living as possible.

He had plenty of that around him now. In addition to old Judge Reynolds and Carter Harrison, who was already a little drunk, beside him on the padded front seat and Carter's wife Amy and their kids in the rear, a line of thirty to forty cars had formed behind the limousine. Everyone knew Miss Bunty.

Two other people had had to leave considerably sooner, however. Young Josh Harrison's trim, silver grey Porsche had already been on the Marsh Creek Road from Berkeley for three quarters of an hour when the procession left Suttman's Funeral Home, and Sonny Boy Higgins had pushed his wheelbarrow to within a mile of the cemetery by the time the Cadillac turned onto the 99 over-pass.

It was a hot summer day, and the heat shimmered above the dry fields of barley and glistened in the irrigation ditches. The line of cars, like a metallic centipede, found its way at last to the man with the wheelbarrow. Sonny Boy made that peculiar gulping sound in the hollow of his throat when Suttman's hearse approached. He dropped the wheelbarrow and pulled off the old felt hat, crushing it against his chest. With an open mouth he inspected each car that moved slowly by. The people looked back at him. No one thought about stopping to see if he wanted a ride. Sonny Boy pushed

---

his wheelbarrow to all the funerals, even the ones out here. He pushed his wheelbarrow everywhere, for that matter, straight down the middle of the road, and wouldn't have it any other way. That was Sonny Boy. So nobody felt sorry for him.

The tiny cemetery waited under the valley sun. In one corner, to the rear, a rainbird was going at the end of a frayed garden hose. The grass was green under the spray, but there were many places where the lawn was dry or uneven in color. When the cemetery had begun, all those years ago, no one had bothered to level the ground, and they had started by burying a few Indians. Then stone and marble slabs with names of people who had settled there began to appear. Someone planted locust trees, sprinkled lawn seed and set up barbed wire fence to prevent the cattle wandering in from the surrounding pastures. And then someone else put up a gate along the road. Whoever mowed the place in summer neglected to trim, and the wild bermuda, which had taken over in time, grew in scruffy beards at the base of all the scattered trees and headstones. There was no order to any of it, except that they did stop burying Indians after a while and moved the ones already there to a row just inside the south fence. Someone said it added to the history. But the Indians weren't given any markers.

Then all the cars were stopped, strung out along the shoulder of the road. Billy Rylands parked the grey hearse in front of the gate, and everyone climbed out of the limousine and waited. The people came out of the cars. Walking slowly along the hot asphalt road, they looked like over-dressed refugees from a fallen village. The thin valley wind turned the hem of a print dress or cooled the brow of a red face pinched by a tight, starched collar. Then they formed in a bunch, and Billy opened the rear of the hearse and the six men, all of whom Bunty had designated as pallbearers, came forward awkwardly and stood like stumps in two rows, just as Arnold Suttman had rehearsed them.

Then they took the weight of Bunty by the handles of the bright aluminum casket and led the people through the gate and across the uncut grass to the square hole that had already dried out beneath the sun. There was a pale green canvas draped over the mound of freshly dug earth. The pallbearers set the casket upon the two hickory boards that lay over the hole, and Arnold Suttman returned the wreath of flowers to that place which he knew, after years of practice, must be situated just under the chin of the deceased. It was at this point that Sonny Boy, who had pushed his wheelbarrow up against the barbed wire fence and wiped his forehead with the cotton plaid sleeve of his workshirt, turned and saw young Josh Harrison's silver machine arrive from the south.

• • •

Judge Reynolds gave the eulogy. Tall, with a grey, balding head and a wisp of moustache, he spoke with pale, trembling lips, and by the time Josh Harrison had covered the distance between the gate and the place where his great aunt was to be buried, the judge had already talked about Bunty's childhood among Indians, the Prince of Waxton, who had sought her hand, the fire that had swept across the hills of barley before The Great War, the first coming of electricity and the big house there on Harrison Road. It was history and everyone knew it anyway, but it always gave the judge pleasure to talk about the past and about old things. The people nodded, remembering other stories, and thought they had been to Bunty Harrison's for tea, though not one of them had ever set foot in the big house.

"Some things in life," the judge said, recognizing Josh as the people let him through, "should never change."

And that brought him to Bunty's virtues.

"Though she never married," the judge said, "this kind and gracious lady loved children. She loved baking for the P.T.A. You all remember the book of her favorite recipes she had printed at her own expense and that she gave away free at each county fair for over forty years. She gave things when people never knew of the giving." He cleared his throat. "Bunty Harrison was that kind of woman."

He paused and looked full into Josh Harrison's eyes, so that the young man lowered them a moment.

"Bunty Harrison loved animals," the judge said. "She loved riding that great white stallion Emperor in the Fourth of July parade. She was grand marshal at the fair six—no, seven—times. She loved cats and dogs. She loved the sound of cattle in the early evening after milking. She never owned a television."

Josh Harrison looked up. He had heard a humming sound, something deep in the throats of all the people around him. He studied them. They had the kind of faces that no one remembers, old photograph faces. They were, in fact, quite plain, simple, hardworking people who, it seemed to him, had little in common with the strange, reclusive maiden aunt he had known as a boy. While he thought of distant places where his work was about to carry him, these people appeared concerned only that such things might one day be said about them beneath a hot valley sky. For this was where they would all be, here or in the big cemetery in town, where his own father was buried.

He thought of the old woman, who pressed her clothes under the mattress while she slept, who had the cardboard figure of a man sitting on the stuffed mohair sofa by her front window and the mannequin made up like a woman, which stood facing the yellowed ivory keys of the grand piano, the first ever shipped around the Horn, and which she played after dark because during the day it disturbed the spider monkey she kept in the tiny wire screen house that Jessie Two Rivers had built out back. He remembered the pictures of Bunty on the white horse that had been killed by a tractor ten years before he was born, and the five other ladies who also rode in the parades, three of whom were still alive and here, huddled to one side like faded ostriches, and who came forward now in turn to place a twig of broken cyprus on the coffin.

He turned to glance at Sonny Boy, who had rested his chin against the fence post, his sleeves rolled to the elbow. Then he looked past the grave to the rolling, dun-colored fields, where white-faced cattle lifted blank faces and stared with mute, but curious, indifference.

When it was over, the judge squeezed into the car, complaining that it was difficult being the only survivor of the mesozoic age.

"It's time to tell you some things, dear boy," the old man said, "and then we're expected at your brother's. I'm afraid he's already hip and thigh in the slough of alcohol."

The others were leaving, too. They shuffled brokenly to their Fords and Chevrolets. The cars scattered slowly in both directions. A blue film hung in the air from their exhausts, and little flurries of dust followed them along the road. Sonny Boy was nowhere in sight.

"Your great aunt was an exceptional woman," the judge said, as the young man slipped in beside him. "You're not really aware how exceptional, I believe. It's about the will, you see."

Josh Harrison looked through the fence to the casket gleaming in the sun. It rested beside the mound of dirt covered with green canvas. The single rainbird spurted a limp stream of water. The air was dry and still. He sat with his hands on the leather steering wheel. The judge leaned forward to look.

"What is it?" Josh asked.

At the far end of the cemetery, beyond the barbed wire fence, the cattle had begun to stir. Then they separated, their heads high and stiff. Two men in work clothes were coming across the field. They were carrying shovels. The judge squinted.

"Indians," he said. "One of them is Frank Rivers, Jessie Two Rivers's grandson. They're here for Bunty." The judge looked straight ahead. "May we go now, please?"

He started the car and drove onto the asphalt. Three quarters of a mile down the road they came upon Sonny Boy pushing the wheelbarrow. Sonny Boy dropped the handles when he heard the sound of the engine. He stood in the middle of the road waiting, his tongue working against the roof of his open mouth. They passed slowly, and Josh Harrison stared into the clear, brown, empty eyes.

"Poor boy," the judge said. "He's never been anywhere he couldn't walk to without that wheelbarrow. He's lived here all his life. Strange, isn't it?" A pointed light gleamed in the old man's eyes. "He'll be by the big house in a while. It's part of the ritual."

"I remember him when I was a kid," Josh said, "pushing that thing up the center of Main Street. He never carried anything in it that I remember."

"It's always empty," the judge said. "It's always been empty."

"What does he use it for, then, Judge? I could never figure it out."

"Nobody knows. Nobody asks anymore. But it holds something, dear boy, you can be sure of that. Something for Sonny Boy." The judge chuckled, but it ended in a strange, furry sound at the back of his throat. "I've lived here all my life too," he said, "and I've often envied him for what he carries in that wheelbarrow."

Josh Harrison thought of his own father, who fell down one day in a field and didn't get up again. His father had never left either.

The car moved down the road. The fields on either side were fluffed with grain. They rolled and dipped with the light wind. The machines would be in them soon, paddlewheel blades sweeping beneath dun-colored stalks, opening arrow-straight paths, then turning again and again to make an island of wheat. Three or four machines at once on one piece of land. The dust and chaff drifting in the air. The big noise and the thrash-thrash-thrash. The rows of sheared stubble. The scent of broken barley in summer heat. The final, empty ground. In the morning darkness as a boy he had waited beside his father to ride the big combines. Green metal. Oil. The smell of cold, black rubber. The sunrise above the blue Sierra.

"You knew that Jessie Two Rivers died three months ago," the judge was saying.

"No, Judge, I didn't." He recalled the copper-skinned man in dungarees and flannel shirt who always lurked about the old house, a silent, ghost Indian who never spoke, not even when he had saved him from drowning that time in the creek out back. "I'm sorry," he said.

The judge sighed and gazed through the side window. "You've been away. Yes, Jessie Two Rivers is dead, and today his grandson is burying your great aunt." He sighed again. "I tell you, dear boy, it's not easy being the sole survivor of the mesozoic age." The judge tried to laugh but ended by clearing his throat. "The rocks and stones are all broken, you see. The work is done, dear boy. Now all we must do is—." The old man turned in the seat of the car that was too small for him. "—I mean you, dear boy," he finished.

He glanced at the judge. The soft face was blotched beneath the eyes. The eyes were watery and crimson. The lips were damp and the color of clay. The judge had always been old, like his great aunt.

A vision of something distant appeared in his mind. It rose to the sky, concrete and steel, and breathed smoke and flame. Large ships came to it and planes and many, many people.

"That fifty acres her father promised Jessie is still fallow," the judge said.

"What was that, Judge?"

"Jessie Two Rivers. The fifty acres. Jessie was an Indian. He could never work land cut by a plow. Bunty specifies that it never be touched. A memorial for all the years. The suffering. The redemption of her father's pledge. It's all in the will, dear boy, which I am leaking to you rather peremptorily, before the pack closes in. Forewarning, as it were. Preparation. There are forces at work. Real estate salesmen. Profiteers. Insipid relations." The judge blinked his wet eyes. "You have been chosen, you see. A clear-sighted judgment on Bunty's part, I must say. The old have a peculiarly keen intuition. I ought to know. Being millions of years past my time, after all." He coughed and laughed.

At that moment he realized how much he had always liked the old man.

"Judge, what are you telling me?"

"She's named you executor for the ranch, dear boy. All of it. Thousands of acres. Everything. Not your older, drunken, adulterous brother Carter or your asinine little sister Martha, who can't even sew a seam straight—you'll pardon me—or even your kind but helpless and distant mother, her sister's only child, but to you. The only one to attend the university. To educate, as it were, his sensibilities. To have any kind of desire to see beyond, shall we say, the Harrison nose. So that, perhaps, he could, after a suitable time in the world, one day return, truly, and claim his birthright."

He was shocked. "Judge, I don't think I'm suited to—"

"In short, dear boy, if not you, then no one. No one at all. And—," he paused, musing, "—the mesozoic age comes to an end."

Then they were there. He hesitated a moment, looking at the place. It was just the same, only a bit more run down. The big house squatted near the road that was named after the great aunt who had lived there so long and was gone.

He got out of the car. The judge struggled after him. He went to the slatted gate whose paint had vanished long ago. The latch was fastened by a piece of bailing wire. To that were twisted other pieces of rusted metal which had been used over the years to keep the gate from swinging open. Even now the latch was ready to fall. He tried to tighten the wire, and it snapped in his fingers.

"Don't bother, dear boy," the judge said. "The whole place is that way. Broken things held together by wire and cracked glue. She never repaired anything. It just crumbled around her. The one failing, you see, in the head of the family. No small matter. The chief argument, in fact, for the brother and sister. Even your mother. Break the dinosaur's bones, they say. Scatter them and tract houses will appear. To our long awaited profit. To hell with Jessie Two Rivers. Avoid madness. The prosecution rests."

"Judge—"

"You haven't heard the defense, dear boy. Part of it is here with me. The rest is there." He pointed. "Here's the key."

"Judge, this is all overdone. I don't think—"

"Exhibit A: it was your great aunt who paid for your schooling."

Astounded, Josh Harrison stared at the old man. "What are you saying? My father—"

"Never had a dime, dear boy. He died flat broke. There in Harrison fields."

"I don't believe it."

"It's true, dear boy. Would you like to see the cancelled checks?"

"But, why—?"

"That's not the end of it." The old man patted the hood of the silver machine. "This automobile, your trips to Europe. All Bunty, dear boy. Your great aunt Bunty."

Josh Harrison leaned against the car. "I—I don't understand, Judge."

"Intuition, dear boy, intuition. Bunty's greatest virtue. Out of all the miserable lot, she recognized the only one—"

"Judge—"

"—who had the strength, as it were."

"But, Judge," he blurted, "I've gone away to school. Now I'm going to New York. You said she wanted me to go."

"Yes, I believe she did."

"I've left."

"Yes, dear boy. And it's hard to leave, isn't it? Small town. Small life. Your relatives are here, living on haunted ground. The past rifles their flesh, strips them of courage. They linger, like ghosts." The judge's eyes filled. "Your great aunt was wise. Those people at the funeral, for example. Proud, but resigned, you see. They go on because . . . they can't do anything else. Bunty understood that about herself as well. She never took care of anything. There were others following, after all. Then time runs out. And there's only the one." The judge looked over the gate. "The barn needs a new roof. The pump has to be fixed. The fences need wire. Everything needs doing. Bunty knew the price you'd pay. Going is one thing. Staying is another. Hardest of all is returning."

"But she sent me away."

"That was so that you wouldn't have missed something when you realized you had to come back. Whenever you do. In time, that is. You have a choice. The rest of us. . . ." The old man looked away and shrugged. "When you have to stay, dear boy, everything runs down. It's a kind of revenge the spirit takes upon the land, you see. For not letting you go. Jessie Two Rivers was the only one who understood."

The judge raised his eyes to a hawk circling high against the clear sky.

"It's fear, dear boy. The fear we all have. That you've lost life, somehow, back in the mesozoic age. You wonder, was there more than the hum-drum days? It has a cumulative effect. A thousand years from now this land will be offered to some other son." The judge looked at him. "Harrisons have always lived on Harrison Road."

Josh Harrison tried to swallow. He stared at the man who never was a judge, only a justice of the peace.

"I still do perform marriages," the judge said. "You'll be married yourself one day, dear boy."

Vaguely he saw himself someplace in the future. A two story Tudor in the suburbs. Square lawns. Bronze-headed sprinklers. Privet hedges. A wife and children.

"I loved Bunty," the judge said. "Oh, not really, but I loved her in a special way, you see. It comes among the old. Does any of this make any sense, dear boy? Jessie Two Rivers saved your life. That also makes you the last Indian, as it were."

"Judge, I—"

"Exhibit B," the judge said. "Go on. Through the gate. Walk about the place. In the name of the defense. Then we'll go to your brother's and the witnesses for the prosecution. The jury of relatives scorned. I'll wait here in your beautiful automobile."

He went through the gate and across the gravel driveway. The great house stood before him. The screens to the porch were torn. There were broken window frames and old furniture stacked against the inside wall. The pyracantha had pushed up against the eaves that were threaded with cobwebs and streaked with dirt. He walked around to the front. The steps had sagged another half inch. Everything needed three coats of paint. He pushed the key into the lock. There was a cold, metallic click and the door opened. He went inside.

There was the smell of dust and old furniture. The mohair sofa. The hand-carved table from Spain and the matching chairs with leather seats. The odd brass floor lamps with printed shades from which drooped little crocheted woolen balls on silver wires. The threadbare Arabian rug. The stained glass. It was all as he remembered.

In the room with the piano was a row of pictures in chipped frames. Bunty's father and mother. The sister, who had died just after the photo was taken. Some uncles and aunts. They confronted him with demanding, unsmiling faces.

In the kitchen there were utensils everywhere. Plates stacked to one side. A porcelain vase and a peculiar metal thing that looked like a scale without numbers. Old cups that had been used so often even the white was faded. The heavy, deep sink. The plain, steel faucets. The torn curtains that covered the tiny window. He went upstairs.

There was the bed in which she had been born and had died. The mahogany dresser that was worn through. The funny oval mirror. On the matching chest of drawers was the faded picture of a foreign young man in military uniform, the hair flat against the head and parted in the middle, a tiny moustache above a thin, unsmiling mouth. In the closet were all her things. The scent of dried wood mixed with old cloth. He touched the pair of riding boots resting on the floor.

He went downstairs and opened the back door. He walked past the monkey house that Jessie Two Rivers had made. The bar still hung in the center of the cage. He went to the crooked steps of the tankhouse and sat down. He could see the judge waiting in the car. His head was back and his eyes were closed.

Then he thought that he himself would be old, like this ancient house, fallen into disrepair, bound together by rusted bones and blood, longing for youth and immortality.

What was it he experienced now? He was too young to know. He leaned back and looked at the screenless windows, the tall, unbending locusts, the quiet, still earth.

Then he stood and walked to the car. He climbed in beside the judge. The old man kept his eyes closed and said nothing. He started the engine.

A half mile away, he came upon Sonny Boy. He drove slowly by, staring again into the dark, empty eyes. Then in the rear view mirror Sonny Boy grew smaller and smaller, going up the center of the road, pushing the wheelbarrow that contained nothing at all. And then he was gone.

Catherine Webster

CHILD OFF HWY 99

because out comes moist starry night to tease and kiss clover and brome
  into bloom
    because hovering starry night pierces hairroots in my San Joaquin
      farmsoil,
    bulbs swell with juices from tiniest white onions to baby blue eyes
      with night's hum. But because
    you, San Joaquin Planning Commission, behind your microphones,
      oblivious to night, vote unanimously
to smother the musk-smell of valley soil   with steel and concrete
because your naive-as-a-child's eyes and mouths are shut to night's
  delicate gift to
San Joaquin Valley, rapt night has touched his lips' wonderment to my lips—

        I, Catherine Webster, dew on my lips, on cheeks, in my hair,
am like urgent summer night, yearning to be born, year after year, a
  burning green oat, wheat, grass leaf,
I flaunt my own dew-sweet fleshy roots, straining for this farmland,
  because I'll be irrigated, hoed, thinned,
picked, tossed into lugs boxes,
my yearnings and feelings for this soil are every San Joaquin farmers' in the
  county—I am this county's bard,
      . . . I sing of my own regal grace, lithe form, I reek of the abundance

($874,620,000 county gross value ag-production) I reap off our county's
  prime farmland—
farmlands of statewide importance, unique farmlands, farmlands of local
  importance,   grazing lands from
Roberts Island East, off Highway 99 to Victor Road, Clements Road,
  Linden Road
    I sing the poem of my incredible SOIL,
      when I squeeze loam in my hand, mash it,
my breaths get hot, come as fast as mating jack rabbits' thump and rut;
    when I squeeze soil, in my heart and blood

I'm wet, underarms and my mouth, sticky with green, I blurt out
the blessing gathered behind my eyes—
I'm the pulse of the Linden farmer's wild
pangs to couple with me—I'm the holiness
in every farmers mating call: "get them goddamn fat-as-tics cows
onto the truck,"
"don't ride the cows so close,"
"heee yyya those pretty cows outa this corral,"
"lookit my cows—sleek and female hips to hocks, glory me,"
Because I feel the stirring when the moon rubs a shine into my
donkey's eyes
and behind my knees, I hear in ten-thousand sweet rye heads
rubbing a song that I Catherine,
am Holy
I'm the Kingdom of HEAVEN . . .   and to any man who has put
down his shovel
the shovel, for this deep deep soil, I'm will come to you,
an unabashed, the ploughing farmer in the field feels good
groin large and getting tighter at the sound of my poem,
bone-strengthening, heroic, wonderous, I do not deny any of them
my flower, I'm   springtime burning
in carrots and sugarbeets
swelling the farmer's thick roots, I'm here, again, that craving in me

for soil's good health sparked
by a single engineflare, pistons pulling crank case,
"Shove the stick into low, ease her, ease her into reverse, kick
the clutch in and get that tractor moving," those that know me seek me
here in loam, I thrive on my body's itch
to breathe and bloom
for the sake of San Joaquin Soil, shove the black gear nob into low,
to chug out across 400 acres of incredible soil,

to luxuriate in Mendez's, Podesta's, or Boggiano's grit caught
between my teeth,
together, with you, dear reader, find loam, scrape in hipholes
then, our fingers interlocking, arm-in-arm,
my lush thighs as spiritual as dew on a Merlot grape cluster
your biceps as visibly masculine as sweet corn,

as the rump muscles flexing as a bull calf struts calf's rump and
testicles swaying
to become myself, I flaunt my want to hold Solari's grandfather's shovel
my want for my children, Sarah and Matthew, to wear their
grandfather's
pocket watch, belt, and Stetson hat,
On farms with barns stacked against the 6th drought year
in Linden, Lockeford, Clements, Galt, Bellota, Peters,
I am prayer at the dinner table . . .
. . . I'm ass-kicking gumption
I want San Joaquin Country's kids east of Hwy 99 to mouth an
almond blossom,
caress the peach blossom's spread petals
I want my children to gently lip the pistil's rigid legs
I want the nectar that sweetens my grandson's first fruit
as his teeth tingle deep in his gums
I want the fruit to drive him wild with hunger, make him desperate
I want to share my granddaughter's early blooming plum,
I want my children to split a ripe peach with their fingers, to feel the
thrilling
likeness of peach, in me

Cherríe Moraga

from HEROES AND SAINTS

Characters

CEREZITA VALLE, *the head*
AMPARO, *the comadre and activista*
ANA PEREZ, *the news reporter*
DOLORES, *the mother*
BONNIE, *a neighbor's child "adopted" by* AMPARO *and* DON GILBERTO
YOLANDA, *the hairdresser sister*
MARIO, *the sometimes-student brother*
FATHER JUAN, *the "half-breed" leftist priest*
DON GILBERTO, *the compadre,* AMPARO*'s husband*
POLICEMAN
EL PUEBLO, *the children and mothers of McLaughlin;* THE PEOPLE/PROTESTORS/ AUDIENCE *participating in the struggle (ideally,* EL PUEBLO *should be made up of an ensemble of people from the local Latino community)*

Notes on *CEREZITA*

CEREZITA *is a head of human dimension, but one who possesses such dignity of bearing and classical Indian beauty she can, at times, assume nearly religious proportions. (The huge head figures of the pre-Columbian Olmecas are an apt comparison.) This image, however, should be contrasted with the very real "humanness" she exhibits on a daily functioning level. Her mobility and its limits are critical aspects of her character. For most of the play,* CEREZITA *is positioned on a rolling, tablelike platform, which will be referred to as her "raite" (ride). It is automated by a button she operates with her chin. The low hum of its motor always anticipates her entrance. The raite can be disengaged at any time by flipping the hold on each wheel and pushing the chin piece out of her reach. At such times,* CEREZITA *has no control and can only be moved by someone manually.*

Setting

The play takes place in McLaughlin, California, a fictional town in the San Joaquin Valley. The year is 1988.

McLaughlin is a one-exit town off Highway 99. On the east side of the highway sits the old part of town, consisting primarily of a main street of three blocks of small businesses—the auto supply store, a small supermarket, the post office, a laundromat, an old central bank with a recently added automatic teller machine, a storefront Iglesia de Dios and, of course, a video movie rental shop. Crossing the two-lane bridge over Highway 99, a new McLaughlin has emerged. From the highest point of the overpass, a large island of single-family stucco houses and apartments can be seen. The tracts were built in the late '70s and reflect a manicured uniformity in appearance, each house with its obligatory crew-cut lawn and one-step front porch. Surrounding the island is an endless sea of agricultural fields which, like the houses, have been perfectly arranged into neatly juxtaposed rectangles.

The hundreds of miles of soil that surround the lives of Valley dwellers should not be confused with land. What was once land has become dirt, overworked dirt, overirrigated dirt, injected with deadly doses of chemicals and violated by every manner of ground- and back-breaking machinery. The people that worked the dirt do not call what was once the land their enemy. They remember what land used to be and await its second coming.

To that end, the grape vineyards, pecan tree orchards and the endless expanse of the Valley's agricultural life should be constant presences in the play and visibly press upon the intimate life of the Valle family home. The relentless fog and sudden dramatic sunbreaks in the Valley sky physically alter the mood of each scene. The Valle family home is modest in furnishing but always neat, and looks onto EL PUEBLO *through a downstage window. Scenes outside the family home can be represented by simple, movable set pieces, e.g. a park bench for the street scenes, a wheelchair for the hospital, a set of steps for the church, etc.*

Act One

Scene One

At rise in the distance, a group of children wearing calavera masks enters the grape vineyard. They carry a small, child-size cross which they erect quickly and exit, leaving its stark silhouetted image against the dawn's light. The barely distinguishable figure of a small child hangs from it. The child's hair and thin clothing flap in the wind. Moments pass. The wind subsides. The sound of squeaking wheels and a low, mechanical hum interrupt the silence. CEREZITA *enters in shadow. She is transfixed by the image of the crucifixion. The sun suddenly explodes out of the horizon, bathing both the child and* CEREZITA. CEREZITA *is awesome and striking in the light. The crucified child glows, Christlike. The sound of a low flying helicopter invades the silence. Its shadow passes over the field. Black out.*

Scene Two

Mexican rancheras can be heard coming from a small radio in the Valle home. ANA PEREZ *is on the street in front of the house. She holds a microphone and is expertly made up.* AMPARO, *a stocky woman in her fifties, is digging holes in the yard next door. She wears heavy-duty rubber gloves.*

ANA PEREZ *(to the "cameraman"):* Bob, is my hair okay? What? . . . I have lipstick? Where? Here? *(She wets her finger with her tongue, rubs the corner of her lip.)* Okay? . . . Good. *(Addressing the "camera.")* Hello, I'm Ana Pérez and this is another edition of our Channel Five news special: "Hispanic California." Today I am speaking to you from the town of McLaughlin in the San Joaquin Valley. McLaughlin is commonly believed to be a cancer cluster area, where a disproportionate number of children have been diagnosed with cancer in the last few years. The town has seen the sudden death of numerous children, as well as a high incidence of birth defects. One of the most alarming recent events which has brought sudden public attention to the McLaughlin situation has been a series of . . . crucifixions, performed in what seems to be a kind of ritualized protest against the dying of McLaughlin children. (DOLORES, *a slender woman nearing fifty, enters. She carries groceries.)* The last three children to die were each found with his corpse hanging from a cross in the middle of a grape vineyard. The Union of Campesinos, an outspoken advocate for pesticide control, is presently under investigation for the crime. *(Spying* DOLORES.*)* We now are approaching the house of Dolores Valle. Her daughter Cerezita is one of McLaughlin's most tragic cases.

(Upon sight of ANA PEREZ *coming toward her with her microphone,* DOLORES *hurries into the house.* AMPARO *intervenes.)*

AMPARO: You should maybe leave her alone; she don' like the telebision cameras too much no more.

ANA PEREZ *(to the "camera"):* Possibly this neighbor can provide us with some sense of the emotional climate prevalent in this small, largely Hispanic farm worker town.

AMPARO: She says es como un circo—

ANA PEREZ *(to the camera):* A circus.

AMPARO: Que la gente . . . the peepo like tha' kina t'ing, to look at somebody else's life like that t'ing coont never happen to them. But Cerezita's big now. She got a lot to say if they give her the chance. It's important for the peepo to reelize what los rancheros—

ANA PEREZ *(overlapping):* The growers.

AMPARO: Are doing to us.

ANA PEREZ: Cerezita. That's an unusual name. Es una fruta ¿qué no?

AMPARO: That's what they call her because she look like tha' . . . a red little round cherry face. I think maybe all the blood tha' was apose to go to the resta her body got squeezed up into her head. I think tha's why she's so smart, too. Mario, her brother, el doctor-to-be, says the blood gots oxygen. Tha's gottu help with the brains. So pink pink pink she turn out.

ANA PEREZ: And how old is Cerezita now?

AMPARO: A big teenager already. Cerezita come out like this before anybody think too much about it. Now there's lotza nuevas because lotza kids are turning out all chuecos and with ugly things growing inside them. So our pueblito, pues it's on the map now. The gabachos, s'cuze me, los americanos are always coming through McLaughlin nowdays. Pero, not too much change. We still can' prove it's those chemcals they put on the plantas. But we know Cere turn out this way because Dolores pick en los files cuando tenía panza.

ANA PEREZ: Uh . . . pregnant, I think.

AMPARO: Dolores tells me que no le importa a la gente and maybe she's right. She says all the publeesty gives peepo somet'ing to do. Peepo que got a lotta free time. It gives them a purpose, she says—like God.

ANA PEREZ: Señora, what about the boy?

AMPARO: ¿Qué boy?

ANA PEREZ: The boy on the cross . . . in the field.

AMPARO: Memo?

ANA PEREZ: Yes. Memo Delgado.

AMPARO: He died a little santito, son angelitos todos.

ANA PEREZ: That's the third one.

AMPARO: Yes.

ANA PEREZ: Why would someone be so cruel, to hang a child up like that? To steal him from his deathbed?

AMPARO: No, he was dead already. Already dead from the poison.

ANA PEREZ: But ma'am . . .

AMPARO: They always dead first. If you put the children in the ground, the world forgets about them. Who's gointu see them, buried in the dirt?

ANA PEREZ: A publicity stunt? But who's—

AMPARO: Señorita, I don' know who. But I know they not my enemy. *(Beat.)* Con su permiso. *(AMPARO walks away.)*

ANA PEREZ *(with false bravado)*: That concludes our Hispanic hour for the week, but watch for next week's show where we will take a five-hour drive north to the heart of San Francisco's Latino Mission District, for an insider's observation of the Day of the Dead, the Mexican Halloween. *(She holds a television smile for three full seconds. To the "cameraman")* Cut! We'll edit her out later.

(BONNIE and a group of small CHILDREN enter wearing calavera masks. They startle her.)

THE CHILDREN: Trick or treat!

ANA PEREZ: No. I mean . . . I don't . . . have anything to give you.

She exits nervously.

Scene Three

Crossfade to the Valle kitchen. It is late afternoon. YOLANDA is breastfeeding her baby. CEREZITA observes.

CEREZITA: I remember the first time I tasted fear, I smelled it in her sweat. It ran like a tiny river down her breast and mixed with her milk. I tasted it on my tongue. It was very bitter. Very bitter.

YOLANDA: That's why I try to keep calm. Lina knows when I'm upset.

CEREZITA: I stopped drinking. I refused to nurse from her again, bit at her breasts when she tried to force me.

YOLANDA: Formula is expensive. Breastfeeding is free. Healthier, too. I'll do it until Lina doesn't want it no more. *(YOLANDA buttons her blouse, puts the infant into her crib, sings to her softly.)* 'Duerme, duerme, negrito' . . . *(Continues singing.)*

CEREZITA: But imagine my sadness, my longing for the once-sweetness of her nipple.

(YOLANDA positions CEREZITA for her weekly beauty treatment. She takes out various beauty supplies from a bag. MARIO enters, towel wrapped around his hips. He is well built, endearingly macho in his manner. He is drying himself briskly.)

MARIO: ¡Hijo! It's freezing! These cold showers suck, man! We should all just get the fuck outta here. I'm gonna move us all the fuck outta here!

CEREZITA: Where to, Mario?

YOLANDA: Go 'head, chulo. You keep taking those showers purty boy and your skin's gonna fall off in sheets. Then who's gonna want you?

MARIO: The water was cold, man. Ice cold.

YOLANDA: I turned the water heater off.

MARIO: Great. My skin's gonna freeze off from the cold sooner than any chemicals. How can you stand it?

CEREZITA: Where you gonna move us to, Mario?

MARIO *(looking out the window)*: What?

CEREZITA: Where we going?

MARIO: I dunno. Just away.

YOLANDA *(has filled up a glass of water from the faucet)*: Here.

MARIO: Chale. The shit stinks.

YOLANDA: C'mon, chulo. Tómalo. Why don't you just throw it down your throat better? It's the same thing. You suck enough of it up through your skin taking those hot baths three times a day.

MARIO. Two.

*(*YOLANDA *starts to spread the beauty mask onto* CEREZITA*'s face.* DOLORES *can be seen coming up the porch steps after her day's work.)*

YOLANDA: You wanna see Lina's nalguitas? They're fried, man. The hot water opens your pores and just sucks up the stuff. She cried all night last night. This shit's getting outta hand! Doña Amparo told me—

DOLORES *(entering):* Es una metiche, Amparo.

YOLANDA: They shot through her windows the last night.

CEREZITA: Who?

YOLANDA: Who knows? The guys in the helicopters . . . God.

DOLORES: Por eso, te digo she better learn to keep her damn mouth shut. Ella siempre gottu be putting la cuchara en la olla. I saw her talking to the TV peepo last week right in front of the house. It scare me.

YOLANDA: What are you scared of?

DOLORES: They come to talk to Amparo on the job yesterday.

MARIO: Who?

DOLORES: The patrones.

MARIO: The owners?

DOLORES: Not the owners, pero their peepo. They give her a warning que they don' like her talking about the rancheros.

YOLANDA: Cabrones.

DOLORES: She gointu lose her job.

MARIO: Got to hand it to Nina Amparo. She's got huevos, man.

DOLORES: She got a husband, not huevos. Who's gointu support Cere if I stop working?

(The room falls silent. CERE*'s face is now covered in a facial mask.)*

MARIO: Well, I better get ready. *(He starts to exit upstage,* DOLORES *stops him.)*

DOLORES: I better see you back el lunes temprano ¿m'oyes? I got the plaster falling down from the front of the house.

MARIO: Okay.

CEREZITA: Where you going, Mario? .

DOLORES *goes to the stove, puts a pot of beans to boil.*

YOLANDA: Don't talk, Cere. You're gonna crack your face.

MARIO: ¡San Pancho, 'manita!

YOLANDA *(running a slab of facial down his cheek, softly):* Better stay away from the jotos, you don't wanna catch nothing.

MARIO *("slabbing" it back, teasing):* I got it covered, hermana.
DOLORES: What are you two whispering about?
MARIO: Nothing, 'amá.
DOLORES: You know secrets kill sometimes.
YOLANDA: It was nothing, 'amá.
DOLORES: You don' believe it, pero tha' place, it's crazy. They got all those crazy peepo que sleep on the street nowdays. You never know one could come up and shoot you right in the head.
YOLANDA: They're shooting us here anyway.
DOLORES: ¿Crees que soy una exagerada? We'll see.
MARIO *(mimicking):* "We'll see." ¡Hijo! I hate when she says that like she knows something we don't.
YOLANDA: I know.
DOLORES: Pues, maybe I do.
MARIO *(coming up behind* DOLORES *and wrapping his arms around her):* I'm fine, 'amá.
DOLORES *(softening):* "I'm fine, 'amá." ¿Qué sabes tú about "fine"?

*(*AMPARO *can be seen coming up onto the porch.* JUAN *trails behind her carrying a five-gallon tank of spring water. He wears jeans and a flannel shirt.)*

AMPARO: ¡Halo! Anybody home? I got a sorprise for you!
DOLORES: Abra la puerta, hijo.
AMPARO *(calling out behind her):* Right here! This is the house!
MARIO (going to the door): What's up, Nina?
AMPARO: ¡Ay! Te vez bien sexy.
DOLORES *(spying* JUAN *at the porch):* ¡Ay, Dios! *(She quickly pushes* CEREZITA *out of sight, drawing a curtain around her.)*
YOLANDA *(whispering):* Why do you do that to her?
DOLORES: Cállete tú.
MARIO *(to* JUAN*, with interest):* Hello.
JUAN: Hello.
AMPARO: This is my godson, Mario. *(*MARIO *takes the bottle from* JUAN.*)*
JUAN: Thanks.
MARIO: No problem.
AMPARO: That's Yolanda y su baby, Evalina.
YOLANDA: Hi.
AMPARO: And this is my comadre, Dolores Valle.
DOLORES: Halo.
JUAN: Mucho gusto.
DOLORES: ¿Habla español?
JUAN: Soy mexicano.

DOLORES: ¿Verdad?

AMPARO *(aside):* Half y half.

MARIO *(suggestively):* Like the cream?

AMPARO: And a priest. Father Juan Cunningham.

DOLORES: Mario, why you standing around sin ropa? Go put some clothes on.

MARIO: All right. I was just helping the man. I mean, the priest.

(He puts the water onto the dispenser, then exits. JUAN*'s eyes follow him.)*

DOLORES: Siéntese, Father.

YOLANDA: So, where'd you get the water, Doña Amparo?

AMPARO: The Arrowhead donated it.

JUAN: Thanks to Doña Amparo. Last week's newscast stirred up everyone!

AMPARO: It wasn' me. It was la crucifixión. That's what brought the newspeepo here.

DOLORES: ¡Es una barbaridad!

AMPARO: The newspeepo, they wanted to talk to Cereza, comadre.

DOLORES: Y ¿por qué?

YOLANDA: Cere knows, 'amá.

DOLORES: Cerezita don' know nothing.

YOLANDA: She sees.

DOLORES: She sees nothing. *(To* JUAN*)* She looks out the window all day, nomás. What can she see?

(The lights crossfade to CEREZITA *at the window.* BONNIE *sits near her, playing with a doll. She prepares bandages for it, tearing a flour-sack cloth into strips and wrapping it around the doll's head.)*

CEREZITA: The sheep drink the same water we do from troughs outside my window. Today it is an orange-yellow color. The mothers dip their heads into the long rusty buckets and drink and drink while their babies deform inside them. Innocent, they sleep inside the same poison water and are born broken like me, their lamb limbs curling under them.

BONNIE *(takes out a thermometer and puts it into her doll's mouth):* ¿Estás malita, mija? *(Checking the temperature.)* Yes. I think you got "it." *(She rubs the top of its head, chanting.)*

'Sana sana colita de rana, si no sanas hoy sanarás mañana.
Sana sana colita de rana, si no sanas hoy sanarás mañana.
Sana sana colita de rana, si no sanas hoy sanarás mañana.'

(She puts the doll into a small box and covers it tenderly with the remains of the cloth.)

CEREZITA: I watch them from my window and weep.

(Fade out.)

C.W. Moulton

ATWATER: WATCHING B-52'S COMING IN TO CASTLE

Twenty five years ago
I flew out of here with a crewcut and watched
the Merced River flow
into the Sierras.

Today I'm walking between the landing lights
at the end of the runway
out Fox Ave. to the corner of Belleview.

On the ditch bank I find a discarded clawfoot
bathtub, turn it up, get in and watch
a gray squirrel pack his cheeks with seed.

In the far distance a green CAT
is plowing a field of sandy loam
raising a tempest of dust where a B-52 is turning
lining up an approach.

It takes a long time for these frankensteins
to land. Nearer a stone frame tankhouse windmill
hails me.

Nearer still half a herd of cows leans on the fence
looking on with red tags in their ears,
and some with yellow ones.

Smell of cowshit is pungent, more vivifying than
the cities. Blackbird croaking on the fence
was eating bashed insects off the road.

Originally appeared in *Piecework: 19 Fresno Poets,* published by Silver Skates (1987). Reprinted by permission of Mike Moulton.

I don't like bombers because they usually
manage to hit some guy who hates his job.
Who thinks the big bosses are a bunch of ass-holes.

Behind me the shredded carcass of a truck tire
has been discharged to the whippoorwill.

VISITING MUSSEL SLOUGH LOOKING AT STONES AND CLOUDS

> The railroad owners make money simply so they can make more money. The ranchers support a way of life in concert with the land.
>
> Land grabs, manipulated sales, political and journalistic maneuvering threaten to rob the people of their land.
>
> —Frank Norris, *The Octopus* (1850)

> We are an explorer, a miner, an insurance agency, a farmer, a pipe company, a builder, a trucking firm, a gold mine, a logging company, a printer, an energy company. We are 18 companies all across America waiting to serve you. We are also a railroad.
>
> —a commercial for Southern Pacific

> Eating is a political act:
> You make the connections.
>
> —George Ballis

The stones here have faces heavy against the cold ground,
skin wrinkles, fingertips, eye sockets. John Henderson,
Archibald McGregor, faces upturned toward choirs of light.
Edwin Haymaker, Iver Knudsun, noses stuffed with dirt.
This folded speckled granite the face of James M. Harris?
Windblown boulder out there the forehead of Dan Kelly?
Clouds swell with light and drift over the green fields.

They build barns, roads (that went "this way" and then
"over that way"), dirt farmers getting in and out of
wagons, with shovels dug canals to drain this that was

a swamp, planted and harvested crop after crop, got up a
school; the men that were shot down here, must have walked
through gates they built with their own hands opening
into gardens luxurious with green leaves and came before
the guns of the law, guns of the railroad, with little
paper promises from the government in their hands, stood
there and got the shit blown out of them.

And the big mouth blind beast beggar of a railroad bribed
the government and after a hundred years they're still here
with Standard Oil, Dole, Chiquita, Bangor-Punta, people
that don't live here, multi-nationals, multimillionaires
getting ready to heist the water off Mount Shasta with
your tax dollar, putting together another monopoly (food
this time) buying up TV stations calling themselves heroes
now that these fathers are in the ground and forgotten.

Roberta Spear

SOME VOICES

In the last days of October,
when clouds made the shapes of the dead,
we stretched our small bodies out
on the levee and sucked the crimson
out of pomegranate flesh.
Flat on our bellies and trying
to make the willows kiss
the gray water, we watched
their feathery shoots scrawl
on the surface the history of
a leaf flipped over or a stone
punted around the world
and back to this very spot.
A history of anything blessed
both with silence and the power
to break the sluggish surface
into light. Soon, something else
caught my eye: across the river,
the glint of a steel track
tracing the curves of an orchard.
And, peering out at that hour
from the streamer of windows,
faces that made us stand and watch
as though the slope of a cheek
or a waving hand could
bring life to the flatness
of this sky, this land.

I always knew where I came from,
but I couldn't believe it.

In childhood dreams, strong tides
carried off the mockingbirds
and the tufts of cotton
exploded in the night,
like Italian stars. At the station
a mile into town, my mother
straightened her stockings
and strangers thrummed on valises
and old uncles bowed, begging
for hugs. The crowd gathered
slowly, all smiles and tears,
until a man in a black cap,
his teeth like a keyboard,
sang out *Leaving for Corcoran.*
And nothing but steam filled the air.

Poor Corcoran.
Poor Shafter. Poor Taft.
The blunt steel nose pushes
its way through those
brief exhalations of dust.

Ain't no place to get off,
the man slumped beside me mumbles,
but I'm going there anyway.
A tumbleweed races us through town
and the ricks of cotton
stretch out like bones
along the rails. And he recalls
with a sudden sweetness
old lady Roosevelt back in the '30's
standing on those crates
in a sun hat wide as an ocean liner,
soothing the Okies with words
that swept the dust
right off their bread.

The women still come here, he says—
aunties, daughters, and the rest

to see their folks
at the new prison. Men who'll
leave *after a spell*, remembering
little but the heat and fog,
the sound of wheels grinding
in the middle of nowhere
like a woman's voice
as you fall off to sleep.

My son kneels down and places
a penny flat on the rail,
then sifts some gravel
into a pyramid. Seconds later,
he adds the torn sleeve of a foxtail
which has blown his way. Anything
that will let him touch
that huge, steamy body which
he has come to believe
is God. The yellow warning line
still two feet behind him,
he turns his head and rests
his ear on the track,
listening for the distant hum.
But when it comes, he reaches
for my hand and stays close.
The engine grows immense,
hissing; the glint of steel
blinds us. And as we start to
mount the silver steps, I can see
that he has already learned
some voices will take you anywhere.

Bill Barich

PRISON VALLEY

THE SAN JOAQUIN VALLEY had a new nickname, Prison Valley. There were prisons in Madera, Corcoran, Delano, Avenal, and Wasco. In other sections of the Central Valley, in Chino, Folsom, Ione, Soledad, Stockton, Tracy, and Vacaville, there were also prisons. More prisons were being built in California than anywhere in the world. Frequently, they were built on farmland stripped of its value, gone to pebbles and hardpan. Corporate farmers, the titans of agribusiness, often sold the dead land to the state for a handsome score.

If you thought of prisoners as a new sort of crop, drought-resistant and growing incrementally, the future in California seemed bright. In the early 1970s, fourteen out of a hundred convicted felons were sentenced to a prison term. The current ratio was thirty-five out of a hundred. Drugs were an excellent fertilizer, and the crop tended to reseed itself. A high rate of recidivism was guaranteed. No wonder, then, that prisons were known as "gray gold."

Avenal was just down the road from Coalinga, so I drove over to take a look at it. Along the way tumbleweeds blew across Highway 41 and caught in fences. A single black cow stood in a huge field that was so dry and rock-strewn that my eye couldn't pick out a hint of green. The temperature kept rising.

Avenal made Coalinga look like a cultural mecca. How poor, small, and isolated the town was—as isolated as some towns in the backwaters of the Far North. The distance from Avenal to a supermarket or a movie theater was about thirty miles. Nobody wanted to live there. Correctional officers earned $2,400 a year in hardship pay just for working at the prison. The warden got as far away as he judiciously could and had a house in Hanford, near Fresno.

Asking around town, I learned that the prison had come to Avenal through the efforts of a local pharmacist, Nick Ivans. He had read an article in the paper about how the state had $495 million to spend on prison construction, so he and some other town leaders began their successful lobbying.

Avenal State Prison, built in 1988, was a Level Two facility for lower- to medium-security inmates. It had 3,034 beds, but 3,289 cons had to jostle for space two months after the cells were opened for business. There were about 4,200 inmates now.

Around Pelican Bay State Prison, I had witnessed a weird euphoria and had seen the impact a prison could have on real estate speculation, but only in the San Joaquin did I hear about the most significant wrinkle in the scheme—prisoners could be counted as residents of the town or city where the institution was located. By kiting its population with bad guys, Avenal had set itself up for an annual bonus of several hundred thousand dollars from various state agencies, earning funds that it would not have been eligible for otherwise.

No other bonus seemed to be accruing to Avenal, though. The vaunted boom was sounding no more loudly than the tap of a spoon against a washtub. A new motel, its parking lot empty, had rooms by the month, the week, and quite probably by the minute. A new apartment complex let the public know that it was "Now Renting," as if the privilege had not been available yesterday.

Hillside Vistas, a proposed subdivision, amounted to an arrow pointing to open fields. Foxborough had fared a little better. A few three- and four-bedroom houses had gone up, but the carpenters had put down their hammers halfway through some others, leaving behind framed walls and two-by-fours in stacks. Tagged stakes marked the borders of many lots where no houses stood. Instead of lawns, weeds sprouted copiously from the annealed and useless earth.

Through Foxborough more tumbleweeds were rolling, great, thatched spheres bowled across the plain by unseen hands. The smattering of luckless owners who'd closed escrow before the demise had a view of migrant shacks, laundry hanging on clotheslines, and scrawny children with even scrawnier dogs.

A Mexican man moved along the semipaved streets of Foxborough, pushing a white cart and crying, *"¡Helados! ¡Helados!"* His ice-cream bars found no takers.

For a long time, I studied the tumbleweeds and heard the vendor's plaintive cry, thinking with a warped brilliance brought on by incipient sunstroke that the California Department of Corrections ought to cut a deal to buy the unsold Foxborough units at a discount and transform them into prison adjuncts for the bedless cons of Avenal. The houses could then be allotted by crime, with the rapists here and the molesters there, all the bug-eyed killers and the narco-creeps thrown together on the same block to create an entire suburb dedicated to casual mayhem and first-rate violence.

The plan had a simple elegance. It would please everybody from developers to reprobates to penologists to homeowners who were concerned about drugs and crime in *their* suburbs. They all came out ahead. Any way you looked at it, from any angle, the plan was a winner.

Corcoran State Prison, where Charlie Manson was taking a long vacation from the streets, was not far from Avenal, so I made an appointment to talk with Warden Bernie Aispuro there. Aispuro had put in forty years of service at California's penal institutions and was the ranking warden in the state. He had worked at Soledad, Tracy, Susanville, and San Quentin before coming to the San Joaquin, and Corcoran was his last stop before retiring.

A sharp, chemical stink infected the air in Corcoran. It came from a plant that processed cotton and alfalfa seeds. The plant belonged to J.G. Boswell Company, an agribusiness titan in the West. Named after a retired colonel from Georgia who'd established the company in 1924, Boswell held more than 140,000 acres of farmland in California alone, the Miller and Lux of its time. Its fortunes had been built on irrigation and on loans made to other farming concerns in the San Joaquin, and its global operations were directed from some high-tech offices in a Los Angeles skyscraper.

In effect, Boswell owned and managed Corcoran, another dry, desolate, unredeemable place. Prison families, black and Hispanic, strolled the streets and cashed their checks at a check-cashing outlet painted a shocking pink. They had a new, pork-barrel YMCA and a theater that alternated Hollywood films with movies in Spanish. There was a boarded-up, rust-colored Santa Fe Station where the trains didn't stop anymore.

When the Department of Corrections had chosen Corcoran for a prison site, they had bought a parcel of land from J.G. Boswell—land that was maybe not quite so useful for farming anymore. It was a few miles from town in a belt of cotton fields furrowed with irrigation ditches. In the standing water, in mosquito-dense clouds, avocets were wading and feeding. The soil was grayish and dotted with cotton bales.

In the distance, behind a chain-link fence topped with concertina wire, loomed the prison complex. It bore a strange resemblance to a power-generating facility. The prison building proper was a concrete block with slits for eyes, and what the eyes looked out on was so much nothing. Even sounds were in limited supply, rationed like every other tactile sensation—cars coming and going, the clank of a flag against a metal pole, birds chirping obliviously under a gnawing sun.

How remote the prison seemed from anything having to do with California, I thought. Few Californians would ever see it, much less know that it existed, and yet, increasingly, our prisons and all that they represented were becoming more and more integral to life in the Golden State.

At a guard station, I identified myself and the purpose of my visit. After that, I sat for a while in a sterile waiting room in the prison's reception area. The room had a hospital feel, an antisepsis meant to disguise all the festering illnesses within. A sickness of the soul was the chief complaint at Corcoran, and those who suffered from it were supposed to be just as invisible as AIDS patients or the victims of terminal cancer. We had given up on trying to heal the soul's complaint.

Correctional officers marched through the waiting room at intervals, going at a military clip. Their combat boots rapped against the clean, unscuffed surface of the glossily waxed floors. They were dressed like commandos in green camouflage uniforms. They looked rugged, solid, and devoid of all sympathy, particularly the younger ones, who seemed years away from making a compassionate gesture. Not a few of them had grown symbolic moustaches, as boys do when they're sent to war.

The guards didn't appear to be the sort of men who'd have much in common with the prisoners. Instead, they projected a soldierly apartness. You could have dropped them into any foreign jungle and put them to work killing guerrillas. Doing time at Corcoran, out in the vast reaches of nowhere, would not be easy.

Jon Veinberg

MOTEL DRIVE

Next door
the room is padlocked
from the outside, inside
the children are ransacking
the cupboards and playing
catch with the empty
Cool Whip containers
and most afternoons
you could've found me
behind any of these blue-
stained doors, my girdles
unhitched, my dusty nylons
flagging the window,
contesting the smoke rings
that peel off these sluggish
lips like clouds being freed
from a sore and blistered earth.
You could've plotted out
any old highway that angled
into the railroad tracks
and found me tucked between,
sleep's week-old spittle
on my cheek, spooning the lumpy
pudding and refusing weight.
You could've tossed your jacket
into the sperm-stained closet
and waited out the weak pulse
of noon, the haggle and choke

of the water cooler, and the sun's
desire to flash against the tin-
foiled courtyard.
        Buttermilk and crosstops
for dinner. The truckers
and three-legged dogs scuffle
beneath the neon's ferocity
and you could've asked for Ginger
or Blossom or Bonnie Blaze
and been steered to me. You
could've tracked the whiff of sour mesh
and sweet sin into the Miracle Inn Club
where men in dark tilted hats
roll the bones for another
round of widowmakers, and heard
through the fractured glow
of the t.v. how sad it was
for the tall blond daughters
of millionaires to have gone
so easily berserk. You could've
followed me to the chain link
fence, high heels dangling
from my purse, at the intersection
of cracked ribs and a thousand winks.
You could've slipped past
the tortilla hawkers dodging
the tow trucks, the cats
pairing off in dumpsters,
and the yellow fist of morning
to take my wrist, thin
as a membrane into your hand.
        You could've
stretched out on the rollaway,
felt the pistol's cool nozzle
hidden in the cushions
and watched me will myself
to sleep. My litheless body,
coiled and snug, could've,
once again, fired the hearth
of your quivering waist.

HARDPAN

It could be the chipped stars out of alignment
Or the sadness of old men spitting through screened porches.

It could be the long crawl of the afternoon
And the sun's spindle going crazy splicing the window shades

Into the unkempt bedrooms of long-legged boys
Who'll crowd the pool halls to ask about the pale lovers

Whom they thought would never desert them.
Here, the trees are burning from the roots up,

The noon sky divides and crackles like cellophane
And the wind is as far off as a postcard.

Here, among the cakes of sweat and easily spooked dogs
He watches the body he had just floated out of:

The face anonymous and cracking like a mirror,
The double-winged fly calloused to his lip,

And wonders what could've tempted it to pick this place to die.
Here, beside the pelts of heat and the shovel

Questioning its strength. It could be the rumpled playgrounds
Or the blown tires and soiled mattresses clogging the canals.

It could be the promise of hardpan: to chip
A sliver off the world's clayed and trampled shell.

Here, where spirits too tired to rise scar the air
He watches the dead swallow their tongues

And waits for heaven's soft scaffold to meet him half-way.
Here, where the earth's metronome swings away in slow motion

And the sheaf of clouds advances one dull shoulder at a time
While the gardens are always on the verge of fire.

Jon Veinberg

TERRA BELLA

Morning,
I would dress you as a child
unclenching its hand
for the first time.

Already
the pre-dawn birds
have ended their windless flight,
their feathers greying
under the raised lid of the sun

and I think of you, gone,
as the last flower
able to nourish itself here.

No one believed
that what mattered here was fields,
barley flooded and docile,
the persistence of gopher,
the ambition of weed.

Nor could they see
that the elms were dying,
unable to bear the weight
of another circle—
bark curled and thirsty
and their sickness
thicker than my own.

I am the last son
of an old testament sharecropper
left to walk the plots
of ploughboys in their teens
who chose not to return,
to wonder at what point
the nerve forgets its quiver

or where the eye wanders
other than back to memory.

In the distance
there is a bus
making its last run
through the rubble of car parts
and abandoned stoves.
As I step towards it
the earth holds its breath
waiting for me to follow the others,
waiting till there is no place left
from which to leave.

Dixie Salazar

THRIFTER'S PARADISE

Here, the clothes have really lived,
pajamas, worn out from sleep,
yellow moons of armpit stain
on bridal gowns, a collapsed girdle.
Here is the matador
and Spanish dancer lamps,
fringed, bordello red,
and the frog-bellied old man
who says, "I'm not much,
but I'm yours,"

his plaster of Paris nose
crumbled half away
like a leper's.
Here at Thrifter's Paradise,
mannequins have the look
of battered wives. One is posed
in a "Too Hot to Handle" t-shirt
and blue puff slippers.
Down polyester aisles,
the trophies drag

chipped silver bowling balls.
In appliances, Arthur Godfrey,
tinged a sinister green,
sells Oxydol, while the child nearby
smears Velveeta fists on Arthur's
mutely moving tropical lips.

With back issues of Fortune,
Money, American Girl
and the salvation
of Meal Vac and Tupperware,

we all gather here.
The size twenty-two lady wrestler
in the daisy shift, tells me
there's some cute stuff
if you can find your size.
The Hmong man in women's slacks
studies fluorescent bathing caps.
And the mother whose
coffee-skinned twins
are named Alfonso and Sacramento,

buys an Easter basket
woven from plastic beads
and safety pins,
and a blow up swan
that will never inflate,
will always stay a flat,
seamed, listless swan,
never to sail
on Avacado Lake, lapping
and soothing Sacramento's tears away.

But we are all here
buying scratched Del Shannon records,
pablum stained t-shirts,
and bride and groom salt shakers,
in line with the cast offs, forgotten
and lost. Some part of us
kneels down in the dust,
turning like the twinned ballerina,
bumping round and round
before the cracked mirror.

DIXIE SALAZAR

HOTEL FRESNO

On Belmont, low riders
give the finger to King Muffler,
the moon, whatever moves in the dark.
Headed for The Flamingo,
La Nueva Tapatia, Moon Dog Inn,
they wait for the light to change

on a street of no change.
In a room, shaded
by dark feathered birds,
the old man waits,
a room that holds its breath
behind the window fan
scissoring the air.
With a sweaty finger,

he dials time slowly.
Time is busy
and so he waits.
He waits at the Hotel Fresno
for the lady with the alligator purse
and foreign accent to collect,
for the man with the cigar-shaped scar.

On the corner of Inyo and Y
a young man waits,
neon slipping down his brow
to where his shades cast back
the Walk—Don't Walk
of a red dissolving hand,
like the lost guide of Yaki winds.

But he doesn't cross
as if frozen between centuries,
waiting for a different light.
At Mayfair, the all night clerk
rings the cigarettes by,
seeing lights of El Dorados,
The Big Spin, and Dollar Machines.

A teenager takes her Salems and change,
steps outside to the curb
where she cocks her hip and thumb,
waits for a ride to anywhere,
away from these people
wearing loneliness
like old bathrobes pulled close.

On Motel Drive,
shadows jump from box cars
into cinders and bottle caps,
wait for the freight's long vowel
to shudder away,
so they can cross the singing ties,
find some way to crack the frozen

core of this winter night.
She has always been in line
even as she waited to be born
while her mother waited
in line after line, as she waits
now for her number
that is never called,

one leg-bound baby,
another low riding her hip.
All day asleep in the park
to the click of checkers,
all night, rocking
in time to the rain.
He has stopped waiting now.

nothing to wait for here
where the leaves don't fall,
and the constellations petrify,
where only watermarks bloom
on the wall, and the mirror
reflects the moon, hooked
around a rusted rosette.

DIXIE SALAZAR

REINCARNATION OF THE COMMONPLACE

A white thumbprint of moon
stamps the blue
behind a sky-sized hand
where Mrs. Day, palm reader of Highway 99,
divines the shape of sunlight

on the grackle's wing,
studies the snail's lifeline
of broken silver wanderings,
breaks the code
of the winter solstice moon.

It winds her on its ancient spool,
threads her like a bone needle
stitching firelight to dawn,
foretells the enchantment
of water, salt and wind.

When they have eaten their fill
of No. 4, Klein's Truck Stop Special,
chosen "White Shoulders" or "my Sin"
from the ladies room,
or colored condoms

from the men's room,
travelers will park Winnebagos
by the rose trellis
and ask Mrs. Day to fathom
their folded up dime store hearts,

to forecast the romance of fingernails
and salt, reveal to them their true
Egyptian midnight selves.
But when neon blue spells CLOSED,
Mrs. Day enters the almond sweet air,

black orchards where
the raw boned moon swallows up
the coyote mother's howl
and the owl digests
his midnight cartilage.

Heavy rigs hum then,
with the buzz of power lines,
and Mrs. Day need not predict
the six car pileup on Interstate Five,
the oilspill off Badger Pass Road,

the unexplained violence
of the bloated moon.
She trails its bloodline then
through wet vineyards; travels backroads
past "Rifles Made to Order,"

"Adventures in Moving With U-Haul."
And beyond the Bluebird Motel,
in the pastel aura of mushrooms and gnomes,
she will trace the cryptic shadows
of her lover's eyelids,

the one who asks nothing
from the moon's light, circling her eyes
and melts with her
in the formulas of pure shadows
and the simple alchemy of love.

FRESNO PET CEMETERY

Behind the toy windmill,
purple swells into black—
trees gather darkness,
and roots drink slowly
from the flooded plots

where their own bones sink now.
An old stump is sticky
where flies lay spumes of eggs.
A nameless cat wanders,
sniffling the miniature markers,
mewling for food
or the weight of a warm hand.

She stops to lift a leg over
"Nicki Smith, Our Punkinheimer,
Stop and Smell the Tulips."
The pet owners leave chrysanthemums,
sprayed blue and glitter dusted,
brush fallen pine needles
from the pet photographs
and whisper the loam sunken names,
calling for a wagging dog
to come bounding up,
licking and sniffing the dusk...
"Nugget Sears, He Loved Everyone."

When all the humans leave,
a flock of thrushes
dips out of the trees,
releases from their tiny graves,
the pets, who come running now
only to the faint call
of birds and light,
moving over "Danny Bushby,
Thanks for Everything, Sweet Boy,"
and "Po Po Stephanian,
He was Like a Son.
May God Bless Him."

Their water-darkened bodies
blend with the atmosphere,
far beyond the star-washed slab
of "Duchess of Emerald,
Life Too Short, Death No Reason,"
And in this waning light,

Duchess of Emerald jumps
the last white-washed fence,
meets the Big Swede
in some forbidden earth-lit field,
and they lie together
in joy and absolute defiance.

C.G. *Hanzlicek*

ADRENALIN

It's best I put down my pen now.
I should give the trees
A deep watering,
I should hunt the undersides of leaves
For the snails
That have been eating the pansies,
I should butter a slice of bread,
I should keep everything low-keyed.

I sat down to write of adrenalin.
In the past month, two climbers
Lost their grip on El Capitan,
A parachutist
Pulled the wrong cord
Just before she hit a field of wheat,
Six bodies of rafters
Were wedged among rocks in the Kings River.

It was all for fun, see?
It was a way to take
A sluggish pond of blood
And make it race like the Kings.
I should put down my pen.
The trees have been dry too long,
The snails must be stopped,
And I keep seeing faces.

UNDER STARS

It's night on the water in the canal,
Darker still in the kingfisher's burrow;
Night glistens among the day lilies,
Walking the track of a garden snail.

It's night in the red eye of a box turtle,
In the stirring of loam
Where a sprout splits the seed,
Night in the thought that is better off dead.

It's night in the doorway of my daughter's dream,
In the hand on my wife's shoulder;
Night stretches out in the shadow
Cast on the lawn by my shadow.

Night quietly searches for itself
Where the moon breaks up in a maple,
But without people under stars
There would be no one to find it.

Philip Kan Gotanda

from FISH HEAD SOUP

Characters

TOGO IWASAKI, "PAPA," Japanese American, mid-sixties in age, seemingly mentally ill

DOROTHY IWASAKI, his wife, in her middle fifties but still attractive, waitress in a Japanese restaurant

VICTOR IWASAKI, their older son, in his mid to late thirties, tour-bus driver for a Japanese hotel, Vietnam veteran

MAT IWASAKI, their younger son, late twenties to early thirties

Voices of THREE YOUTHS

VIET CONG SOLDIER

The voice of a DRILL SERGEANT

Time

Late fall, into winter, 1989.

Place

Town in the San Joaquin Valley, California.

Settings

ACT ONE. The Iwasaki house, Friday evening to Saturday morning.

ACT TWO. The Iwasaki house and various locales around the town, Saturday evening.

ACT THREE. The Iwasaki house and various locales around the town, Sunday.

Act One

Lights up on VICTOR *and* PAPA. *It's very wet and foggy outside. They've just returned from a church memorial service.* VICTOR *is helping* PAPA *into his overstuffed chair.* VICTOR *has a pager in his belt.* PAPA *starts to jerk his hands up and down in front of him, mouthing "Mat."* VICTOR *notices and gently grabs his*

---

hands, stopping him. He lets go and PAPA *starts again.* VICTOR *stops him.* VICTOR *and* PAPA *look at each other.* VICTOR *slowly releases his hands. He watches.* PAPA *doesn't move his hands.*

DOROTHY *enters through the front door with a dripping umbrella and crosses to the upstage bathroom.*

DOROTHY: How is he? Is he all right now?

VICTOR *(nodding):* Un-huh.

DOROTHY: I can't believe he acted like that . . . *(She goes into the bathroom and puts the umbrella into the shower stall.)*

DOROTHY *(calling from inside):* The toilet seems fine, Victor, it's not running, I don't see why you're making such a fuss about it . . .

*(*VICTOR *helps* PAPA *out of his wet coat.* DOROTHY *reenters.)*

DOROTHY *(taking off coat):* And in front of all those people—Mat's baseball coach, what's his name, big tall fellow . . .

VICTOR: Mr. Guilfoyle.

*(*DOROTHY *moves to the corner table with a small* obutsudan *[Buddhist shrine] sitting on top of it.)*

DOROTHY: . . . Yes, yes, he was there, and Mr. Sanders the principal, Mrs. Thompson his debate teacher—they all came up and talked to me so nicely you know, before the memorial service began. After all this time they still remember . . . *(Fingering* MAT*'s photo)* and then he jumps up and starts screaming and yelling . . .

VICTOR: I was watching him, okay, Mama. *(*VICTOR *begins wiping* PAPA*'s face.)*

DOROTHY: I was so embarrassed, all those folks there, wanting to show how much they loved and respected Mat and he gets up and starts to do that. I think he does it on purpose. Did you do it on purpose, Papa? Huh? To embarrass me in front of all those people . . .

VICTOR: He didn't, Mama.

DOROTHY: How do you know? Who knows what goes on inside of there? Besides, you said you'd keep an eye on him.

VICTOR: I was, I said I was watching . . .

DOROTHY: It's bad enough they have to sit there in a Buddhist church, incense smelling up the place, all that mumbo jumbo sutra chanting. I mean, it's okay for us Japanese, we're used to it. But they have to sit there breathing in that stuff and then they have to go up there in front of everyone and gassho . . . *(Starts to laugh)* Did you see Mr.—what did you say his coach's name was . . .

VICTOR: Guilfoyle.

DOROTHY: Right, right, he's so tall he nearly bumped his head on that lantern near the shrine and then he didn't know what to do with the incense . . . *(Stops laughing, looking at* PAPA. VICTOR *is combing* PAPA*'s hair.)* And then he starts doing that thing with his hands. *(Upset)*

DOROTHY: You have to go in today?

VICTOR *(shaking head):* Un-uh. I called in.

*(*DOROTHY *watches* PAPA *for a beat.)*

DOROTHY: Victor, the toilet's fine. Okay, Victor? Victor?

VICTOR *(nodding):* Un-huh.

DOROTHY: I have to get dressed.

*(*DOROTHY *turns and exits into the back hallway.)*

*(*VICTOR *stands there for a moment, then pulls out a large wrench from his back pocket. Grips it, staring down at* PAPA*. Then turns and disappears into the upstage bathroom door. We begin to hear the banging of pipes coming from within.)*

(Lights fade to black. VICTOR*'s banging changes into a watery, echoing din. We hear a droning sound.* PAPA *lit in a pool of light standing on top of his chair. Staring out, he begins to jerk his hands up and down wildly as if snagging with an imaginary fishing pole.)*

PAPA *(calling in a raspy voice):* Mat. Mat. Mat . . .

(Upstage, the silhouette of a man is brought up in a bluish watery shaft of light. Drowning in very slow motion—grasping upwards mouth and eyes gaping.)

DOROTHY *(off):* Victor!

(Lights immediately up on the house. Sound cuts away. PAPA *stands, disoriented, on his chair, perched precariously.* VICTOR *is emerging from the bathroom carrying a toilet in his arms, struggling to make his way carefully to the hallway.* PAPA *starts to fall.)*

DOROTHY *(off):* Victor! Victor!

*(*VICTOR *notices* PAPA *falling, hurriedly puts down the toilet. Starts to move towards* PAPA *when his pager goes off. Unable to make up his mind—should he phone in or catch his father? The pager continues to beep and his father continues to totter. He rushes over and catches* PAPA *just before he falls.* DOROTHY *speaks the following during the above described action.)*

DOROTHY *(off):* Know what I hate most about going to work this time of year? Know what? The fog. Know what I hate second most? Driving in this fog. It's so scary, can't see more than a few feet in front of you. And you can't slow down 'cause those damn semis will run you right off the road. So what do you do? You can't go forward 'cause you can't see but you have to keep going forward or you'll get run over—so what do you do, huh? What do you do?

(VICTOR puts PAPA back in his chair, then hurries to the phone, checking the phone number on the beeper.)

VICTOR *(phoning)*: Hello, Mr. Toyama? Mr. Toyama-san? . . . Yes, moshi-moshi, yes it's . . . Yes, moshi- moshi, Victor desu . . . Hai, hai . . . I'm, uh, uh . . . boku wa very busy desu . . . Not suppose to be on call today, desu, call the office desu . . . Yes, yes, sayonara desu . . .

(VICTOR hangs up and goes back to his toilet moving.)

DOROTHY *(off)*: Victor!

(VICTOR, struggling with the toilet, ignores her calls. PAPA is climbing back up on his chair, teetering.)

DOROTHY *(off)*: Victor! Victor!

(VICTOR sees PAPA starting to fall again. His pager starts beeping. Momentary confusion, then rushes over just as PAPA begins to slip off the chair. They tumble to the ground in a heap this time. DOROTHY, dressed in a kimono, enters from the back hallway, awkwardly walking in ministeps owing to the constraints of the kimono.)

DOROTHY: Victor, you have to keep an eye on him, you just can't . . . there's no telling what he'll do . . .

VICTOR *(helping PAPA up)*: Yeah, but I'm trying to . . . *(Continues.)*

DOROTHY *(overlapping)*: What's that sound?

VICTOR: . . . fix the toilet and then I'm supposed to be watching Papa . . .

DOROTHY *(interrupts)*: It's not broken, Victor . . . *(Spots the ripped-out toilet sitting on the floor)*

VICTOR *(shutting off beeper)*: And my clients are calling me . . .

DOROTHY *(staring)*: What the hell did you do . . . Victor, didn't I tell you . . .

(DOROTHY turns back to VICTOR and sees PAPA moving his hands again in an agitated fashion.)

DOROTHY *(unnerved)*: Victor, he's doing it again. He's doing it again, Victor . . . *(VICTOR hurriedly stops PAPA. DOROTHY tries to compose herself.)*

DOROTHY: I have to go to work. *(Turns to leave)*

VICTOR: You coming home late again?

DOROTHY: Look, we just need the money, okay? I can't help it. Ice cream executives. A special party or something.

PAPA: Mama. I want Mama. I want Mama. Mama?

(DOROTHY looks at PAPA for a beat. Something in her softens. Crosses to him, smiling.)

DOROTHY *(reaching in to stroke his cheek)*: Here I am, silly, Mama's right . . .

PAPA *(knocks her hand away)*: I want Mama. I want Mama.

(They stare at each other for a beat. She starts to adjust his clothing. He resists and tries to push her hands away, but she overpowers him, forcing

him to sit still while she adjusts his clothing. She speaks during the above struggle.)

DOROTHY: He started walking when he was just eight months old. I mean, really walking. Everyone was amazed. And when I took him to swimming lessons, like a minnow, just like a little minnow. Everyone right away started talking about the Olympics. And they were always trying to skip him—second grade, fifth grade—but Daddy wouldn't let them. Just wouldn't let them. Said it was showing off, "What would other people think, putting our kid ahead like that?" *(Finishes. Beat. Catching her breath.)* Mat would've been something very important by now. Something high up, like one of those international business brokers between Japan and America . . .

*(*DOROTHY *hikes up her kimono between her legs and turns to leave.)*

DOROTHY *(muttering to herself as she exits awkwardly):* Why do they make them like this, you can barely move. Damn thing . . .

*(*VICTOR *watches her leave, then turns back to* PAPA*. Notices he's upset.)*

PAPA: Mat's not my son, Mat's not my son . . .

VICTOR *(comforting him):* Papa? We were climbing this tree. You know, Mat and me? And Mat kept going higher and higher. "Stop, Mat, you crazy? Come back down." I was scared 'cause we were way up there you know, and if anything happened to him, Mama always blamed me, always. I kept telling Mat, "Come back, come back down," but you know Mat, he just kept going up and up. And then I fell. I broke my arm. Mat came down and looked at me. He started crying. Mama got so mad at me. "What if Mat broke his neck and killed himself?" *(Beat)* Sometimes I dream about him. Mat. He's somewhere far away, Papa. He's happy, too. And he's so high up, the world could never pull him back. Never pull him under . . .

PAPA *(sadly shakes his head back and forth):* Victor's house, Victor's house . . .

VICTOR: No, no, Papa's house. Papa's house . . .

(Dim to darkness on VICTOR *and* PAPA*.)*

(Night. Iwasaki house. Foggy outside. Shadowy figure enters through the window, flashlight and carry bag in hand. Late 20s to early 30s dressed like a cross between a Japanese gangster and a sleazy Hollywood producer. Appears as if he's been up for several days, unshaven, hair unkempt. Flashes light around. Begins picking up objects, touching them, inhaling their smells. Picks up a vase, quietly studies it for a beat, then playfully tosses it from one hand to the other hand. Crosses to PAPA*'s chair. Notices* PAPA*'s hat sitting on the seat. Picks it up, examining it, and then puts it on. He sits in the chair. Lights come up abruptly.* VICTOR *is standing by the light.)*

MAT: Hi, Victor.

(Silence. VICTOR *stares, wide-eyed.)*

MAT: It's me.

(More silence. VICTOR *continues to stare.)*

MAT *(moving towards* VICTOR*)*: It really is, Victor, it's . . .

VICTOR *(raising wrench threateningly):* Stay back . . .

*(*MAT *stops.)*

MAT: I know this is a bit of a shock—I mean, after all this time and everything. But it's me.

*(*VICTOR*, desperately confused, begins to make high-pitched whimpering sounds.)*

MAT *(worried):* No, no, don't do that, Victor, don't do that . . . you're all right, you're all right, you're going to be just fine . . .

*(*VICTOR *starts to move around the house with* MAT *following him.)*

VICTOR: See, see, I'm fixing the house, I am 'cause it's all broken down and rotting and Mat can't see up, 'cause he's so high up and it's all beneath the surface, inside the walls under the floors and I'm ripping it out, ripping . . . *(Continues.)*

MAT *(overlapping):* Calm down, calm down. It looks great, you're doing a great job, you are . . .

VICTOR: . . . it out, the pipes all leaky, all the wiring, making it better, making it better. See, see inside, inside here . . . *(Continues.)*

MAT *(overlapping, moving closer):* It is me, Victor. It's me, Mat. I'm back, Mat's back . . .

VICTOR: . . . all copper pipes, all copper pipes, all copper . . . NO!

*(*VICTOR *swings his wrench wildly at* MAT.*)*

MAT *(ducking):* Victor, what the hell you . . .

*(*PAPA *enters from the hallway.)*

PAPA: Mat? Mat, is that you?

*(*MAT *notices* PAPA*'s strange appearance.)*

MAT: Yeah, it's me, Papa . . .

*(*PAPA *stares at* MAT.*)*

PAPA: You kids better go to bed now. I want to get an early start tomorrow. Victor, tell Mama I want fried chicken and onigiri [rice balls]. More salt, too, on the rice—not enough salt last time. *(Beat)* Who knows? Maybe we'll get lucky this time and catch him. Yeah. Haul him in kicking and screaming like some big striped bass.

MAT: Papa, it's me. Mat. I'm back. *(*PAPA *stares at* MAT.*)*

PAPA: Mama? Mama! Where the hell is she? It's 1:30 in the morning, where the hell is she, Victor?

VICTOR: She has to work late sometimes.

PAPA: Victor, make sure all the equipment is loaded into the car. And don't forget the chum I got frozen in the icebox. Oh, and pack my old pole, too. *(To* MAT*)* It's yours now. *(*PAPA *turns and walks to the hallway. Stops and turns to* MAT.*)* Aren't you dead?

VICTOR: Namu Amida Butsu, Namu Amida Butsu . . .

*(*PAPA *turns and exits. Lights begin a slow fade.* VICTOR *withdraws into the shadows, continuing to watch* MAT. MAT *is isolated in a pool of light.* VICTOR *fades to black.)*

*(*MAT *lit alone.)*

MAT: I didn't like myself. I didn't like anything about myself. No. I hated myself. Yes. Hated. I hated my yellow face. My flat nose. The name that no one could ever pronounce correctly. So I got rid of them. Got rid of them and started all over. From the ground up. This time it was going to be my way, my creation.

(Fade in) I became Italian. I went on a fruit and water diet, worked out religiously on various contraptions of self-torture. My butt was so taut and my pants so tight you could have skied off the drop. But the rest of me? No curves, no soft lines. Only angles. Sleek angles and sharp edges. And I would purposely walk down the darkest alleys in the most dangerous parts of town knowing not a soul would touch me because I was like a shiny stiletto cutting this way and that way . . . My name was Paolo. I loved saying it. It was like some exotic bird taking flight off my tongue. And when other people said it, "Paolo? Paolo?"—"Yes?" . . . I was a million miles from here.

(Looks back at PAPA*'s hat on his chair)*

But then, after a while, it began to sound like my old name. Feel like my old name. Someone would say, "Paolo?" and I would hear your voice calling me through it. So I changed it again. I became Joaquim. Joaquim dressed in black exclusively, drank only espresso, and spoke with an accent that could have been Peruvian, that could have been Panamanian, but that could not have been, nor ever be, mistaken for Japanese . . . *(Moves towards chair)* But then I started to hear you again. Every time someone said my new name—"Joaquim?" "Joaquim?"—I could almost feel your voice. Your voice, like a big hand reaching right out through it and grabbing me by the scruff of the neck—"Mat, Mat" . . .

(Pause.)

Then one night as I lay in bed, I began to change . . . *(Lighting change)* My head pushes out forward, grows long, and my eyes move to the sides of my head. My body elongates like a shiny torpedo. And the trick to the whole thing is my tail. And the two fins on the sides of my head. *(Starts to move around)* I begin to move with a knowing. I mean, I don't know but I know. My body just seems to be . . . *(Continues.)*

(PAPA *and* VICTOR *lit upstage in a shadowy light.* PAPA *stands shivering in a wide-rimmed bucket, filed with water.* PAPA *is naked.* VICTOR *dips a small container into the bucket to get water, then pours it over the body of* PAPA. *As it touches his skin he makes a mournful high-pitched wailing sound. It intermingles with the sound of the water splashing into the bucket.)*

MAT: . . . leading me, taking me somewhere. Turn here, turn there, go straight here . . . Piece of cake, I'm on automatic pilot, I've been doing this forever—Look, Ma, no hands! Wait, wait, what's this? Up ahead, I hear a growing roar. Then I see it . . . The rapids . . . The current picks up, I'm being swept along . . . Suddenly I'm pulled under, tumbling and rolling along the bottom, pebbles like tiny razors scraping at my skin. *(Breaks through surface gasping for air)* Ahhh . . . What's this . . . Rocks! Big, small, jagged—leap out like flying car fenders! Ouch! *(Bangs into furniture, knocking things over)* Ahh! Ahhh! Leap high, high out of the water! Then deep, deep, deep . . . Finally, I surface. Over there, a quiet pool of water beneath a shady tree. *(Pulls himself along floor now)* My left side has a deep gash, my dorsal fin is ripped. But I must keep going . . . Then I see something . . . yes, yes, I know this place . . . There, I recognize that street, it's where I used to play as a kid. And, and that gas station, I used to fill up the air in my bicycle tires there . . . And those maple trees . . . And there, on the corner, I see it . . .

*(*MAT *passes out on the ground.* PAPA *and* VICTOR *dim to darkness.)*

(DOROTHY *enters house in semi-darkness, trying not to make noise. All dressed up, carrying shoes. She stumbles on* MAT*'s sleeping body.)*

MAT *(waking up):* Mom?

(DOROTHY *sees him.)*

DOROTHY: AAHHH! *(Silence.* DOROTHY *stares as* MAT *gets up.)*

MAT: Mom?

(MAT *slowly approaches* DOROTHY. *He reaches out and embraces her. She is rigid with shock. She slowly comprehends the situation.)*

DOROTHY *(embracing him):* Mat, Mat, you're alive. My baby's . . . (Continues.*)*

MAT *(overlapping):* It's all right, Mama, it's all right. I'm back. Mat's home.

DOROTHY: . . . alive. My baby's come back to me. Mat, Mat, Mat . . .

(DOROTHY *stops, suddenly angry. She starts to attack him, striking at him.* MAT *has to grab her flailing arms as they whack at him.)*

DOROTHY: How could you? Do you know what you put us through? Do you know? Do you? He just stood by the river, by the river all night . . . *(Can't take it, breaks down, reaching out and embracing him again)* Mat, Mat, Mat . . .

(As MAT *stands there being held by* DOROTHY, *lights fade to half.* PAPA *enters, dressed in his worn robe and holding a bowl of mush. He makes his way across the room towards his chair. His slow movements and blank expression give his appearance an eerie quality.* MAT *watches this strange apparition that is his father move across the room. A weeping* DOROTHY *continues to cling to him.)*

MAT: Mama? Mama?

*(*DOROTHY *looks up.)*

MAT: It's the middle of the night. Where were you?

*(*They stare at each other. Fade to black.*)*

Clark Brown

CÉZANNE'S FINGERS

THEY LEFT ABRUPTLY, piling what they could in the U-Haul trailer and abandoning the rest. Two thousand miles on baloney, bread, jug wine and benzedrine. Stopped only at gas stations. Then in Nebraska they hit a blizzard and the wipers quit. He doubled up his fist and punched out the glass. In Utah the trailer broke an axle. He unhooked it and left it for the Mormons. "You should have seen the crap I got from the U-Haul people," he would say later, when they had arrived in abundant time. (There had been no need to hurry.) Then he would lean forward, his gravelly voice low, gleeful and triumphant. "I used to be in real estate," he would confide. "I was good, damn good, but I couldn't hack the pressure."

This was probably true. There was a snapshot he used to display: himself, two-hundred and forty pounds; ballooning from a light seersucker suit, dark shirt and white tie (as though in photographic negative)—a man from whom you would not readily buy anything, and yet, no doubt as he said, "damn good."

When he came to the town, however, he was a chunky one hundred and eighty, dressed in tweeds and guards ties and silent about real estate. He taught American Studies at the college while Ruth tended Howie and Jennifer and meditated dinner parties, and they both fretted in a fierce enlightened way over the rotten schools. Soon they were absorbed: P.T.A., Women's League, Pop Warner, Cub Scouts, Youth Soccer, fights for curbstones and streetlights—all this in the California Sudan where the winter rains are torrential and the summer light like powdered glass. You could, without much trouble, have imagined him in Rotary or Kiwanis (he talked cautiously now of "leverage" and "carrying paper"), her running for schoolboard. You could have, I say, and you would have been a damn fool.

For suddenly, you see, during those lost, sad, drunken years they too discovered they were artists. It was now that they fled to the foothills, and Harris began his sculpture while Ruth attempted ceramics. Now he appeared on campus in lederhosen, tanktop and sandles, thick dark hair sprouting in whorls on his legs and chest and peeking in tufts from beneath his arms. A drooping moustache and the lank, shoulder-length black hair gave

---

him the air of a Mexican bandido. Oh and Ruth! *Her* long dark hair, formerly so wavy and soft, was yanked in a painful pony tail, turning her sharp-boned face to a blade. Saks and I. Magnin vanished, and she padded about in paint-speckled jeans and a floppy man's shirt, cheap "go-aheads" flapping on her dirty feet. Bitterly too and repeatedly she announced that she was sick, sick unto death of "this fucking hole!"

He collected a coterie, not the brightest students, but disciples all the same to nourish on cocaine and Jung. Soon he brought a young woman into his house and bed, then nagged Ruth into affairs and bragged of *them*. There was the legend of the Reclyner chair in his office, there were stories, which he encouraged, and several times he shoved beneath ambivalent eyes more photographs: himself, Ruth and various unidentified persons tightjawed and gloomily naked. *With us,* he announced, *anything goes!*

Meanwhile he published—or threatened to. "Read that!" he would command. "This baby is coming out in *Artyfuss: a journal of the smarts!* (by which he meant *Artifice: a journal of the arts.)* "That," he would tell you, "is the way you gotta write. You gotta zing it in there!" The prose hardly zang, but you got only a glimpse before he snatched away the cloudy Xerox sheets:

> *. . . here it may be proper to remark that the "lost sheep" unquestionably represent an aspect of the self, often depicted in such symbology as friendly animals. The imperative to leave them alone and the assurance that they will faithfully return verbalizes an insistent plea for trust in the subconscious . . .*

A colleague was moved to parody.

> *The Greedy Man is he who sits*
> *And bites bits out of plates,*
> *Or else takes up an almanac*
> *And gobbles all the dates.*

> *Obviously* [the satirist wrote], *the plates represent the* Tai-git-tu, *the Chinese symbol of eternity, and the biting is an attempt to reduce the instantaneous whole of eternity to time which measures change (v.* Summa Theologiae, *1A, 10,4.) The third and fourth lines reverse the process. Thus, the Greedy Man gobbling all the dates is ravenous for both immortality and the gratifications of this life . . .*

And so on.

Copies distributed in the Department coffee room provoked wry amusement, except in the case of one phlegmatic graduate student who seemed to miss the point.

"It's a put-on," the author explained. "You see, Harris is publishing this impossible thing about 'Little Bo Peep'—I'm not making this up—and he's running around bragging about it."

"You mean," the young man asked, "the one about archetypes and reliance on the subconscious?"

"Right. It goes on and on. So I dreamed this up for the hell of it. I couldn't stop myself. There's something about Charlie that brings out the worst in me." And the wit permitted himself a chuckle.

The young man nodded. "But the thing is," he said tonelessly, "is that Harris didn't write it. I did—in his seminar."

This is the part which, related at cocktail parties or on the cooling twilight patios, produces the shrillest laughter. It never fails, and why should it? You can almost see it: Abbott-and-Costello Harris opening a closet and bringing flatirons, teddy bears, girdles, washboards and roller skates crashing down upon him. A limp inner tube settles about his neck. A lampshade crowns him briefly. A bowling ball conks him cross-eyed and rubber-legged.

"Why?" someone will ask. "Why did he do it? That's what I want to know."

"Because," someone else will offer, "he's Charlie. Why else?"

"No," the first will protest. "I don't mean why did he steal it. That I can understand. What I don't get is what possessed him to tell everybody. He must have known what would happen."

"Well," the second will reply, "that's Charlie too."

Old Philip Moore—if I may now speak of myself as having acquired the third person—was delegated to treat with him. Authorized to flourish the Education Code and to mutter darkly about the Chancellor's Office and "legal action," he was to induce Harris to resign, which, it was agreed, would be best for all. This commission Moore accepted and set forth one burning day—a tall grave man near retirement who wore tweeds and neckties all year round, without affectation.

East of the highway the land begins to rise. The flat tawny fields give way to forest, creek and canyon: first the sparse Digger pine, then cedar, black oak, then tall columns of plated yellow pine. Bounding over the twisting road, jolted into chuckholes and abrupt depressions, Moore flew over sudden rises and yanked the car around sharp turns, blinking at sunbursts through shadowy leaves. His head throbbed. Pine dust itched his nose and throat. Now and then a glare seared his eyes—creek flashes like molten tin.

It had been ages since he had come this way, yet when he and Melissa (dead these nine years) had first arrived in town—long before the westside freeway was done—when the children had been truly that (flopping like chimpanzees across the front seat) and he a brainy young historian certain (he thought) to make his mark—why then they would take this route, racing for a weekend in San Francisco: the river, the rice fields, the live oaks, the scored volcanic rock, the shining mountains far away. Sometimes, on an autumn evening, he recalled, you would see the rice stubble on fire, the flames silky in the dusk, and sometimes after dark, zipping through the small towns, you would start at the looming shadow of a sawmill cone and the bright flurries of sparks.

But that was long ago. And since? Spouses traded round. Children passed about like the furniture. Friends, enemies, colleagues shovelled out of sight. The sleepy town a humming city. And now? Widower, grandfather, elder statesman, more than a little ridiculous with his old-fashioned courtesy and impractical clothes, he would do for fools' errands like this. Too removed, too dispassionate to care greatly, he would lend to the whole silly proceeding an air of judicial earnestness and dignified fairness. *Hell, that crazy Harris can't blow up at old Phil!* He was, of course, being used—precisely because he would allow it, because it seemed a thing of no importance. He had dutifully rendered service. He could do so once again. Really, it wasn't much to ask.

He turned west and the country grew rugged: steep walls of basalt—dark, craggy, pitted—carved for twenty million years by the tumbling creek. Then he was higher—pine and fir and cedar and along the creeks aspens and willows. Sometimes he glimpsed slushy meadows—an unearthly emerald dotted with wildflowers. The road narrowed. He rumbled across a wooden bridge. Below, dark water spilled over granite boulders and swirled into pools. Off in the dimness loomed a tent—or something. Now his throat burned; he was parched and lightheaded. Then he emerged from the trees and dusty dappled sunlight. Here he had come eons ago with Melissa to cut Christmas boughs. Or was that someplace else? His head pounded. His nostrils itched and his eyes stung. He couldn't recall.

A wooden box with a steep tin roof, the house sat well back from the road. It snowed here (it never snowed in town), and sometimes drifts reached the second story. Maybe, he thought, Harris rising of a frozen morning contemplated a Christmas card world: the meadow gleaming with powder, tree limbs stark as lace.

Climbing out, he was staggered by the sun. He sucked in the hot dusty air—like breathing through gauze! Then he moved toward the weathered garage. From it there now issued a furious hissing, then a burst of light in the open door. Sparks rose and fell in showers, and a foul smell reached his nose, mingling with pine dust and soiled air.

A silhouette—booted, armored, helmeted—sat upon an upturned crate, squeezing fire from its crotch. The hiss grew louder. The smell pricked Moore's nose and throat—a rank sulfur aroma. Now he could make out the tall tanks chained together, the hoses and gauges. Lost in the fierce light, the manic rushing sound, the craftsman toiled on.

Legs weak, Moore pressed forward. The helmet jerked. The clumsy gloves fumbled the torch. The flame edge diminished to a spear of blinding light, then popped and vanished. The last orange sparks bounced about the booted feet.

"Harris?"

The black shield regarded Moore, and in the small glass rectangle he saw a blurry image: a tall, skinny man in suit and tie staring from a sunlit doorway—himself! Then the helmet shifted and the picture fled. A memory: eight years old, kneeling before a shadowy grille.

Moore cleared his throat and entered. "Look, Charlie," he said, a little too loudly, "you probably know what's up. What it comes down to is, you can resign or fight it. If you fight it the college will too, and so will the Chancellor's Office—if they have to. The more public the thing becomes—if you ask me—the worse for everybody." He paused. His words seemed to clatter, and the sudden stillness was oppressive. Limply, he said: "I'm sorry."

The helmet stared.

Uneasily, Moore looked around: tools, anvil, sander, brazing rods, burnt-out water heaters lying on their sides, crumpled five gallon cans, a heap of small junk: bent forks, stiffened paint brushes, burst flashlight batteries, buttons, cups . . . Some of this trash would be glued to a board and coated with resin, garter snaps and fanbelt pulleys coyly juxtaposed. The smashed cans would be sanded raw and exposed on floodlit turntables. Through it all you felt some rage against the smooth and simple and pleasing, some fury to compel the fastidious eye to wallow in crud.

A gloved hand cocked up the helmet, and peering beneath this visor, this *lid*, Moore saw the dirty sheepskin pad across the forehead, the drooping black moustache, the ragged fringe of beard. The face studied him, the throat speckled with tiny scars, spark holes in the limp shirt as though tiny teeth had nibbled there.

Moore cleared his throat again. "Listen," he said—more quietly now—"there are nine causes for dismissal—as you may know. The first three are immoral conduct, unprofessional conduct and dishonesty." He thought all at once: had it ever really happened?

Harris grunted and spat. Glaring, passionate, he said: "Know what I did one time?"

"What, Charlie?"

"We were coming out here, see." (The voice was hoarse, wild.) "And I threw everything I could in this U-Haul trailer. . . ." Then it all poured out again, Harris, bard and hero both, celebrating his own saga, while Moore fidgeted and pointedly drew breath. But suddenly, in the midst of so much nonsense—The Charlie Harris Story crashing about you like so much cheap crockery—suddenly in the open door a figure appeared, blocking the light. Both men turned. Harris looked up.

Slim and bony in the dirty jeans and loose shirt, she approached, the ratty hair falling over the shoulders. Now Moore saw the narrow, highcheeked face—bitter and sullen. The dark eyes flashed.

Reaching Harris, she turned and glared. In the proud head tilt, the silent sneer, you caught an absurd queenly belligerence, something both protective and childish. Literally now, she stood beside her man. Harris extended a gloved hand—an extravagant gesture that might have included the world.

"You know something?" he said softly. "Language doesn't make it."

"What?"

"I can't get off on anything verbal anymore."

And suddenly Moore understood. With the force of a vision it struck him: it had happened before, would happen again. Harris would cheat, run into debt, lose jobs, infuriate anyone on whose good will he depended, and she would endure, victim and consort. Here was one more humiliation, one more defeat, one more excuse for her to leave him finally. In his elliptical, self-lacerating way (language didn't make it) he insisted: *you can't love me. Look what I've done now!* Then would follow—what? Charlie on his knees? Tears? Ecstacy? Frantic sex? It would start all over: another place, a different job, a fresh crisis. Like gypsies the Harrises packed up and moved on, abandoning along the road every impediment to their crazed love.

And Moore was moved—abruptly, mindlessly—shoved to the brink of stinging tears and wished, absurdly, to lift his hand in blessing. All at once he felt himself as old as the earth itself, but it wasn't for *him* to offer gestures. Now Harris cleared *his* throat.

"The painter Paul Cézanne," he intoned, shucking off the gloves, "used to link his fingers together like this, to show how everything is locked together. You know?" Harris looked up.

Moore nodded, wincing at the pedantry. Then Harris said:

"It's all right."

"What's all right?"

"You don't have to be sorry."

"Who said I was sorry?"

Moore blinked, and again a ridiculous memory seized him: kneeling; the patient knowing face beyond the shadowy grate.

Harris shrugged. "I'll send in a letter."

"That would be best."

"Yeah." Then he rose—a short, squat, roundshouldered man wearing a great tin hat, a gnome, a troll surely, hunched and stealthy with that glittering eye and sallow skin. Didn't he live beneath some bridge? Grandly now, he placed his arm around Ruth, who pouted and sagged against him. Then her chin lifted. Her eyes blazed. His free hand—dirt-caked, nails rimmed—Harris extended.

"Thanks, Phil."

"For what?" With distaste Moore shook hands.

"Listen," Harris said, dropping his hand. "You want something?"

"Want something?"

"I got ten dollar twenty-four grams of Mex, twenty-four dollar full ounce—"

"No, no," Moore said. "I don't want anything."

"Quarter gram of coke? Hardly cut?"

"No," Moore said. "I don't want it."

Harris shrugged, the face-shield tilted above his head like an enormous beak. Then he lifted his hand.

"Well," he said, "see you, huh?"

Below, the valley spread—flat, dun, field upon field stretching toward the misty Coast range. Down there grazed dozens of nearly sheared sheep—scrawny, dust colored, heads down, rumps to the south (often depicted in such symbology as friendly animals). North, where the freeway began, you could see the dark shapes of cattle under scraggly oaks. Further on, the town trembled in the haze, sunken among trees. Here shone a water tower; there glinted a radio spire; and nearer now there quivered a man-made lake, flat and polished like old silver, reflecting the milky sky. And all of it—trees, lake, spire, town—swam and shimmered in the heat, some fragment of a dream.

Moore could breathe now, though the air was thick and soiled and smelled of exhaust, tar and burning pavement, and though his heart pained him still. "Heartache," he thought, a silly phrase, a metaphor, but now, eyes stinging, he felt it again—a dull throb, a gnawing—and blindly, he plunged on.

Gary Thompson

HARD PAN

Six inches
below this city is another world
protected by hard pan.
The trees here live on the surface.
Only a few brave houses
have basements; their owners store
half their memories
in another, deeper life.

At least here the earth
is solid. People born in this city
are very sure of themselves.

We could not dig up our dead if we wanted.

IN THE WILD RICE

In the wild rice fields two rivers
meet:
one from the north bears soil
so fertile that the dead
grow
in their graves;
the other carries the cleaned bones
and empty skins
of animals that once lived
inside the mountain snows. When the sun goes,
these old friends stay outside

---

and exchange
stories of the past and the silent days
when being a river
was something to be proud of.
For warmth, they drink up their own gifts.
Now because they are drunk and tired
of the journey,
they lie down in their beds.
Friendship falls through the heavy water of sleep
like a stone.

CHINESE WALLS

The wall esses
  like a sinuous New Year's dragon
    up old Humboldt Road

into Sierra foothills
  where the last Yahi hid
    with Ishi growing up

lonely through the decades.
  The dragon is fashioned
    from fieldstones spewn

(some claim) from Lassen
  in an ancient eruption
    that shattered California.

Today, lovers in spring art-
  collage spandex
    lean their neon bikes

against the century-
  weathered fence.
    They pick and poke

among lupin, endangered
    meadowfoam, buttercup
        in blazing vernal pools.

The young man whistles
    the obvious, catchy
        tune of their lives

while the woman nestles
    California poppies
        in her tight, brown hair,

points toward Mt. Shasta
    (the other visible
        volcano), wonders

is that Lassen?
    Sun brightens into gold
        the saffron moss

clinging in niches
    of crumbling wall
        that follows contours

of grazing land
    once run by a rich man,
        Christopher Lemm.

Last century, Chinese crews
    (used and turned loose
        by Central Pacific)

lugged and stacked each rock,
    mile after grunt mile,
        15 cents a rod,

while their ancestral
    Great Wall was crumbling
        and mandarin China

fell, inevitable brick
    upon brick, around
        the last emperor's throne.

That oft-dreamed wall,
    thousands of crenelated miles
        of spring vitality, stood

as safety for sacred ways
    against the heathen
        hordes from the north.

For the corvee, the poor
    who built the dragon
        sometimes in its belly,

the wall was deadly.
    Mengjiang (the story goes)
        widowed, walked

a hundred lonely nights
    along the sprawling wall
        until it split

her husband's dirt
    and forgotten tomb
        open to Ming world,

legend, love,
    and a simple temple
        that articulates grief.

Centuries west,
    some codger, a pooped-out
        forty-niner, spun

a yarn about the cheap
    California rancher's cheap,
        dastardly deed:

"Them squint-eyed
coolies built Sam
the prettiest dang fence

in Cal-i-for-nee.
Sam schemes up
a grand who-ee,

invites them Chinee
to grub down too.
Poisoned 'em all,

buried them right there
beneath their own damn
wall."

Later, Ishi hid
in caves of Deer Creek
on the waters of Lassen,

while the brooding
gods of Mt. Shasta
held the northmost breach

of summer valley
against northern hordes
of settlers and clouds.

John Muir, late in spring
'75, got caught
on Shasta, watched a hundred

"square miles of white
cumuli boiling
dreamily" turn quickly

into storm so violent
the massive volcano
seemed "rent to foundations."

Muir survived
bedding in the hissing,
sputtering mud of the fumaroles,

the lingering fires
of ancient eruptions.
He hiked out frostbit

and stunned by the "novel
misery," still a devotee
of "mountain-cloud flowering."

CHICO ENTERPRISE—
16, March, 1877:
"About a mile up

Humboldt road, six
Chinese were grubbing
and clearing land by contract.

They had built a cabin of plank.
A lighted candle
upon a small table

aided the bloody
murderers
in their hellish work.

Blood and brains
formed a pool. Kerosene
was poured and a match

applied, but the wounded
man rose
and put out the flames.

The assassins were six—
five *Melican* men
and one boy, or little man.

Is it because they are Chinamen
    the law is lax
        to ferret out perpetrators

of bloodshed, incendiarism
    and rapine?
        The Order of Caucasians

is opposed to Chinese labor.
    Mr. Lemm informs
        they were good working Chinamen."

Today air is early-
    afternoon lazy
        as the lovers pledge

a future, plucking
    petals from a buttercup.
        They love Humboldt Road

in the '90's first spring.
    They love their bodies
        and the surge of pain

they feel pedalling
    against Ishi's hill.
        They love their new vows.

Two springs back
    Cal-Trans tore down
        a hundred feet of Humboldt wall

to complete a new road.
    That proud work done,
        a scruffy crew of unskilled

labor lugged and restacked,
    rock by rock
        and curse by word,

the scrambled wall
    so untold tales
        might scrabble

into the settling hills
    of always-shattering
        California.

Alan Chong Lau

2 STOPS ON THE WAY HOME

for jeff chan and the tofu store in marysville

1 bean curd

once
maybe two times a year
when returning home
stop in marysville

past orchards, fruit stands
sweet musk of rotting peaches
collide with death smells of skunks
in the valley's heat

dust
hurting up eyelids
fruitpickers sun brown
a speck of red bandana
their only shout

this japanese
with a store of woodplanks
lined with fishing poles
immigrant groceries
wading boots slumped over the counter

coca cola vies with kimchee
scallions and mudstained fish
spotted tails still flop
over in a tin
                    bucket

i cry for nehi
momma rushes into a toilet

---

lined with newspapers
she can only half-read

but in a backroom
the lardcan
sliced in half
tucked between round
baskets of beansprouts

cubes of white tundra
in clear icy water

2 wild potatoes

jump a stone fence
ancestors built
years before

muddy currents oozing silt
water skipper characters
on a glimmer

take off the shoes
dig dig deep
deep underneath rushes
you smell
whenever they burn the delta
out

even without light
grandma can still pick out
the tenderest green shoots

water just soaks cuffs
alone in darkness
we move to the hum of mosquitoes

the occasional moan of a train
in the distance

Greg Pape

ON OUR WAY

for Linda

Delano, MacFarland, Famosa...

I name them and they vanish.
Little towns that have listened
so long to fenders crumpling,
flying glass, to tires unraveling
and men catching fire.

My black shepherd and my woman
sleep beside me on the seat
as boxcars loaded with sugar beets drift by—
their frail root hairs burning
on the broad San Joaquin.

Is this what I wanted, dreamed of
propped against the doorpost
watching the big moon
sink into the quiet leaves of the fig orchard?
Is this it: to be swinging through the lanes of dense heat,
looking up into the glazed eyes of truckers,
or down on a red dog twisted like a rag?

Yes. This is what I wanted;
to be here,
knowing my destination,
knowing it's a long way off.

Lebec, Gorman, Castaic...
Gas stations moored to the dead hills—

"Fill it" the man asks,
greasy fingers fumbling through khaki pockets
for a smoke.

Beaumont, Cabazon, Indio…
This is what you wanted,
and here we are:
pilgrims under the late sun,
a would-be evangelist
bad-mouthing the silence,
a river of white water and a barrel
disappearing over the falls…

Let's close our eyes
and press into this grill of light
until the dead road
opens like a fern.

Thiphavanh Louangrath

OLD MAID

THE NIGHT of April 10, 1988 was a big night for me, and I could not wait to get out of work earlier than usual. I had dreamed of that night for months, and it was finally coming true. I had been planning for one particular night for so long it seemed like infinity to me while I waited anxiously. Who could blame me; every Laotian girl in Fresno had dreamed of this night. They even went out to buy the most formal and expensive outfits they could find for the night's preparation. I also had bought the most beautiful and formal dress I could find in Fashion Fair. Every one of us wanted to look our best, because the night was going to be the biggest Laotian New Year's party ever held in Fresno. People from all over California would gather there to see who would be crowned as Miss Laos of 1988 in Fresno at midnight. I never thought that night would be my turning point from "party" girl to an "old maid" girl.

A couple of years ago, many friends of mine called me a "party" girl, because I went to every single party that existed in Fresno. They could always count on me to be at every single party, dressed in wild styles like Madonna and the New Wave. I had a lot of friends and was very well known among my peers because my friends would cluster around me and we would dance in the latest style as a big group. Those people who saw us at the party would say we were a bunch of wild party kids who could think of nothing except having fun all the time. Also, they said we were carefree with no responsibility and worry. They were right. I admitted it. I was a party addict. I always had fun going to parties. I could not recall a time I did not have fun at a party. Furthermore, I had not missed one single party

---

From *Passages: An Anthology of the Southeast Asian Refugee Experience* compiled by Katsuyo Howard, California State University, Fresno (1990). Reprinted with permission.

because I would not let anything on this earth stop me from going to a party; not even the sound of my grandmother's pleading. But one particular night changed all of that.

That night, I was overwhelmed with happiness and excitement; I wanted to burst with joy. But the thought of having to work at Fashion Town on that night drained all the energy from my body. I had begged my boss for the night off, but she stubbornly insisted that I should work, no matter what. Like a bull attacking the waving red flag in the stadium, I furiously argued for the night off. As a result, I lost the argument. Therefore, I was asked to work the long and dreadful hours until the store closed. Without another word, I began to work. The hours seemed to be longer than the Mississippi River. I could not concentrate on my work and my thoughts wandered off. My fantasy about the party broke off when I was interrupted by my boss. She disturbed my day-dreaming to tell me she had decided to let me off work earlier than usual, for which I was very thankful. I was so delighted; I jammed my car home without wasting any more minutes.

The minute I got home, I rushed to the bathroom for the hot shower for which I had waited a long time. The feeling of cleanliness and freshness made me feel like a whole new person when I stepped out of the shower. Not being able to control my feelings of happiness, I hummed as I sat in front of the long narrow mirror to fix my hair. While applying makeup and fixing my hair, I noticed that the person in front of me kept smiling at me with a delightful expression. Slowly and carefully, I checked to see if my hairdo and makeup were perfect. Satisfied with what I saw, I went to put on the new dress I had bought recently. As I pulled up my dress, I saw a most pretty girl in front of me, dressed in a soft pink formal dress with beautifully designed laces. To my surprise, she was my twin sister, whom I admired greatly. It seemed as though she was looking forward to a hot night on the town. I envied her looks, because she looked so wonderful and outstanding that she could be compared to Cinderella. I kept staring at her and was very pleased with what I saw. I said to myself that if those guys at the party could only see her, they would follow her around like hungry tigers would follow a piece of big, red, juicy beef. Suddenly, my thought was interrupted by my grandmother's voice. She had been watching me the whole time.

Shaken by the unnoticed interruption, I asked my grandmother to repeat what she just said to me. Promptly, she said, "When are you going to get sick and tired of going to parties? You always go, and you never once stay home with me. Just stay home with me tonight. Haven't you had enough fun yet?" As always, I ignored her questions by saying, "Grandmother, tonight is a big night for me, and I can't miss it. You know all my friends are counting on me to be there."

To my amazement, she remained silent. I was so shocked to see that because I was expecting her to give me the same old lecture she always gave me before I went out. But this time, it was different! She no longer gave me her speech. "Why?" was all I could ask myself over and over again. Suddenly, my curiosity burst out when I asked her, "Grandmother, aren't you going to give me your lecture tonight?" "No," she replied carelessly. "Why? I don't mind listening to it again. As a matter of fact, I am used to it already. Your lecture is like a song to my ears," I teased. "I know that; that's why I stopped. What is the point if you don't listen or care. I can say it over a hundred times until my teeth fall out, and it still won't make any difference to you," she answered softly. "It might this time, just try it," I teased again. She simply said, "Granddaughter, if you are mature, you will understand what I have been trying to tell you all along." "I am mature!" I protested. "Mature persons take responsibility and make decisions wisely, which you haven't done yet," she explained. Without another word, I stormed out of the house to my car.

While I sat in the car waiting for the engine to warm up, I felt emotion and pain rush through my body. I felt ashamed of myself and could not forget what my grandmother had said to me. The words she said kept popping up in my head. "What did she mean, that I had not taken any responsibility or made any wise decisions yet? How could she say that to me? If that's what she thinks of me, then what do others think of me?" I kept asking myself. The thought of it made me so furious and upset I wanted to scream out loud. Wanting to find out what she meant, I ran home.

Inside the house, she sat on the pink sofa in our living room. As I approached her, I saw a sad expression on her aging face. Slowly and quietly, I proceeded toward her. Without hesitation, I apologized to her, knowing that she was also hurt by our earlier conversation. As before, she remained silent. Feeling her silence made me want to get down on my knees and beg for her forgiveness.

Not knowing what to do next, I began to ask her many questions to loosen the tension between us. To my astonishment, I found out she was scared to be left alone for some reason and she did not want me to go out at night. I asked her why but she hesitated to tell me. Because she did not answer my question, I thought it was one of her plans to make me stay home with her. But I was proven wrong when she decided to tell me of the incident which frightened her when I was out.

She told me that a couple of nights before there was somebody knocking at our door. They knocked so loudly she thought they would break down the door. When she went to peep to see who it was, they covered the door hole with a big hand. Frightened and terrified, she asked them who they were. With no answer, she ran back to the living room to call my cousin who lived next door. Because she was so scared, she asked my cousin to peep through her door hole to see who was knocking at our door. As she had suspected, they were two American men. More scared than ever, she locked herself in the bedroom and kept talking on the phone until the knocking stopped, which indicated that they had already left our front door.

After hearing her story, my heart softened. I could see that she was very scared and frightened because her face turned pale and her eyes were red after she finished telling the story. Those men must have scared her to death. "Poor grandmother," I felt so sorry she had experienced such an incident. I wished I had stayed home with her that night. If I had, she probably would not be that scared. The thought of it made me sad and ashamed. How could I go out and have fun while my grandmother was terrified to death at home? How could I have done that to her? Do the parties and friends mean more to me than my dear grandmother? Why have I not taken any responsibility?

Uncontrollably, my tears ran down my hot cheeks. I remembered clearly what my mother told me before she died. She told me to take good care of my grandmother and to obey her. If my mother could know that I had not been a good granddaughter, she would be very sad and disappointed in me. I felt so guilty and ashamed that I wished I could go back and change the past.

For a moment, I thought it was impossible for me to do something nice for my grandmother. But suddenly, something inside told me it was not too late to do so. I could start by not going to that night's party even though it might kill me not to go. So I did; I chose to stay home with my grandmother. This was a turning point from childhood to adulthood.

From then on, I have always stayed home with her, no matter what names my friends call me. They called me an "old maid" because I no longer went to the parties with them when they asked me. No matter what they might call me besides an "old maid," it doesn't bother me because I strongly believe what I did was right, and I will continue doing it. As long as my grandmother is proud of me, I am very happy. I just hope that someday my friends will understand what I have done.

Bill Barich

HMONG TEMPLE

In the *Fresno Bee*, I read that some Hmong Buddhists wanted to build a fifty-seven-foot-tall pagoda at their compound on North Valentine Avenue. They were scheduled to go before the County Planning Commission for a hearing at the Hall of Records, an Art Deco marvel in the city center.

On the morning of the hearing, I found about sixty Hmongs waiting in a second-floor hallway to be called before the commission. Only two women were among them, a toothless crone and a very young wife, maybe fifteen, with her little son. The Hmong men were a nut-brown color and rather small and compact. They didn't seem to be put out by the wait. It was as if they'd had considerable training, as if waiting were a skill that they had mastered over the centuries and then had inbred.

They were perfect at it, really. Some men stood patiently without moving a muscle, while others slid down to a squatting position, elbows resting on their knees and the small of their backs pressed against a wall. When they tired of squatting, they sat on the floor and crossed their legs. They were comfortable touching one other. They grinned and grabbed at each other's arms and tapped each other on the shoulders. There was even a little friendly Hmong goosing going on. They might not have behaved any differently in the fields at home.

I watched the Hmongs and realized that some of them were watching me, staring directly and intently, making no bones about it. I knew what they must be thinking: He's about the right age; he could have been there.

From doorways and office windows, the clerks and typists of Fresno were watching the Hmongs, too. In the diversity of faces, I saw again the new face of California being formed.

Although the Hmong people were usually from either Laos, Thailand,

---

Vietnam, or China, this group claimed to be from Cambodia. Of the eighty different ethnic groups in Fresno County, they were the latest to arrive and the lowest on the totem pole. Some farmers resented them for working so hard and relying solely on their families and relatives for labor. Already they'd cornered the cherry-tomato market in the San Joaquin and were moving in on other crops.

Some people in Fresno complained about how the Hmongs lived, with as many as five families sharing a single tract house. They complained that the Hmongs grew backyard poppies for personal-use opium and dealt in child brides. They complained about how the Hmongs drove. Every Hmong neighborhood had a designated driver who'd managed to get a car and a license, and these drivers were supposed to be a menace. I had seen some evidence of this—a Hmong behind the wheel in white-knuckled terror over the horsepower at his command.

A story was going around about a Hmong who got a speeding ticket and killed himself, both from shame and because he was afraid of the punishment. The Hmongs were frequently accused of eating dogs, a bogus charge that had been levied at every Asian immigrant group in California since the Chinese.

When the doors to a hearing room opened, the Hmongs rose as a body to go inside. For the first time, I noticed their clothes. It looked as though they'd been set free on a shopping spree at K mart. As newcomers to the state, they had no mastery over the language of apparel, so they had gone wild and crazy, mixing and matching whatever clothes appealed to them. A curious element of Hmong cool was on display. One flashy dude was got up in a wrinkled trench coat and a pair of expensive cowboy boots, like a private eye who worked part time on a ranch.

Once the Hmongs were seated, the Planning Commission members came in. They were all white and middle-aged or older. A couple of them appeared to be perplexed, not out of any negative feelings toward the Hmongs but from a general disbelief at what was happening, Buddhists from maybe Cambodia wanting to build a pagoda on ag-land in Fresno County. It just didn't compute.

The hearing went smoothly. The Hmongs' attorney, a white Fresno man, presented an impressive array of documents and architectural drawings. He lectured the panel about Angkor Wat and explained that the height of the pagoda was an homage to Buddha. In response to an obvious question, he said, "We would appreciate the nonrepeating of things over and over again."

The opposition to the pagoda was minimal—two calls, one letter, and one petition. Only one person showed up to protest in the flesh. He was mean-faced and uncompromising and seemed to see himself as the last sane man in the San Joaquin fighting for the flag, God, and country.

The Hmongs made noise during their celebrations, he said, and they used loudspeakers. Their drivers might run down children. The pagoda would have an adverse effect on property values. And what about the impact of the Hmongs' "lifestyle"? He wondered why they couldn't scale down the project.

"They should Americanize their ideas a little bit," he suggested.

He was no match for the Hmong, who responded through an interpreter.

"When the government of Cambodia fell down to Communist regimes," the Hmong speaker began, "thousands were killed or tortured. We lost lives, children, family, everything. The Communists took over and banned all religions. Nobody can have a religion except for Communism. All pagodas, even Christian churches, were demolished.

"*This* country give us freedom. Our temple will be both for worship and education. We want our children born in Fresno to be Americans—but to preserve our own culture, too. We want to live in peace in this community.

"All kind of volunteers will work on the pagoda," he continued, his voice rising. "Thank you to government of U.S.A. for letting us have this freedom! We want to help you, to share all kind of responsibility and lifestyle with you. We want to calm and educate our people. Many thank you!"

Wherever these Hmongs were really from, they knew their way around. We listened to some further discussion about such fine points as sewage disposal and the groundwater situation on North Valentine, but everybody in the hearing room knew that Fresno would soon have the only Buddhist temple between Stockton and Bakersfield.

Luis Omar Salinas

MY FIFTY-PLUS YEARS CELEBRATE SPRING

On the road, the mountains
in the distance are at rest
in a wild blue silence.
On the sides of the highway
the grape orchards unfurl
deep and green again
like a pregnant woman
gathering strength
for the time to come.
And with the passing
of each season
human life knows little
change. Forty years
in this valley,
the wind, the sun
building its altars
of salt, the rain that
holds nothing back,
and with the crop
at its peak
packing houses burn
into morning,
their many diligent
Mexican workers stacking up
the trays and hard hours
that equal their living.

I've heard it said
hard work ennobles
the spirit—
If that is the case,
the road to heaven
must be crowded
beyond belief.

Philip Levine

SNAILS

The leaves rusted in the late winds
of September, the ash trees bowed
to no one I could see. Finches
quarrelled among the orange groves.

I was about to say something final
that would capture the meaning
of autumn's arrival, something
suitable for bronzing,

something immediately recognizable
as so large a truth it's totally untrue,
when one small white cloud—not much
more than the merest fragment of mist—

passed between me and the pale
thin cuticle of the mid-day moon
come out to see the traffic and dust
of Central California. I kept quiet.

The wind stilled, and I could hear
the even steady ticking of the leaves,
the lawn's burned hay gasping
for breath, the pale soil rising

only to fall between earth and heaven,
if heaven's there. The world would escape
to become all it's never been
if only we would let it go

streaming toward a future without
purpose or voice. In shade the ground
darkens, and now the silver trails
stretch from leaf to chewed off leaf

---

of the runners of pumpkin to disappear
in the cover of sheaves and bowed grass.
On the fence blue trumpets of glory
almost closed—music to the moon,

laughter to us, they blared all day
though no one answered, no one
could score their sense or harmony
before they faded in the wind and sun.

MAGPIETY

You pull over to the shoulder
    of the two-lane
road and sit for a moment wondering
    where you were going
in such a hurry. The valley is burned
    out, the oaks
dream day and night of rain
    which never comes.
At noon or just before noon
    the short shadows
are gray and hold what little
    life survives.
In the still heat the engine
    clicks, although
the real heat is hours ahead.
    You get out and step
cautiously over a low wire
    fence and begin
the climb up the yellowed hill.
    A hundred feet
ahead the trunks of two
    fallen oaks
rust; something passes over
    them, a lizard
perhaps or a trick of sight.

The next tree
you pass is unfamiliar,
the trunk dark,
as black as an olive's; the low
branches stab
out, gnarled and dull: a carob
or a Joshua tree.
A sudden flaring up ahead,
a black-winged
bird rises from nowhere,
white patches
underneath its wings, and is gone.
You hear your own
breath catching in your ears,
a roaring, a sea
sound that goes on and on
until you lean
forward to place both hands
—fingers spread—
into the bleached grasses
and let your knees
slowly down. Your breath slows
and you know
you're back in central
California
on your way to San Francisco
or the coastal towns
with their damp sea breezes
you haven't
even a hint of. But first
you must cross
the Pacheco Pass. People
expect you, and yet
you remain, still leaning forward
into the grasses
that if you could hear them
would tell you
all you need to know about
the life ahead.

Out of a sense of modesty
  or to avoid the truth
I've been writing in the second
  person, but in truth
it was I, not you, who pulled
  the green Ford
over to the side of the road
  and decided to get
up that last hill to look
  back at the valley
he'd come to call home.
  I can't believe
that man, only thirty-two,
  less than half
my age, could be the person
  fashioning these lines.
That was late July of '60.
  I had heard
all about magpies, how they
  snooped and meddled
in the affairs of others, not
  birds so much
as people. If you dared
  to remove a wedding
ring as you washed away
  the stickiness of love
or the cherished odors of another
  man or woman,
as you turned away
  from the mirror
having admired your newfound
  potency—humming
"My Funny Valentine" or
  "Body and Soul"—
to reach for a rough towel
  or some garment
on which to dry yourself,
  he would enter
the open window behind you
  that gave gratefully

onto the fields and the roads
    bathed in dawn—
he, the magpie—and snatch
    up the ring
in his hard beak and shoulder
    his way back
into the currents of the world
    on his way
to the only person who could
    change your life:
a king or a bride or an old woman
    asleep on her porch.

Can you believe the bird
    stood beside you
just long enough, though far
    smaller than you
but fearless in a way
    a man or woman
never could be? An apparition
    with two dark
and urgent eyes and motions
    so quick and precise
they were barely motions at all?
    When he was gone
you turned, alarmed by the rustling
    of oily feathers
and the curious pungency
    and were sure
you'd heard him say the words
    that could explain
the meaning of blond grasses
    burning on a hillside
beneath the hands of a man
    in the middle of
his life caught in the posture
    of prayer. I'd
heard that a magpie could talk,
    so I waited
for the words, knowing without

the least doubt
what he'd do, for up ahead
an old woman
waited on her wide front porch.
My children
behind her house played
in a silted pond
poking sticks at the slow
carp that flashed
in the fallen sunlight. You
are thirty-two
only once in your life, and though
July comes
too quickly, you pray for
the overbearing
heat to pass. It does, and
the year turns
before it holds still for
even a moment.
Beyond the last carob
or Joshua tree
the magpie flashes his sudden
wings, a second
flames and vanishes into the pale
blue air.
July 23, 1960.
I lean down
closer to hear the burned grasses
whisper all I
need to know. The words rise
around me, separate
and finite. A yellow dust
rises and stops
caught in the noon's driving light.
Three ants pass
across the back of my reddened
right hand.
Everything is speaking or singing.
We're still here.

Wilma Elizabeth McDaniel

SELF-KNOWLEDGE

I look out on the upper
  branches
of a knowing old tree
  and realize that I could step
  out this window
and walk on top of the
    universe
but I prefer flat farmland
  and the dry and tedious
  for my brief stay

BIOGRAPHICAL INFORMATION

MARY AUSTIN (1868–1934), a native of Illinois, moved with her family to the area south of Bakersfield in 1888. This region inspired many of the novels and stories which earned her a reputation as a major author of the Southwest. She later was a prominent member of the artists' colonies at Carmel and Santa Fe. The author of thirty-one books, numerous short stories, poems and essays, Austin was also a playwright and wrote an autobiography, *Earth Horizon*, which was published two years before her death.

BILL BARICH is the author of *Laughing in the Hills* (1980), *Traveling Light* (1984), and *Hard to Be Good* (1987). His writing has appeared frequently in *The New Yorker*. He lives in the San Francisco area.

WILLIAM HENRY BREWER (1828–1910), a scientist and geologist, was born in Poughkeepsie and raised in Ithaca, New York. Between 1862 and 1864, he was first assistant to Josiah Whitney in a geological survey of California. From 1864 to 1903, he held the chair of agriculture at the Sheffield Scientific School at Yale. He was active in public health and science issues in Connecticut and helped to establish the Yale Forest School in 1900.

CLARK BROWN has lived in the Sacramento Valley for thirty years. He has published a novel *The Disciple* (1968), and his fiction and essays have appeared in *The Pushcart Prize Anthology, The Interior Country: Stories of the Modern West, California Childhood*, and *Where Coyotes Howl and Wind Blows Free* as well as in journals such as *Writers' Forum, Sou'wester, The South Carolina Review*, and *Occident*. He has also published criticism, poetry, and translations. He currently teaches at Chico State University.

CARLOS BULOSAN (1914–1956), born in the Philippine village of Mangusmana, immigrated to the U.S. in 1931. He worked throughout California, Washington, and Alaska as a laborer and later became a labor leader. He wrote several autobiographical books and poems, and is best known for his autobiographical novel *America is in the Heart* (1943), which chronicles his life in the Philippines and his early years working in the U.S.

ART COELHO is a poet, short-story writer, novelist, and painter. He is the grandson of Azorean-Portuguese immigrants from the Azores Islands. Through his Seven Buffaloes Press, he has published numerous poetry and story collections by Central Valley writers. He grew up on two family farms in Fresno County, in the Riverdale and Five Points area, and currently lives in Montana.

MIKE COLE grew up in Fresno and now lives near Oakhurst in the Sierra Nevada foothills. He graduated from Fresno State University and worked as a laborer in his father's tunnel construction business, drilling holes under roads and highways up and down the Valley. He went on to be a printer and a bookmobile driver. He currently teaches high school Spanish and English. His work has appeared in numerous magazines including *Antioch Review, Laurel Review, Home Planet News,* and *The San Joaquin Review*.

GEORGE DERBY (aka Squibob and John Phoenix) (1823–1861), a humorist and career military man, was born in Dedham, Massachusetts and graduated from the United States Military Academy in 1846. In 1849, he was sent to California and conducted three expeditions, including a trip to the Tulare Valley in 1850. Writing under his pseudonyms, however, Derby was a popular satirist, a master of puns, grotesque exaggeration, ridiculous understatement, and irony. A collection of his writing entitled *Phoenixian; or Sketches and Burlesques*, published in 1856, was an immediate success and remained popular for a generation.

JOAN DIDION is considered one of America's leading prose stylists. Born in 1934 to a family that had lived in the Central Valley for five generations, Didion received her BA from UC Berkeley. She won *Vogue* magazine's Prix de Paris in 1956 and spent eight years working for *Vogue* in New York before returning to California. Although her first novel *River Run* (1963) received moderate attention, it was her non-fiction prose in the collection *Slouching Towards Bethlehem* (1968) that brought her significant public notoriety. In addition to writing novels and essays, she has collaborated on several screenplays with her husband, John Gregory Dunne, a writer.

CHITRA BANERJEE DIVAKARUNI was born in India and spent the first nineteen years of her life there. She came to the U.S. and earned an MA in English from Wright State University in Ohio and a Ph.D. in English from UC Berkeley. She has written three books of poetry *Dark Like the River* (1987), *The Reason for Nasturtiums* (1990), and *Black Candle* (1991), and edited *Multitude* (1992), an anthology of multicultural writing. Her latest book is a fiction collection, *Arranged Marriage* (1995). Among her numerous honors are an Allen Ginsberg poetry prize, a Pushcart Prize, and awards from the PEN Syndicated Fiction Project, the Gerbode Foundation, and the Santa Clara Arts Council. She lives in the San Francisco area and teaches creative writing at Foothill College.

RICHARD DOKEY's short fiction has been published widely around the country, and he has three collections of stories, including *August Heat* (1982), published by Story Press in Chicago. His novel *The Hollow Man* will soon be published by Delta Press, which will also publish a collection of his stories. He lives in Lodi.

WILLIAM EVERSON (aka Brother Antoninus) (1912–1994) was born in Sacramento and raised in Selma. He is recognized as a major American poet of the 20th century and the first great poet to emerge from the Central Valley. Inspired by the poetry of Robinson Jeffers, Everson began writing poems in the 1930s. He converted to Catholicism and joined the Dominican Order of the Catholic church as a monk in 1951. Although still a monk, he was associated with the San Francisco

Beat Poets, and he left the order in 1969 to marry. Everson was the author of over forty volumes of poetry; the most recent, *The Blood of the Poet* (1994), was published the year Everson died.

PETER EVERWINE was born in Michigan and grew up in Western Pennsylvania. He has taught at Iowa University, Haifa University in Israel, and Fresno State University, from which he recently retired. He received the 1972 Lamont Prize from the Academy of American Poets for his book *Collecting the Animals*. He currently lives in Fresno.

PEDRO FAGES (1730–1794) left his native Catalonia in 1767 for infantry service in Mexico. Between 1770 and 1791, he served as governor of Alta California, where he led expeditions of the Salinas, Santa Clara, and San Joaquin Valleys.

ERNEST J. FINNEY lives in the San Joaquin Valley and writes stories and novels. His first story collection, *Birds Landing,* won the California Book Awards Gold Medal for Fiction in 1986. "Peacocks," reprinted in this anthology, from his novel *Winterchill,* received an O. Henry Awards First Prize. He is the author of *Words of My Roaring* (1993), *Lady with the Alligator Purse* (1992), and a new story collection, *Flights in the Heavenlies* (1996).

ANDREW FREEMAN (c. 1881–Date of death unknown) was a Nomlaki shaman. Living on the Paskenta rancheria and working as a sheepherder in 1936, he conveyed his account of European invasion (titled in this collection as "The Arrival of the Whites") to anthropologist Walter Goldschmidt.

ERNESTO GALARZA (1905–1984) was born in Jalcocotán, Mexico, and left with his mother and two uncles to escape the violence of the Mexican Revolution. They eventually settled in Sacramento. Galarza received a BA from Occidental College, an MA from Stanford University, and a Ph.D. from Columbia University. A union leader and writer, Galarza fought tirelessly for farm workers' rights. After retiring from union work, he devoted himself to community service, elementary school teaching (he was a pioneer in bilingual education) and writing.

LEONARD GARDNER was born in Stockton in 1936. A brief career as a boxer provided him the background for *Fat City*, his 1969 novel, which was nominated for a National Book Award and later adapted into a film directed by John Houston. He graduated from the creative writing program at San Francisco State University and is the recipient of a James D. Phelan Award and a Guggenheim Fellowship. He lives in Mill Valley.

PHILIP KAN GOTANDA is a Japanese American playwright and filmmaker who was born in Stockton. His plays have been produced extensively throughout the country. He has been a strong proponent of the Asian Pacific American Arts Movement and has provided a sustained creative voice in the Asian American theater scene and successfully fought to bring this under-represented perspective to the stages of larger mainstream venues. Among his numerous theater honors are a Guggenheim Fellowship, the Lila Wallace-Readers' Digest Writer's Award, and the PEW-TCG National Theater Artist Award. Gotanda holds a law degree from Hastings College of Law, studied pottery in Japan, and worked as a musician before settling into a career in theater and film. He currently lives in San Francisco.

C.G. HANZLICEK is professor of English and director of creative writing at Fresno State University. He is the author of seven collections of poetry, the most recent of which is *Against Dreaming* (1994), published by the University of Missouri Press.

GERALD HASLAM was born in Bakersfield in 1937 and raised in nearby Oildale. Haslam's background includes farm labor and oil-field work, and he has received degrees from San Francisco State University, Washington State University, and Union Graduate School. The author of eighteen books, he is noted for short fiction and personal essays. He has received the Benjamin Franklin Award, the Josephine Miles Prize for Fiction, the Bay Area Book Reviewers' Award, and the Commonwealth Club Medal for non-fiction. He co-edited with James Houston *California Heartland: Writing from the Great Central Valley* (1978). He lives in rural Penngrove and teaches at Sonoma State University.

JAMES D. HOUSTON is the award-winning author of five novels, including *Love Life* (1987) and *Continental Drift* (1978). Among his several non-fiction works are *Californians: Searching for the Golden State* (1982), *The Men in My Life and Other More or Less True Recollections of Kinship* (1987), and *Farewell to Manzanar* (1973), co-authored with his wife, Jeanne Wakatsuki Houston. He also co-wrote a screenplay based on the latter book for a film broadcast on NBC. He lives in Santa Cruz, where he offers workshops at UC Santa Cruz.

LAWSON FUSAO INADA was born in 1938 in Fresno and was interned in Japanese American concentration camps in Arkansas and Colorado during World War II. His book *Before the War: poems as they happened* (1970) was the first volume of poetry by an Asian American published by a major press. His second book, *Legends from Camp* (1992), won the American Book Award. He co-edited the seminal book, *AIIIEEEEE!: An Anthology of Asian American Writers* and the follow-up collection *The Big AIIIEEEEE*. He has taught at Southern Oregon State University in Ashland since 1966, where he is a Professor of English

JEAN JANZEN has lived in Fresno for thirty-five years, raising four children with her pediatrician husband. In 1980, she studied with Philip Levine and Peter Everwine at Fresno State University. A recipient of a National Endowment for the Arts grant in 1995, she has published four poetry collections: *Snake in the Parsonage* (1995), *The Upside-Down Tree* (1992), *Words for the Silence* (1984), and *Three Mennonite Poets* (1986). She teaches poetry writing at Fresno Pacific College and Eastern Mennonite University.

GEORGE KEITHLEY is the author of six books of poetry, two plays, and numerous short stories. His award-winning epic, *The Donner Party* (1972), a Book-of-the-Month Club selection, has been adapted as a stage play and an opera. His latest book of poems is *Earth's Eye* (1994). He lives in Chico with his wife, Carol Gardner.

SUSAN KELLY-DEWITT has lived in the Sacramento Valley for thirty-five years. Her work has appeared in anthologies and journals such as *Poetry*, *Prairie Schooner*, and *Yankee*, among others. Her awards include a Wallace Stegner Fellowship. Her first book, *The Woman Living Inside Me*, is currently in search of a publisher.

DAVID KHERDIAN has published fourteen novels, poetry collections, memoirs, translations, and children's books. He co-edited *Down at the Santa Fe Depot: 20 Fresno Poets* (1970), the first collection of literature by Central Valley authors. His recent retelling of the Chinese legend of the Monkey King, *Monkey: A Journey to the West* (1992), was selected by the Quality Paperback division of the Book-of-the-Month Club. His awards include a Newbery Honor Book, the Jane Addams Peace Award, the Friends of American Writers Award, and a nomination for the American Book Award. He currently edits *Forkroads*: A *Journal of Ethnic-American Literature* in Spencertown, New York.

MAXINE HONG KINGSTON was born in Stockton in 1940. Recognized as a major American writer of her generation, Kingston first received national attention with the publication of *The Woman Warrior* (1976), which won the National Book Critics Circle Award for Non-fiction and has been dramatized throughout the country. Her second book, *China Men* (1980) won the National Book Award and was a runner-up for the Pulitzer Prize. *Tripmaster Monkey: His Fake Book* (1989), her first novel, won the PEN USA West Award in Fiction. Among her numerous honors are the Lila Wallace-Reader's Digest Writers Award and fellowships from the National Endowment for the Arts and the Guggenheim Foundation. She currently teaches at UC Berkeley.

ALAN CHONG LAU was born in Oroville and raised in nearby Paradise, California. He is the author of *Songs for Jadina* (1980) and the co-author (with Garrett Hongo and Lawson Inada) of *Buddha Bandits Down Highway 99* (1978). His poems appear in numerous anthologies including *The Open Boat* (1993) and *The Before Columbus Foundation Poetry Anthology* (1992). He is also a visual artist, and his work appears in *On a Bed of Rice* (1995) and *Tricycle Magazine*. He is represented by Francine Seders Gallery in Seattle, where he lives.

PHILIP LEVINE was born in Detroit in 1928 and was educated there in public schools and at Wayne State University. After a succession of industrial jobs, he left Detroit and lived in various parts of the country before settling in Fresno. His book *The Names of the Lost* won the Lenone Marshall Award for best book of poetry published by an American in 1976. *Ashes* and *7 Years from Somewhere* received National Book Critics Circle Awards. *Ashes* also won the American Book Award in 1980. In 1987, he received the Ruth Lilly Poetry Prize, awarded by *Poetry* magazine and The American Council for the Arts. *What Work Is* won the 1991 National Book Award for Poetry. *The Simple Truth* was awarded the Pulitzer Prize for Poetry in 1995. He currently lives in Fresno and New York.

LARRY LEVIS (1946–1996), born in Fresno, received a BA from Fresno State University, an MA from Syracuse University, and a Ph.D. from the University of Iowa. His first book, *Wrecking Crew*, won the United States Award of the International Poetry Forum. His second book, *The Afterlife*, won the Lamont Award of the American Academy of Poets in 1976. *The Dollmaker's Ghost* was the 1981 winner of the Open Competition of the National Poetry Series. He was awarded three fellowships from the National Endowment for the Arts, a Fulbright Fellowship, and a Guggenheim Fellowship. Before his death, he taught at Virginia Commonwealth University in Richmond.

THIPHAVANH LOUANGRATH immigrated to the United States in 1979. She graduated from Fresno State University in 1993 and currently lives in the San Francisco Bay Area with her grandmother, whom she wrote about in the story "The Old Maid," included here.

DAVID MAS MASUMOTO is a Sansei, third-generation Japanese American, who grows peaches and grapes on an 80-acre family farm in Del Rey. He is the author of *Epitaph for a Peach* (1995) and has written for major periodicals including *USA Today*, *Los Angeles Times*, and *Glamour* magazine. His other books include *Silent Strength* (1984), *Home Bound* (1987) and *Country Voices: The Oral History of a Japanese American Family Farm Community* (1987).

THOMAS JEFFERSON MAYFIELD (c. 1843–1928), the son of a former military officer, was raised by and lived among the Choinumne Yokuts Indians of the Kings River area for about ten years spanning his childhood and adolescence. He left the Choinumne in the early 1860s and worked herding sheep and prospecting, making trips to Utah, Oregon, the high Sierras, and the deserts of California in search of gold and other precious minerals. When not traveling in his later life, he lived (and finally retired) in a small mining town in the southern San Joaquin Valley called Tailholt (later renamed White River) where he was affectionately known as "Uncle Jeff."

WILMA ELIZABETH McDANIEL was born in 1918 to a family of Oklahoma sharecroppers. In the 1930s, they joined thousands of other "Okies" in journeys to the Central Valley, where McDaniel worked as a farm laborer. Although she wrote poetry from the age of eight, her first book was published forty-seven years later. Called the "biscuits and gravy poet," McDaniel is the author of over fifteen collections of poetry and prose, the most recent of which is *The Ketchup Bottle* (1996), a collection of stories.

ROBERT MEZEY was born in Philadelphia in 1935. He taught at Fresno State University between 1966 and 1968 and now teaches at Pomona College where he is a professor of English and poet-in-residence. The author of nine books of poetry and several works of translation, he won the Lamont Award from the Academy of American Poets in 1960 for his collection *The Lovemaker*. Among his honors are Stegner, Ingram-Merrill, National Endowment for the Arts, and Guggenheim Fellowships; Academy of American Poets prizes; and the Poetry Award from the American Academy and Institute of Arts and Letters.

ARCHIE (KHATCHIK) MINASIAN (1913–1985) was born in Fresno and attended Fresno State College between 1932 and 1934. He moved to San Francisco in 1935, where he worked a variety of jobs before becoming a professional house painter. He began writing poetry when he was fifteen and from that time regularly contributed to various Armenian American publications, earning him the reputation as the preeminent Armenian American poet of his generation. His *Selected Poems* (1986) was published posthumously.

JOSE MONTOYA was born in Escoboza, New Mexico, and raised there, Albuquerque, and California. A multidisciplinary artist—poet, painter, writer, and musician—he is a founding member of the Rebel Chicano Art Front, also known as the Royal Chicano Air Force, a group of artists and writers in Sacramento. He is currently a professor of art at Sacramento State University.

CHERRÍE MORAGA is a playwright, poet, and writer. She has written the plays *The Shadow of a Man* and *Giving Up the Ghost*, the latter of which was published by West End Press in 1986. She won the American Book Award for *This Bridge Called My Back: Writings by Radical Women of Color*, a collection she co-edited with Gloria Anzaldúa. She has taught at UC Berkeley and is a co-founder of Kitchen Table Press.

C.W. MOULTON (1936–1995), born and raised in the San Joaquin Valley, was the founder and director of the Fresno Poets Association, through which he organized readings by some of America's leading poets and encouraged valley poets of all ages. He taught poetry and received a Contemporary Lifetime Achievement Award in Literature from the Fresno Area Council of Teachers of English for his efforts to promote poetry in the schools. In addition to writing and teaching, he was a sculptor, painter, fire fighter, and teacher's aide for severely handicapped children. A collection of his poetry will be issued by the Fresno Poets Association in 1996.

JOHN MUIR (1838–1914), perhaps best known as a naturalist and environmentalist, also wrote several evocative books about California's landscape including *My First Summer in the Sierra* (1911) and *The Yosemite* (1912). Born in Scotland, Muir immigrated with his family to Wisconsin in 1849. Largely self-educated once in the U.S., he studied at the University of Wisconsin but left without taking a degree. After accidentally injuring an eye while working at an Indianapolis wagon factory, he devoted his life to the study of nature. He was influential in establishing Yosemite as a national park; and during the 1870s, he wrote a series of articles on western forestry for *Scribner's Monthly* and *The Century* which had a profound impact in educating the American public about conservation.

AMADO MURO (aka Chester Seltzer) (1915–1971), the son of an influential newspaper editor in Cleveland, majored in journalism at the University of Virginia and studied creative writing at Kenyon College. Soon after starting a career in journalism, he was imprisoned for refusing to fight in World War II. Upon his release, he married and began a family in El Paso. He resumed writing and worked for various newspapers in the West, including the *Bakersfield Californian*, but he never lost his wanderlust and roamed the Southwest and Mexico on freight trains. He wrote fiction under the pseudonym Amado Muro (a variation of his wife's maiden name), and he is noted for his stories about migrant laborers.

LEE NICHOLSON was born in Hanford in 1936 and has lived in the San Joaquin Valley most of his life. He graduated from Millsaps College in 1957 and received an MA from the University of Mississippi in 1959. He received a National Endowment for the Humanities scholarship to study T.S. Eliot at Princeton University, and he continued his study of Eliot at Oxford University. He currently teaches at Modesto Junior College.

FRANK NORRIS (1870–1902) was born in Chicago and moved to California with his family when he was fourteen. While a student at UC Berkeley, Norris discovered the work of the French naturalist writer Émile Zola, whose novels influenced Norris's literary philosophy. Norris worked as a journalist in South Africa for the *San Francisco Chronicle* and covered the Spanish-American War for *McClure's Magazine* before a short but prolific career as a novelist. *The Octopus* (1901) was the first novel of a planned trilogy dealing with wheat. *The Pit* (1903), the second novel in the trilogy, is set in Chicago. Norris died suddenly of peritonitis before the final novel in the series was published.

GREG PAPE is the author of *Border Crossings* (1978), *Black Branches* (1984), and *Storm Pattern* (1992), all published in the Pitt Poetry Series. His book *Sunflower Facing the Sun* (1992) was co-winner of the Edwin Ford Piper Award. He has received a Discovery/*The Nation* Award and two fellowships from the National Endowment for the Arts. He teaches at the University of Montana.

DEWAYNE RAIL was born in Round Prairie, Oklahoma, in 1944. He first moved to Fresno in 1952, and moved back and forth between Oklahoma and Fresno until 1957 when he settled in Fresno. He attended Fresno State University and received an MFA from UC Irvine, where he studied with Galway Kinnell and Charles Wright. His poems have appeared in numerous anthologies and magazines including *Poetry Now, Southern Poetry Review, Epic, Poetry Northwest,* and *Western Humanities Review*. He has taught at Fresno City College since 1969.

WILLIAM RINTOUL is a freelance writer and the author of two short story collections, *Rig Nine* (1983) and *Roustabout* (1985), and four non-fiction books about California oil history. A native of Taft in the southwestern area of the Valley, he currently lives in Bakersfield.

RICHARD RODRIGUEZ was born in San Francisco in 1944 and raised in Sacramento. He was educated at Stanford University, Columbia University, and the University of California and first received public attention with the appearance of his memoir *Hunger of Memory: The Education of Richard Rodriguez* (1982). He is a contributing editor to *Harper's Magazine,* and his essays regularly appear in major American magazines and journals. He is an Associate Editor for the Pacific News Service and is a regular commentator on "The News Hour with Jim Lehrer." His honors include an Emmy Award, the Frankel Medal from the National Endowment for the Humanities, and the Christoper Prize for Autobiography. His second book, *Days of Obligation* (1992), was a finalist for the Pulitzer Prize.

WENDY ROSE was born in Oakland in 1948. Her tribal affiliations are Hopi and Miwok. She is the author of ten books of poetry. Since 1984, she has been coordinator and an instructor of American Indian Studies at Fresno City College. She currently lives in Coarsegold.

LUIS OMAR SALINAS was born in Robstown, Texas, in 1937 and moved to California when he was nine. He attended Fresno State University and has been awarded the Stanley Kunitz Poetry Prize, the Earl Lyon Award, and the General Electric Foundation Literary Award. The author of six collections of poetry, his work is widely anthologized. He currently lives in Sanger.

DIXIE SALAZAR is a visual artist and writer, who has lived in Fresno for about the last twenty years. She has published a chapbook, *Hotel Fresno,* and a novel, *Limbo.* She currently teaches writing at Fresno State University and also at the Corcoran State Prison.

WILLIAM SAROYAN (1908–1981), born in Fresno, was a leading literary figure of the 20th century, winning the Pulitzer Prize for Drama (for *The Time of Your Life*), the Academy Award (for his script adapted from his novel *The Human Comedy*) among other honors. His first success was the collection *The Daring Young Man on the Flying Trapeze and Other Stories* (1934), the title story of which won an O. Henry Award. He wrote over 400 short stories and also authored memoirs, novels, plays, and songs.

DENNIS SCHMITZ has published six collections of poetry; the most recent is *About Night: Selected and New Poems* (1994). He won the 1988 Shelley Memorial Award from the Poetry Society of America. A past winner of the Poetry Center's Discovery Award, he has also received National Endowment for the Arts and Guggenheim Fellowships. He teaches at Sacramento State University.

ARTHUR SMITH was born in Stockton in 1948 and received his BA and MA from San Francisco State University and his Ph.D. from the University of Houston. He won a 1981 Discovery/*The Nation* Award in Poetry, the 1982 Southwest Discovery Prize in Poetry from International P.E.N., a writing grant from the National Endowment for the Arts, and the Norma Farber First Book Award from the Poetry Society of America for *Elegy on Independence Day* (1985).

JEDEDIAH SMITH (1798–1831), an explorer and trader, was born in Bainbridge, New York. Between 1826 and 1827, he traveled throughout California to explore the practicality of entering Oregon country from California. He was killed near Cimarron while involved in the Santa Fe trade.

GARY SNYDER was born in San Francisco in 1930 and moved to a small farm in Seattle soon after his birth. He studied at Reed College and Indiana University before entering a doctoral program in Japanese at UC Berkeley. In the early 1950s, he was part of the Beat movement and spent twelve years in Japan, living mainly in Buddhist monasteries. He won the 1975 Pulitzer Prize for Poetry for his book *Turtle Island*. He currently teaches at UC Davis. He is a member of the American Academy of Arts and Letters and lives with his wife and family in the Sierra foothills.

GARY SOTO was born in 1952 in Fresno and received his BA from Fresno State University and an MFA from UC Irvine. He is the author of over fourteen collections of poetry and prose, the most recent of which is *New and Selected Poems* (1995). His awards include an Academy of American Poets Prize, the U.S. Award from the International Poetry Forum, the Bess Hokin Prize from *Poetry*, an American Book Award, and a Guggenheim Fellowship. He lives with his wife and daughter in Berkeley.

ROBERTA SPEAR was born in Hanford and received BA and MA degrees from Fresno State University. She won the 1985 Literary Award from P.E.N. Los Angeles Center for her book *Talking to Walter*, and her collection, *Silks*, was selected by Philip Levine for the National Poetry Series. She won the 1979 James D. Phelan Award and is the recipient of writing grants from the Guggenheim Foundation and the Ingram Merrill Foundation.

DAVID ST. JOHN's most recent collection of poetry, *Study for the World's Body: New and Selected Poems* (1994), was nominated for a National Book Award in Poetry. His latest publication is *Where the Angels Come Toward Us: Selected Essays, Reviews, and Interviews* (1995). He teaches in the department of English at the University of Southern California.

JOHN STEINBECK (1902–1968), a native of Salinas, won the Nobel Prize for Literature in 1962. Throughout the 1930s, in a series of novels including *Tortilla Flat, In Dubious Battle*, and *Of Mice and Men*, Steinbeck brought California alive for an international readership. He is perhaps best known for the novel *The Grapes of Wrath*, which won the Pulitzer Prize, and was both widely praised and vilified. In addition to writing novels, he was also a playwright (winning the New York Drama Critics Circle Award in 1938 for an adaptation of his novel *Of Mice and Men*), a screenwriter (receiving Academy Award nominations for *Lifeboat* and *A Medal for Benny*) and a Vietnam correspondent for *Newsday* just prior to his death.

JOYCE CAROL THOMAS was born in Ponca City, Oklahoma, and later moved to the San Joaquin Valley. She received an MA in education from Stanford University. Honored with a National Book Award and an American Book Award for *Marked By Fire*, her first novel, she has also edited *A Gathering of Flowers: Stories About Being Young in America* and is the author of five books of poetry and five other novels. She currently lives in Berkeley near her family.

DON THOMPSON was born in Bakersfield in 1942. He holds an MFA from the University of British Columbia, and his fiction and poetry have appeared in such journals as *Kayak*, *Carolina Quarterly*, and *Fiction International*. He is the author of three poetry collections, *Toys of Death* (1970), *Granite Station* (1977), and *Praise from a Stone* (1985).

GARY THOMPSON moved to California in 1962 when he was fifteen and except for a period earning an MFA in Montana, he has lived in the Sacramento Valley ever since. His poems have been published in numerous magazines, anthologies, and have been featured in *Northern California Handbook*, a travel guide for Moon Publications. *Hold Fast*, a collection of early poems, was published by Confluence Press in 1984. *As For Living*, a chapbook, was recently published by Red Wing Press. He teaches in the Creative Writing Program at Chico State University.

ERNESTO TREJO was born in 1950 in Fresnillo, in central Mexico, and grew up in Mexico City and Mexicali. He received an MA in Economics from Fresno State University and attended the Writers' Workshop at the University of Iowa. He co-edited *Piecework: 19 Fresno Poets* (1987) and is the author of three chapbooks, three volumes of translation, and the poetry collection *Entering a Life* (1990). Before his untimely death in 1990, he taught at Fresno City College.

LUIS VALDEZ was born in Delano in 1940. After graduating from San Jose State University, he returned to Kern County and organized *El Teatro Campesino* and used drama as a tool in the United Farm Workers' strikes of the 1960s. He is recognized as an important American playwright for more recent plays such as *Zoot Suit*. Valdez has also enjoyed success in Hollywood with the movie *La Bamba* (1988). Valdez and *El Teatro Campesino* are now located in San Juan Bautista.

LILLIAN VALLEE is a freelance writer and award-winning translator. She has lived in Modesto since 1985 and is interested in the natural and American Indian history of the Central Valley. She is an instructor of English at Modesto Junior College.

ROBERT VASQUEZ was born in 1955 in Madera and grew up in Fresno. He is the winner of the James D. Phelan Award, a Stegner Fellowship, and was a finalist for the AWP Award Series in Poetry. He is the author of *At the Rainbow* (1995). He teaches at the College of the Sequoias in Visalia.

JON VEINBERG was born in Germany after his family fled Estonia. He has lived in Fresno since 1960 and is a graduate of Fresno State University and UC Irvine. He has received two National Endowment for the Arts fellowships and is the author of the collection of poetry *An Owl's Landscape* (1987). He co-edited *Piecework: 19 Fresno Poets* (1987).

CATHERINE WEBSTER, a third-generation California cattlewoman, has her first book forthcoming from the Black Rock Press, University of Nevada, Reno. The editor of *Hardspan of Red Earth: An Anthology of American Farm Poems* (1991), she teaches creative writing at the University of the Pacific in Stockton. She is editor of the forthcoming anthology, *World Farm Poems*, published by the University of Iowa Press.

FRANZ WEINSCHENK was born in Mainz, Germany, in 1925. Being Jewish, his family was forced to leave Nazi Germany in the mid-1930s. They lived for several years in New York City and then moved to Madera. After graduating from Fresno State University, he became an instructor and administrator at Fresno City College, where he still teaches part-time.

SHERLEY ANNE WILLIAMS was born and raised in the Central Valley and received a BA from Fresno State University. She received an MA from Brown University. Her book *Peacock Poems* (1975) was nominated for a National Book Award. She is the author of the poetry collection *Some One Sweet Angel Chile* (1982) and a novel, *Dessa Rose* (1986). She is a professor of literature at UC San Diego.

YOIMUT (c.1845–1933) was the last full-blood survivor of the Chunut Yokuts people who lived on the northeast shore of Tulare Lake. She was born in the *Telumne* village of *Watot Shulul*, about a mile southeast of present-day Visalia. The speaker of seven languages, Yoimut was the historian of her people. Her lament, published here as "I Am The Last," is taken from a longer account dictated to San Joaquin Valley historian and ethnographer Frank Latta in 1930. She died in Hanford.

AUTHOR INDEX

BIBLIOGRAPHY

The following bibliography contains works by authors included in this anthology as well as other authors who have lived in the Central Valley.

ANTHOLOGIES & COLLECTIONS

Baloian, James and David Kherdian, eds. *Down at the Santa Fe Depot: Twenty Fresno Poets*. Fresno: Giligia Press, 1970.

Coelho, Art, ed. *At the Rainbow's End: Dustbowl Okie Anthology*. Big Timber, MT: Seven Buffaloes Press, 1982.

_________. *Blackjack #7: A Collection of Western Writing from the Central Valley of California*. Big Timber, MT: Seven Buffaloes Press, 1978.

_________. *Father Me Home, Winds*. Riverdale, CA: Seven Buffaloes Press, 1975.

_________. *Home Valley: Five Heartland Prose Writers*. Big Timber, MT: Seven Buffaloes Press, 1980.

_________. *99 Vintage: Nine Heartland Poets*. Big Timber, MT: Seven Buffaloes Press, 1980.

_________. *Proud Harvest: Writing from the San Joaquin* . Big Timber, MT: Seven Buffaloes Press, 1979.

Haslam, Gerald and James Houston, eds. *California Heartland: Writing from the Great Central Valley*. Santa Barbara: Capra Press, 1978.

Hongo, Garrett, Alan Chong Lau, and Lawson Fusao Inada. *Buddha Bandits Down Highway 99*. Mountain View, CA: Buddhahead Press, 1978.

Howard, Katsuyo K., ed. *Passages: An Anthology of the Southeast Asian Refugee Experience*. Fresno: Southeast Asian Student Services, CSU Fresno, 1990.

Soto, Gary, ed. *California Childhood*. Berkeley: Creative Arts Book Co., 1988. (Includes a section of writers from the Central Valley)

Veinberg, Jon and Ernesto Trejo, ed.. *Piecework: 19 Fresno Poets*. Albany, CA: Silver Skates, 1987.

Watt, Jane. *Valley Light: Writers of the San Joaquin*. N.p.: Poet & Printer Press, 1978.

POETRY

Adame, Leonard. *Entrance: Four Chicano Poets*. Greenfield Center New York: Greenfield Review Press, 1975. (Also includes poems by Luis Omar Salinas, Gary Soto, and Ernesto Trejo.)

Bidart, Frank. *The Book of the Body*. New York: Farrar, Straus & Giroux, 1977.

_________. *Golden State*. New York: George Braziller, 1973.

_________. *In the Western Night: Collected Poems, 1965-1990*. New York: Farrar, Straus & Giroux, 1990.

_________. *The Sacrifice*. New York: Random House, 1983.

Buckley, Christopher. *Blue Autumn*. Providence, RI: Copper Beach Press, 1990.

_________. *Blue Hooks in Weather*. Santa Cruz: Moving Parts Press, 1983.

_________. *Dark Matter*. Providence, RI: Copper Beach Press, 1993.

__________. *Dust Light, Leaves*. Nashville: Vanderbilt University Press, 1986.
__________. *Last Rites*. Ithaca, NY: Ithaca House, 1980.
__________. *Other Lives*. Ithaca, NY: Ithaca House, 1985.
__________. *Pentimento*. Madison: Bieler Press, 1980.
Coelho, Art. *Evening Comes Slow to a Fieldhand: Selected Valley Poems*. Big Timber, MT: Seven Buffaloes Press, 1982.
__________. *The Man and His Model-M*. Big Timber, MT: Seven Buffaloes Press, 1993.
__________. *My Okie Rose*. Big Timber, MT: Seven Buffaloes Press, 1976.
Divakaruni, Chitra Banerjee. *Black Candle*. Corvalis, OR: Calyx Books, 1991.
__________. *Dark Like the River*. India: Writers Workshop, 1987.
__________. *The Reason for Nasturtiums*. Berkeley: Berkeley Poets Press, 1990.
Everson, William (aka Brother Antoninus). *The Achievement of Brother Antoninus*. Glenview, IL: Scott, Foresman, 1967.
__________. *The Blood of the Poet*. Seattle: Broken Moon Press, 1994.
__________. *The Crooked Lines of God: Poems 1949-1954*. Detroit: University of Detroit Press, 1959.
__________. *The Hazards of Holiness: Poems 1957-1960*. Garden City, NJ: Doubleday, 1962.
__________. *Man-Fate: The Swan Song of Brother Antoninus*. New York: New Directions, 1974.
__________. *The Masculine Dead: Poems 1938-1940*. Prairie City, IL: Press of J.A. Decker, 1942.
__________. *The Residual Years: Poems 1934-1948—The Pre-Catholic Poetry of Brother Antoninus*. New York: New Directions, 1968.
__________. *River-Root: A Syzygy*. Seattle: Broken Moon Press, 1990.
__________. *The Rose of Solitude*. Garden City, NJ: Doubleday, 1967.
__________. *San Joaquin*. Los Angeles: Ward Ritchie Press, 1939.
__________. *These Are the Ravens*. San Leandro, CA: Greater West, 1935.
__________. *The Veritable Years: 1946-1966*. Santa Barbara: Black Sparrow Press, 1978.
Everwine, Peter. *The Broken Frieze*. Mt. Vernon, IA: Cornell College, 1958.
__________. *Collecting the Animals*. New York: Atheneum, 1973.
__________. *From the Meadow*. West Chester, PA: Aralia Press, 1991.
__________. *Keeping the Night*. New York: Atheneum, 1977.
Fagan, Kathy. *The Raft*. New York: Dutton, 1984.
Galarza, Ernesto. *Kodachromes in Rhyme*. Notre Dame: University of Notre Dame Press, 1982.
Hanzlicek, C.G. *Against Dreaming*. Columbia: University of Missouri Press, 1994.
__________. *Calling the Dead*. Pittsburgh: Carnegie-Mellon University Press, 1982.
__________. *A Dozen for Leah*. Santa Cruz: Brandenburg, 1982.
__________. *Living In It*. Iowa City: Stone Wall Press, 1971.
__________. *Night Game* West Chester, PA: Aralia Press, 1988.
__________. *Stars*. Columbia: University of Missouri Press, 1977.
__________. *When There Are No Secrets*. Pittsburgh: Carnegie-Mellon University Press, 1986.

Herrera, Juan Felipe. *Akrilica*. Santa Cruz: Akatraz Editions, 1989.

__________. *Exiles of Desire*. Fresno: Lalo Press, 1983. Houston: Arte Público Press, 1985.

__________. *Facegames*. San Francisco: As Is/So & So Press, 1987.

__________. *Memoria(s) from an Exile's Notebook of the Future*. Santa Monica: Santa Monica College Press, 1993.

__________. *Night Train to Tuxtla*. Tucson: University of Arizona Press, 1994.

__________. *Rebozos of Love*. San Diego: Toltecas En Aztlan, 1974.

__________. *The Roots of a Thousand Embraces*. San Francisco: Manic D Press, 1994.

Inada, Lawson Fusao. *Before the War: poems as they happened*. New York: William Morrow, 1971.

__________. *Legends from Camp*. Minneapolis: Coffee House Press, 1992.

Janzen, Jean. *Snake in the Parsonage*. Intercourse, PA: Good Books, 1996.

__________. *Three Mennonite Poets*. Intercourse, PA: Good Books, 1986.

__________. *The Upside-Down Tree*. Winnipeg: Windflower Communications, 1992.

__________. *Words for the Silence*. Center for Mennonite Brethren Studies, 1984.

Keithley, George. *The Burning Bear*. Sunderland, MA: Heatherstone Press, 1991.

__________. *The Donner Party*. New York: George Braziller, 1972.

__________. *Earth's Eye*. Brownsville, OR: Story Line Press, 1994.

__________. *Scenes from Childhood*. Ventura, CA: Futharc Press, 1987.

__________. *Song in a Strange Land*. New York: George Braziller, 1974.

__________. *To Bring Spring*. Whitehorn, CA: Holmgangers Press, 1987.

Kherdian, David. *On the Death of My Father and Other Poems*. Fresno: Giligia Press, 1970.

__________. *The Nonny Poems*. New York: Macmillan, 1974.

Lau, Alan Chong. *Songs for Jadina*. Greenfield Center, NY: Greenfield Review Press, 1980.

Levine, Philip. *Ashes*. New York: Atheneum, 1979.

__________. *The Names of the Lost*. New York: Atheneum, 1976.

__________. *New Season*. Port Townsend, WA: Graywolf Press, 1975.

__________. *New Selected Poems*. New York: Knopf, 1995.

__________. *1933*. New York: Atheneum, 1974.

__________. *Not This Pig*. Middletown, CT: Wesleyan University Press, 1968.

__________. *On the Edge*. Iowa City: Stone Wall Press, 1963.

__________. *One for the Rose*. New York: Atheneum, 1981.

__________. *Pili's Wall*. Santa Barbara: Unicorn Press, 1971.

__________. *Red Dust*. Santa Cruz: Kayak Books, 1971.

__________. *Selected Poems*. New York: Knopf, 1984.

__________. *7 Years From Somewhere*. New York: Atheneum, 1979.

__________. *The Simple Truth*. New York: Knopf, 1994.

__________. *Sweet Will*. New York: Atheneum , 1985.

__________. *They Feed They Lion*. New York: Atheneum, 1972.

__________. *A Walk with Tom Jefferson*. New York: Knopf, 1988.

__________. *What Work Is*. New York: Knopf, 1991.

Levis, Larry. *The Afterlife*. Iowa City: University of Iowa Press, 1977.

__________. *The Dollmaker's Ghost*. New York: Dutton, 1981.

__________. *Sensationalism*. Iowa City: Corycian Press, 1982.

__________. *The Widening Spell of Leaves*. Pittsburgh: University of Pittsburgh Press, 1991.
__________. *Winter Stars*. Pittsburgh: University of Pittsburgh Press, 1985.
__________. *Wrecking Crew*. Pittsburgh: University of Pittsburgh Press, 1972.
Lopes, Michael. *Mr. & Mrs. Mephistopheles & Son*. Paradise, CA: Dustbooks, 1975.
McDaniel, Wilma Elizabeth. *The Carousel Would Haunt Me*. Tulare: Carl & Irvings Printers, 1973.
__________. *The Fish Hook: Okie and Valley Prose and Poems*. Big Timber, MT: Seven Buffaloes Press, 1978.
__________. *A Girl from Buttonwillow*. Stockton: Wormwood Books, 1990.
__________. *Going Steady With R.C. Boley*. Portlandville, NY: M.A.F. Press, 1984.
__________. *I Killed a Bee for You*. Marvin, SD: Blue Cloud Quarterly, 1987.
__________. *The Last Dust Storm*. New York: Hanging Loose Press, 1995.
__________. *The Peddlers Loved Almira*. Tulare: Stone Woman Press, 1977.
__________. *A Primer for Buford*. New York: Hanging Loose Press, 1990.
__________. *A Prince Albert Wind*. Albuquerque: Mother Road Publications, 1994.
__________. *The Red Coffee Can*. Fresno: Valley Publishers, 1974.
__________. *Sand in My Bed*. Tulare: Stone Woman Press, 1980.
__________. *Sister Vayda's Song*. New York: Hanging Loose Press, 1982.
__________. *This is Leonard's Alley*. Tulare: Stone Woman Press, 1979.
__________. *Toll Bridge*. New York: Contact II, 1980.
__________. *Who is San Andreas? Poems to Survive Earthquakes*. Marvin, SD: Blue Cloud Quarterly, 1984.
McPherson, Sandra. *Elegies for the Hot Season*. Bloomington: Indiana University Press, 1970. New York: Ecco Press, 1982.
__________. *The God of Indeterminacy*. Urbana: University of Illinois Press, 1993.
__________. *Patron Happiness*. New York: Ecco Press, 1983.
__________. *Pheasant Flower*. Missoula, MT: Owl Creek, 1983.
__________. *Radiation*. New York: Ecco Press, 1973.
__________. *Sensing*. San Francisco: Meadow, 1980.
__________. *Streamers*. New York: Ecco Press, 1988
__________. *The Year of Our Birth*. New York: Ecco Press, 1978.
Mezey, Robert. *A Book of Dying*. Santa Cruz: Kayak Books, 1970.
__________. *The Door Standing Open: New and Selected Poems*. Boston: Houghton Mifflin, 1970.
__________. *Evening Wind*. Middletown, CT: Wesleyan University Press, 1987.
__________. *Kenny's: Twenty Poems for a Lost Tavern*. Iowa City: Winhover Press, University of Iowa, 1970.
__________. *The Lovemaker*. Iowa City: Cummington Press, 1961.
__________. *The Wandering Jew*. Mt. Vernon, IA: Hillside Press, Cornell College, 1960.
__________. *White Blossom*. West Branch, IA: Cummington Press, 1965.
Minasian, Khatchik. *Five Poems*. Mt. Horeb, WI: Perishable Press, 1971.
__________. *Selected Poems*. New York: Ashod Press, 1986.
__________. *The Simple Songs of Khatchik Minasian*. San Francisco: Colt Press, 1950. Fresno: Giligia Press, 1969.
__________. *A World of Questions and Things*. Prairie City, IL; Decker Press, 1950.

Montoya, Jose. *El Soy y Los Abajo and Other R.C.A.F. Poems*. San Francisco: Ediciones Pocho-che, 1972. Sherman Oaks: Ninja Pres, 1992.
_________. *In Formation: 20 Years of Joda*. San Jose: Chusma House, 1992.
Page, Greg. *Black Branches*. Pittsburgh: University of Pittsburgh Press, 1984.
_________. *Border Crossings*. Pittsburgh: University of Pittsburgh Press, 1978.
_________. *Little America*. Tucson: Maguey Press, 1976.
_________. *Storm Pattern*. Pittsburgh: University of Pittsburgh Press, 1992.
_________. *Sunflower Facing the Sun*. Iowa: University of Iowa Press, 1992.
Pereira, Sam. *Brittle Water*. Omaha: The Penumbra Press, The University of Nebraska at Omaha, 1987.
_________. *The Marriage of the Portuguese*. Fort Collins, CO: L'Epervier Press, 1978.
Raborg, Frederick A. *Tule*. Bakersfield: Amelia, 1986.
Rail, DeWayne. *Going Home Again*. Mt. Horeb, WI: Perishable Press, 1971.
_________. *The Water Witch*. Fresno: Blue Moon Press, 1988.
Rose, Dorothy. *Dustbowl Okie Exodus*. Big Timber, MT: Seven Buffaloes Press, 1987.
_________. *Dustbowl: Thorns and Roses*. Big Timber, MT: Seven Buffaloes Press, 1981.
Rose, Wendy. *Academic Squaw: Reports to the World from the Ivory Tower*. Marvin, SD: Blue Cloud Quarterly, 1977.
_________. *Bone Dance: New and Selected Poems, 1965-1993*. Tucson: University of Arizona Press, 1994.
_________. *Builder Kachina: A Home-Going Cycle*. Marvin, SD: Blue Cloud Quarterly, 1979.
_________. *Going to War with All My Relations*. Flagstaff, AZ: Entrada Books, 1993.
_________. *The Halfbreed Chronicles*. Albuquerque: West End Press, 1985.
_________. *Hopi Roadrunner Dancing*. Greenfield Center, NY: Greenfield Review Press, 1973.
_________. *Long Division: A Tribal History*. New York: Strawberry Press, 1976.
_________. *Lost Copper*. Banning, CA: Malki Museum, 1980.
_________. *Now Poof She Is Gone*. Ithaca, NY: Firebrand Books, 1994.
_________. *What Happened When the Hopi Hit the Streets of New York*. New York: Contact II, 1982.
St. John, David. *Hush*. Boston: Houghton Mifflin, 1976.
_________. *No Heaven*. Boston: Houghton Mifflin, 1985.
_________. *The Olive Grove*. Syracuse: W.D. Hoffstadt & Sons, 1980.
_________. *The Shore*. Boston: Houghton Mifflin, 1980.
_________. *Study for the World's Body: New and Selected Poems*. New York: HarperPerennial, 1994.
_________. *Terraces of Rain: An Italian Sketchbook*. Santa Fe: Recursos Press, 1991.
Saleh, Dennis. *A Guide to Familiar American Incest*. Oberlin: Triskelion Press, 1972.
_________. *100 Chameleons*. Cathedral Station, NY: New Rivers Press, 1977.
_________. *Palmway*. Ithaca, NY: Ithaca House Books, 1975.
_________. *First Z Poems*. Madison: Bieler Press, 1980.
Salinas, Luis Omar. *Afternoon of the Unreal*. Fresno: Abramas Publications, 1980.
_________. *Crazy Gypsy*. Fresno: Orígenes Publication, 1970.
_________. *Darkness Under the Trees/Walking Behind the Spanish*. Berkeley: Chicano Studies Library Publications, 1982.

__________. *I Go Dreaming Serenades*. San Jose: Mango Publications, 1979.
__________. *Prelude to Darkness*. San Jose: Mango Publications, 1981.
__________. *The Sadness of Days: Selected and New Poems*. Houston: Arte Público Press, 1987.
Schmitz, Dennis. *About Night: Selected and New Poems*. Oberlin: Field Press, Oberlin College Press, 1993.
__________. *Double Exposures*. Oberlin: Triskelion Press, 1971.
__________. *Eden*. Urbana: University of Illinois Press, 1989.
__________. *Goodwill, Inc.*. New York: Ecco Press, 1975.
__________. *Singing*. New York: Ecco Press, 1985
__________. *String*. New York: Ecco Press, 1980.
__________. *We Weep for Our Strangeness*. Chicago: Big Table Publishing Co., 1969.
Smith, Arthur. *Elegy on Independence Day*. Pittsburgh: University of Pittsburgh Press, 1985.
Snyder, Gary. *Axe Handles*. New York: North Point Press, 1983.
__________. *The Back Country*. New York: New Directions, 1968.
__________. *Left Out in the Rain: New Poems 1947-1985*. New York: North Point Press, 1986.
__________. *Mountains and Rivers Without End*. Washington, DC: Counterpoint, 1996.
__________. *Myths & Texts*. New York: Totem Press, 1960. New York: New Directions, 1978.
__________. *No Nature: New and Selected Poems*. New York: Pantheon, 1992.
__________. *A Range of Poems*. London: Fulcrum Press, 1966.
__________. *Regarding Wave*. Iowa City: Windhover Press, 1969. New York: New Directions, 1970.
__________. *Riprap & Cold Mountain Poems*. San Francisco: Four Seasons Foundation, 1969.
__________. *Songs for Gaia*. Port Townsend, WA: Copper Canyon Press, 1979.
__________. *Turtle Island*. New York: New Directions. 1975.
Soto, Gary. *Black Hair*. Pittsburgh: University of Pittsburgh Press, 1985.
__________. *Elements of San Joaquin*. Pittsburgh: Univ. of Pittsburgh Press, 1977.
__________. *Father is a Pillow Tied to a Broom*. Pittsburgh: Slow Loris Press, 1980.
__________. *Home Course in Religion*. San Francisco: Chronicle Books, 1991.
__________. *New and Selected Poems*. San Francisco: Chronicle Books, 1995.
__________. *The Sparrows Move South: Early Poems*. Berkeley: Bancroft Library, 1995.
__________. *The Tale of Sunlight*. Pittsburgh: University of Pittsburgh Press, 1978.
__________. *Where Sparrows Work Hard*. Pittsburgh: Univ. of Pittsburgh Press, 1981.
__________. *Who Will Know Us?* San Francisco: Chronicle Books, 1990.
Spear, Roberta. *The Pilgrim Among Us*. Hanover: University Press of New England for Wesleyan University Press, 1991.
__________. *Silks*. New York: Holt, Rinehart & Winston, 1980.
__________. *Talking to Water*. New York: Holt, Rinehart & Winston, 1984.
Tanaka, Ronald. *The Shino Suite*. Greenfield Center, NY: Greenfield Review Press, 1981.
Thomas, Joyce Carol. *Bittersweet*. San Jose: Firesign Press, 1973.

_________. *Blessings*. Berkeley: Jocato Press, 1975.
_________. *Brown Honey in Broomwheat Tea*. New York: HarperCollins, 1993. (Poems for children)
_________. *Crystal Breezes*. Berkeley: Firesign Press, 1974.
_________. *Inside the Rainbow*. Palo Alto: Penny Press/Zikawuna, 1982.
Thompson, Don. *Granite Station*. Bakersfield: Paper Boat Press, 1977.
_________. *Praise from a Stone*. Bakersfield: Paper Boat Press, 1985.
_________. *Toys of Death*. Surrey, B.C.: Sono Nis Press, 1970.
Thompson, Gary. *Hold Fast*. Lewistown, ID: Confluence Press, 1984.
Trejo, Ernesto. *El Día Entre Las Hojas*. Mexico, D.F.: Fondo de Cultura Económica, 1984.
_________. *The Day of the Vendors*. Fresno: Calavera Press, n.d.
_________. *Entering a Life*. Houston: Arte Público Press, 1990.
_________. *Los Nombres Propios*. Mexico: Editorial Latitudes, 1978.
Vasquez, Robert. *At the Rainbow*. Albuquerque: University of New Mexico Press, 1995.
Veinberg, Jon. *An Owl's Landscape*. Nashville: Vanderbilt University Press, 1987.
Walker, Ardis. *Further Venturings*. Kernville: Sierra Trails Press, 1982.
_________. *The Pagaent*. Kernville: Sierra Trails Press, 1976.
_________. *Quatrains*. London: Arthur H. Stockwell, 1930.
_________. *Sierra Nevada Sequence*. Kernville: Sierra Trails Press, 1968.
_________. *West from Manhattan*. Kernville: Sierra Trails Press, 1979.
Williams, Sherley Anne. *The Peacock Poems*. Middletown, CT: Wesleyan University Press, 1975.
_________. *Some One Sweet Angel Chile*. New York: William Morrow & Co., Inc. 1982.

PROSE

Austin, Mary. *Earth Horizon*. New York: The Literary Guild, 1932.
_________. *The Flock*. Boston: Houghton Mifflin, 1906.
_________. *The Ford*. Boston: Houghton Mifflin, 1917.
_________. *Isidoro*. Boston: Houghton Mifflin, 1905.
_________. *The Land of Little Rain*. Boston: Houghton Mifflin, 1903.
_________. *Lost Borders*. New York: Harper, 1909.
Barich, Bill. *Big Dreams Into the Heart of California*. New York: Pantheon, 1994.
Brewer, William Henry. *Up and Down California, 1860-1864*. New Haven: Yale University Press, 1930.
Brown, Clark. *The Disciple*. New York: Viking, 1968.
Bulosan, Carlos. *America Is In the Heart*. New York: Harcourt Brace & Co., 1943. Seattle: University of Washington Press, 1973.
Coelho, Art. *The Blake Boys Revenge*. Big Timber, MT: Seven Buffaloes Press, 1978.
_________. *Faces of the San Joaquin*. Big Timber, MT: Seven Buffaloes Press, 1978.
_________. *Fresno County Tales: Eleven Short Stories from the San Joaquin Valley*. Big Timber, MT: Seven Buffaloes Press, 1979.
_________. *Papa's Naturalization*. Big Timber, MT: Seven Buffaloes Press, 1991.
_________. *The Valley Oaks*. Big Timber, MT: Seven Buffaloes Press, 1979.

Didion, Joan. *After Henry*. New York: Simon & Schuster, 1992.
__________. *Book of Common Prayer*. New York: Simon & Schuster, 1977.
__________. *Democracy*. New York: Simon & Schuster, 1984.
__________. *Miami*. New York: Simon & Schuster, 1987.
__________. *Play It As It Lays*. New York: Farrar, Straus & Giroux, 1970.
__________. *River Run*. New York: I. Obolensky, 1963.
__________. *Salvador*. New York: Simon & Schuster, 1983.
__________. *Slouching Towards Bethlehem*. New York: Farrar, Straus & Giroux, 1968.
__________. *The White Album*. New York: Simon & Schuster, 1979.
Divakaruni, Chitra Banerjee. *Arranged Marriage*. New York: Anchor/Doubleday, 1995.
Dokey, Richard. *August Heat*. Chicago: Story Press, 1982.
__________. *Birthright: Stories of the San Joaquin*. Big Timber, MT: Seven Buffaloes Press, 1981.
__________. *Sundown*. Big Timber, MT: Seven Buffaloes Press, 1983.
__________. *Two Beer Sun: A Novel of the San Joaquin Valley*. Big Timber, MT: Seven Buffaloes Press, 1979.
Finney, Ernest. *Bird's Landing*. Urbana: University of Illinois Press, 1986.
__________. *Flights in the Heavenlies*. Urbana: University of Illinois Press, 1996.
__________. *The Lady with the Alligator Purse*. Livingston, MT: Clark City Press, 1992.
__________. *Winterchill*. New York: Morrow, 1989. Reno: University of Nevada Press, 1995.
__________. *Words of My Roaring*. New York: Crown, 1993.
Galarza, Ernesto. *Barrio Boy*. Notre Dame: University of Notre Dame Press, 1971.
Gardner, Leonard. *Fat City*. New York: Farrar, Straus & Giroux, 1969.
Haslam, Gerald. *Coming of Age in California: Personal Essays*. Walnut Creek, CA: Devil Mountain Books, 1990.
__________. *Condor Dreams and Other Fictions*. Reno: University of Nevada Press, 1994.
__________. *That Constant Coyote: California Stories*. Reno: University of Nevada Press, 1990.
__________. *The Great Tejon Club Jubilee*. Walnut Creek, CA: Devil Mountain Books, 1996.
__________. *Hawks Flight: Visions of the West*. Big Timber, MT: Seven Buffaloes Press, 1983.
__________. *The Man Who Cultivated Fire and Other Stories*. Santa Barbara: Capra Press, 1987.
__________. *Masks*. Penngrove, CA: Old Adobe Press, 1976.
__________. *Okies*. San Rafael: New West Publishers, 1973.
__________. *The Other California: The Great Central Valley in Life and Letters*. Reno: University of Nevada Press, 1994.
__________. *Snapshots: Glimpses of the Other California*. Walnut Creek, CA: Devil Mountain Books, 1985.
__________. *Voices of a Place: Social and Literary Essays from the Other California*. Walnut Creek, CA: Devil Mountain Books, 1987.
__________. *The Wages of Sin: Collected Stories*. Fallon, NV: Duck Down, 1980.

Houston, James D. *Between Battles*. New York: Dial Press, 1968.
_________. *Californians: Searching for the Golden State*. New York: Knopf, 1982.
_________. *Continental Drift*. New York: Knopf, 1978.
_________. *Farewell to Manzanar*. (with Jeanne Wakatsuki Houston). Boston: Houghton Mifflin, 1973.
_________. *Gasoline: The Automotive Adventures of Charlie Bates*. Santa Barbara: Capra Press, 1980.
_________. *Gig*. New York: Dial Press, 1969.
_________. *Love Life*. New York: Knopf, 1985.
_________. *The Men in My Life and Other More or Less True Recollections of Kinship*. Berkeley: Creative Arts Book Co., 1987.
_________. *A Native Son of the Golden West*. New York: Dial Press, 1971.
_________. *Open Field*. (with John R. Brodie). Boston: Houghton Mifflin, 1974.
_________. *One Can Think About Life After the Fish is in the Canoe and Other Coastal Sketches*. Santa Barbara: Capra Press, 1985.
_________. *Three Songs for My Father*. Santa Barbara: Capra Press, 1974.
Kherdian, David. *Asking the River*. New York: Orchard, 1993.
_________. *Finding Home*. New York: Greenwillow, 1981.
_________. *Monkey: A Journey to the West: A Retelling of the Chinese Folk Novel by Wu Ch'eng-en*. Boston: Shambala, 1992.
_________. *The Mystery of the Diamond in the Woods*. New York: Knopf, 1983.
_________. *The Road Home: The Story of an Armenian Girl*. New York: Greenwillow, 1979.
_________. *Root River Run*. Minneapolis: Carolrhoda, 1984.
_________. *The Song in the Walnut Grove*. New York: Knopf, 1983.
Kingston, Maxine Hong. *China Men*. New York: Knopf, 1980.
_________. *Tripmaster Monkey: His Fake Book*. New York: Knopf, 1989.
_________. *The Woman Warrior: Memoirs of a Girlhood Among Ghosts*. New York: Knopf, 1976.
Kikumura, Akemi. *Promises Kept: The Life of an Issei Man*. Novato, CA: Chandler & Sharp, 1992.
_________. *Through Harsh Winters: The Life of a Japanese Immigrant Woman*. Novato, CA: Chandler & Sharp, 1981.
Latta, Frank F. *Tailholt Tales*. Santa Cruz, CA: Bear State Books, 1976.
Levine, Philip. *The Bread of Time: Toward an Autobiography*. New York: Knopf, 1994.
_________. *Don't Ask*. Ann Arbor: University of Michigan Press, 1981. (Interviews)
Levis, Larry. *Black Freckles: Stories*. Salt Lake City: G. Smith, 1992.
Martin, Larry J. *Rush to Destiny*. New York: Bantam, 1992.
Masumoto, David Mas. *Country Voices: The Oral History of a Japanese American Family Farm Community*. Del Rey: Inaka Press, 1987.
_________. *Distant Voices: A Sansei's Journey to the Gila River Relocation Center*. Del Rey: Inaka Press, 1982.
_________. *Epitaph for a Peach: Four Seasons on My Family Farm*. San Francisco: HarperSan Francisco, 1995.
_________. *Home Bound*. Big Timber, MT: Seven Buffaloes Press, 1989. (Also includes poems and stories by Art Coelho.)

__________. *Silent Strength*. Tokyo: New Currents International, 1984.

Mayfield, Thomas Jefferson. *Indian Summer: Traditional Life Among the Choinumne Indians of California's San Joaquin Valley*. Berkeley: Heyday Books & California Historical Society, 1993. (See also: Latta, Frank F.)

McDaniel, Wilma Elizabeth. *Day Tonight/Night Today Presents Wilma Elizabeth McDaniel: A Special Issue of Selected Short Stories*. Hull, MA: Day Tonight/Night Today, 1982.

__________. *The Ketchup Bottle*. St. John, KS: Chiron Review Press, 1996.

Muir, John. *The Mountains of California*. New York: The Century Company, 1894.

Muro, Amado (aka Chester Seltzer). *The Collected Stories of Amado Muro*. Austin, TX: Thorp Spring Press, 1979.

Norris, Frank. *Blix*. New York: Doubleday, 1899.

__________. *A Deal in Wheat and Other Stories of the New and Old West*. New York: Doubleday, 1903.

__________. *A Man's Woman*. New York: Doubleday, 1900.

__________. *McTeague*. New York: Doubleday, 1899.

__________. *Moran of the Lady Letty*. New York: Doubleday, 1898.

__________. *The Octopus*. New York: Doubleday, 1901.

__________. *The Pit*. New York: Doubleday, 1903.

Rintoul, William. *Rig Nine*. Big Timber, MT: Seven Buffaloes Press, 1983.

__________. *Roustabout: Nine Oilfield Tales from the San Joaquin Valley*. Big Timber, MT: Seven Buffaloes Press, 1985.

Rodriguez, Richard. *Days of Obligation: An Argument With My Mexican Father*. New York: Viking, 1992.

__________. *Hunger of Memory: The Education of Richard Rodriguez*. Boston: D.R. Godine, 1982.

St. John, David. *Where the Angels Come Toward Us: Selected Essays, Reviews and Interviews*. Fredonia, NY: White Pine Press, 1995.

Saroyan, William. *The Bicycle Rider in Beverly Hills*. New York: Scribners, 1952.

__________. *The Daring Young Man on the Flying Trapeze and Other Stories*. New York: Random House, 1934.

__________. *The Human Comedy*. New York: Harcourt Brace & Co., 1943.

__________. *Inhale and Exhale*. New York: Random House, 1936.

__________. *My Name is Aram*. New York: Harcourt Brace & Co., 1940.

__________. *The Saroyan Special*. New York, Harcourt Brace & Co., 1948.

__________. *The William Saroyan Reader*. New York: George Braziller, 1958.

Smith, Jedediah. *The Southwest Expedition of Jedediah Smith: His Personal Account of the Journey to California*. Glendale, CA: Arthur H. Clark Co., 1977.

Snyder, Gary. *The Old Ways: Six Essays*. San Francisco: City Lights Books, 1977.

__________. *A Place in Space: Ethics, Aesthetics and Watersheds: New and Selected Prose*. Washington, DC: Counterpoint, 1995.

__________. *The Practice of the Wild*. San Francisco: North Point Press, 1990.

__________. *The Real Work: Interviews and Talks, 1964-1979*. New York: New Directions, 1980.

Soto, Gary. *Baseball in April and Other Stories*. San Diego: Harcourt, Brace, Jovanovich, 1990.

__________. *Lesser Evils: Ten Quartets*. Houston: Arte Público Press, 1988.
__________. *Living Up the Street: Narrative Recollections*. San Francisco: Strawberry Hill Press, 1985.
__________. *Small Faces*. Houston: Arte Público Press, 1986.
__________. *A Summer Life*. Hanover: University Press of New England, 1990.
Steinbeck, John. *Cannery Row*. New York: Viking, 1945.
__________. *East of Eden*. New York: Viking, 1952.
__________. *The Grapes of Wrath*. New York: Viking, 1939.
__________. *The Harvest Gypsies: On the Road to the Grapes of Wrath*. Berkeley: Heyday Books, 1988.
__________. *In Dubious Battle*. New York: Viking, 1936.
__________. *Of Mice and Men*. New York: Viking, 1937.
__________. *The Pearl*. New York: Viking, 1947.
__________. *Tortilla Flat*. New York: Viking, 1935.
Taylor, Ronald. *The Long Road Home*. New York: Henry Holt, 1988.
Thomas, Joyce Carol. *Bright Shadow*. New York: Avon, 1983.
__________. *The Golden Pasture*. New York: Scholastic, 1986. (Young Adult Fiction)
__________. *Journey*. New York: Scholastic, 1988. (Young Adult Fiction)
__________. *Marked By Fire*. New York: Avon Books, 1982.
__________. *Water Girl*. New York: Avon Books, 1986.
__________. *When the Nightingale Sings*. New York: HarperCollins, 1992. (Young Adult Fiction)
Williams, Sherley Anne. *Dessa Rose*. New York: William Morrow, 1986.

DRAMA

Gotanda, Philip Kan. *Fish Head Soup and Other Plays*. Seattle: University of Washington Press, 1995.
Keithley, George. *The Best Blood of the Country*. Lewiston, NY: Mellen Poetry Press, 1993.
Moraga, Cherríe. *Giving Up the Ghost*. Albuquerque: West End Press, 1986.
__________. *Heroes and Saints and Other Plays*. Albuquerque: West End Press, 1994.
Saroyan, William. *Three Plays: My Heart's in the Highlands, The Time of Your Life, Love's Old Sweet Song*. New York: Harcourt Brace & Co., 1940.
__________. *Three Plays by William Saroyan: The Beautiful People, Sweeney in the Trees, Across the Board on Tomorrow Morning*. New York: Harcourt Brace & Co., 1941.
Valdez, Luis. *Actos*. Fresno: Cucaracha Press, 1971.
__________. *Zoot Suit and Other Plays*. Houston: Arte Público Press, 1992.